Exploring the Role of Universities in Entrepreneurship Education

Exploring the Role of Universities in Entrepreneurship Education

Editors

Valentina Ndou
Otilia Manta
Vera Ndrecaj
Eglantina Hysa

MDPI

Basel • Beijing • Wuhan • Barcelona • Belgrade • Novi Sad • Cluj • Manchester

Editors

Valentina Ndou
Università del Salento Piazza
Tancredi
Lecce
Italy

Otilia Manta
Romanian Academy
Bucharest
Romania

Vera Ndrecaj
Cardiff Metropolitan
University
Cardiff
UK

Eglantina Hysa
Epoka University
Tirana
Albania

Editorial Office
MDPI AG
Grosspeteranlage 5
4052 Basel, Switzerland

This is a reprint of articles from the Special Issue published online in the open access journal *Administrative Sciences* (ISSN 2076-3387) (available at: https://www.mdpi.com/journal/admsci/special_issues/entrepreneurship_universities).

For citation purposes, cite each article independently as indicated on the article page online and as indicated below:

Lastname, Firstname, Firstname Lastname, and Firstname Lastname. Article Title. *Journal Name* **Year,** *Volume Number*, Page Range.

ISBN 978-3-7258-1411-4 (Hbk)
ISBN 978-3-7258-1412-1 (PDF)
doi.org/10.3390/books978-3-7258-1412-1

Contents

About the Editors

Valentina Ndou

Valentina Ndou is a certified Associate Professor Management Science and Engineering at the Department of Engineering for Innovation—University of Salento (Lecce, Italy), with more than 20 years of experience with European, international, and national research projects. She has extensive experience in teaching and education in subjects related to e-business, digital transformation, entrepreneurship education, project management, the innovative management of tourism systems, and developing countries. She has been a project manager, responsible for different international projects such as Interreg Italia-Albanna-Montenegro (2018–2021). She serves as an expert for project evaluation for different science and research funds (e.g., Lithuania council of science, Serbia Science fund). She served as a senior international consultant of the UN IOM organization, assisting the Albanian Ministry of Education and Sports and the Office of Negotiation in the EU approximation process She has authored more than 50 papers in leading international journals such as *Technology Forecasting and Social Change*, *Management Science*, and *Current Issues in Tourism*, as well as being the co-author and co-editor of different books. She is appointed as an advisory expert of the International Academic Council of World Tourism Forum Institute. She is a member of the Scientific Committee of the Contamination Lab of the University of Salento, the extra-curricular laboratory that aims to spread the culture of entrepreneurship in young students. She served as a mentor, developing different innovative entrepreneurial projects, and leading them toward obtaining awards (such as Startcup Puglia award 2018). She has been scientifically responsible for collaborations between the Department of Engineering for Innovation—University of Salento and some Universities in Albania (e.g., Epoka University). She has also been the supervisor of many PhD and MSc students for their thesis.

Otilia Manta

I have been an active member of the International Engineering and Technology Institute (IETI), and has served on various international academic boards, holding the esteemed titles of Professor PhD, Doctor of Economics (Finance), and Scientific Researcher of the Romanian Academy. With over 25 years of experience in financial and banking consulting, EU project management, and scientific research across multidisciplinary fields, I have established myself as a leading authority in International Financial Relations, FinTech, and Entrepreneurship. My expertise extends to being an Evaluation Expert and Rapporteur for EU Projects, specializing in investment projects, capacity building, and sustainable development on both local and global scales. Additionally, I have founded companies and NGOs, demonstrating a commitment to advancing societal wellbeing through innovative financial instruments and projects. Throughout my career, I have authored books, scientific papers, and articles published in international journals, and have served as both publisher and editor. I am also recognized as an international reviewer, contributing to the advancement of knowledge in various fields. My academic and professional endeavors have been focused on the development of innovative financial tools and interactive learning methodologies, particularly in the realm of FinTech. I have also been instrumental in the creation of intelligent learning environments aimed at enhancing financial literacy and fostering societal wellbeing through projects addressing financial technologies and insurance, innovative financial instruments, and other critical areas. I have been the supervisor of many MSc students for their thesis.

Vera Ndrecaj

Dr. Vera Ndrecaj holds the esteemed position of Associate Professor at the Cardiff School of Management, Cardiff Metropolitan University. Her professional accolades extend beyond academia, as she is a distinguished public speaker, author, and recipient of Excellence Awards. Her expertise has garnered recognition from prestigious publications and professional bodies.

Eglantina Hysa

Professor Eglantina Hysa, as a seasoned senior lecturer and experienced researcher of development economics, has been actively contributing to the field of higher education for around twenty years. Her body of work includes numerous scholarly publications, such as book chapters and technical reports, that have been featured in esteemed international journals and publishing houses like the *Journal of Applied Economics*, *Cogent Economics and Finance*, and Routledge. In addition to her research endeavors, she has been an integral part of various EU projects, providing her expertise as a Bologna System expert for higher education. Furthermore, she has been serving as an expert evaluator in prestigious programs such as Horizon 2020/Europe and Erasmus+, while also being an International Expert in several European Accreditation Agencies for Higher Education. Furthermore, she is a highly experienced and certified professional, with more than 15 years in innovation and business consultancy, with a comprehensive background in business models, and a special focus on strategy and business development, KPIs analysis, lifecycle management, digital economy, and sustainability. Additionally, she has seasoned management experience, with extensive experience in directing cross-functional teams, research groups, NGOs, and international projects.

Preface

Universities have played and continue to play an essential role in the training of future entrepreneurs by offering specialized programs and tools which are specific to entrepreneurial education. This reprint, *Exploring the Role of Universities in Entrepreneurship Education*, delves into how universities combine theory with practice, enabling students to develop critical skills such as creative thinking, problem solving, and resource management, especially in the context of current challenges.

The scope of this work encompasses the multifaceted contributions of universities to economic development through innovation. By promoting and facilitating technological research and development, universities create new business opportunities and jobs, thereby stimulating economic growth. The collaboration between universities, industry, and the government is pivotal in supporting technology transfer and the commercialization of inventions.

The aim of this reprint is to highlight the importance of effective management within higher education systems, which is crucial for the successful implementation of entrepreneurship education. Institutional policies and strategies must support innovation and entrepreneurial development through allocating necessary resources and creating a favorable environment for students and educators.

Furthermore, the quality of higher education is significantly enhanced by integrating innovation. Higher education institutions must adopt modern pedagogical methods, invest in technological infrastructure, and promote a culture of continuous improvement. This approach ensures that students are adequately prepared for the dynamic challenges of the business environment.

In a knowledge-based economy, universities serve as primary sources of knowledge generation and dissemination. Entrepreneurship education plays a vital role in shaping individuals who can innovate and leverage this knowledge in order to create new products and services, thus enhancing global economic competitiveness.

Lastly, this reprint addresses the role of universities as essential components of the entrepreneurial ecosystem, which also includes governments, the private sector, investors, and non-profit organizations. By creating incubators, accelerators, and strategic partnerships, universities facilitate the development of startups and support entrepreneurs through every stage of their development.

Through this work, we aim to contribute as a supporting pillar to the education–research–entrepreneurship bridge, reinforcing the vital connections that drive innovation and economic growth.

This reprint includes the following key papers: **"The Impact of COVID-19 on Curriculum and Employability in Lebanon"** (Boustani, N.M., 2023)—This paper explores how the COVID-19 pandemic has forced universities to innovate and adapt, impacting curriculum and employability. **"An Enneagram Approach to Strategy"** (Tlemsani, I. et al., 2023)—This conceptual paper examines how the Enneagram personality typing system can be adopted as a meta-model across multiple domains. **"'Entrepreneurizing' College Programs to Increase Entrepreneurial Intentions: A Mediation Framework"** (Fallatah, M.I.; Ayed, T.L., 2023)—This research examines the impact of entrepreneurial college programs on students' entrepreneurial intentions. **"Entrepreneurship Education: A Challenging Learning Process towards Entrepreneurial Competence in Education"** (Miço, H.; Cungu, J., 2023)—This paper discusses the development of entrepreneurship competence among students and the necessary teacher training. **"Examining the Impact of COVID-19 on**

Entrepreneurial Intention through a Stimulus–Organism–Response Perspective" (Çera, G. et al., 2022)—This study explores how the COVID-19 pandemic influences individuals' entrepreneurial intentions. **"Personality Traits and Business Environment for Entrepreneurial Motivation"** (Zarnadze, G. et al., 2022)—This paper examines the joint effects of personality traits and business barriers on entrepreneurial motivation. **"Factors Influencing Public Higher Education Institutions' Performance Reporting in the Romanian Context"** (Tiron-Tudor, A. et al., 2022)—This study investigates the factors influencing performance reporting in Romanian higher education institutions. **"Economic Growth through the Lenses of Education, Entrepreneurship, and Innovation"** (Apostu, S.A. et al., 2022)—This research analyzes how education, entrepreneurship, and innovation contribute to economic growth. **"Entrepreneurial Education at Universities: A Bibliometric Analysis"** (Dissanayake, H. et al., 2022)—This study identifies the primary topics and trends in entrepreneurship education at universities. **"Emergent Strategy in Higher Education: Postmodern Digital and the Future?"** (Hashim, M.A.M. et al., 2022)—This conceptual paper proposes strategies for developing agile, short-term solutions in higher education.

We believe this collection of works will serve as a valuable resource for academics, policymakers, industry professionals, and anyone interested in the intersection of education, research, and entrepreneurship.

The fellow professors who served as editors for this Special Issue and the authors express their sincere gratitude to all those who have contributed to this research and supported its publication. We extend our thanks to the *Administrative Sciences* editorial team and the "**International Entrepreneurship**" section for their exceptional support. We appreciate the valuable contributions made by the reviewers to the reviewed papers and the authors for their significant role in enhancing the specialized scientific literature. Many special thanks are due to our fellow professors who served as editors for this Special Issue.

Valentina Ndou, Otilia Manta, Vera Ndrecaj, and Eglantina Hysa
Editors

administrative sciences

MDPI

Article

The Impact of COVID-19 on Curriculum and Employability in Lebanon

Nada Mallah Boustani

Faculty of Business and Administration, Saint Joseph University, Beirut 1104 2020, Lebanon; nada.mallahboustany@usj.edu.lb

Abstract: The COVID-19 pandemic negatively impacting the whole service sector, including higher education, has forced universities to quickly innovate and recreate. The sanitary crisis resulted in the greatest disruption to educational systems in human history, as well as a shift in the nature, qualifications, and mix of the workforce. The reopening of higher education institutions is another concern, with numerous new operational procedures in place, new opportunities, and prospective curriculum adjustments based on labour market realities. Due to the development of technology, businesses' requirements for human resource credentials and job types underwent several modifications. This study's goal is to investigate how COVID-19 has influenced curriculum revisions and employability requirements. The research used a mixed methodology, with quantitative analysis of changes in enrolled students by major and a qualitative study including two different sets of surveys based on the innovation and employability theories addressed to five human resource (HR) directors from institutions and organizations in Lebanon, as well as three universities. The objective was to answer the following questions: How can universities adapt to the changing demands of the labour market specifically in times of crisis? Should university curricula place greater emphasis on students' personal growth than on technical and conventional learning? The results lead to re-thinking about what higher education systems and institutions can do to redesign their curricula in accordance with the job market and the expectations of the students in this challenging context, where employment security and job market stability issues are more urgent due to the economic crisis and advanced technologies. According to the findings, the research implications include boosting the implementation of the new curriculum through improved HR practices from the Ministry of Education. This will also encourage innovative performance, which will necessitate realistic, swift technical procedures to be unbeatable, creative, and competitive. This study adds significantly to the literature by suggesting curriculum adjustments for online courses and e-training.

Keywords: innovation theory; education theory; employability; curriculum; skills and job market

Citation: Boustani, Nada Mallah. 2023. The Impact of COVID-19 on Curriculum and Employability in Lebanon. *Administrative Sciences* 13: 128. https://doi.org/10.3390/admsci13050128

Received: 31 March 2023
Revised: 4 May 2023
Accepted: 8 May 2023
Published: 10 May 2023

1. Introduction

Due to the current sanitary and economic crisis, higher education institutions are being compelled to reconsider their curricula and deepen the connections between tertiary education and the labour market. This has made the problem of employability more urgent. The COVID-19 pandemic has had far-reaching consequences for higher education, with a focus on shifting educational results toward online and digital curricula to permit education during lockdowns and emergency remote teaching. These rapid system-level changes have resulted in lower levels of student and staff well-being.

The coronavirus pandemic (COVID-19) crisis caused unforeseen issues around the world in 2020, which have been particularly difficult for the service industry (Suneson 2020). According to Tuzovic and Kabadayi (2021), this unprecedented worldwide pandemic has disrupted the economy in a variety of ways that not only affect service companies but also change how business services are provided (Finsterwalder and Kuppelwieser 2020).

Educational institutions were also closed and have evolved into virtual organizations (Duraku and Hoxha 2020; Carnevale and Hatak 2020).

The disruptive effects of the pandemic affect instructors' motivation and performance (Onyema et al. 2020). Although the conversion of traditional educational institutions into virtual organizations may simplify teaching procedures, it also presents difficulties (Kohntopp and McCann 2020). Pandemic-induced work transformations forced educational institutions to re-evaluate their leadership practices toward staff (Wiradendi Wolor et al. 2020). They also raised concerns about adequate training to address the challenges of technological emergence (UNESCO 2020) and uncertainty in maintaining secured employment (Onyema et al. 2020).

Education is frequently defined by traditional learning as opposed to thinking critically, emphasizing problem-solving techniques, and dealing with real-world problems. Additionally, jobs involving routine tasks that are simple to automate are disappearing because of the impact of rapidly developing technologies, which are frequently ICT-related (information and communications technology), and changes in the organization of work, while new employment creation involves tasks requiring non-routine skills (such as analytical, creative, and interpersonal) that humans are still better at than machines.

New industries are quickly transforming the labour market and altering the nature of employment. Graduates will need to be adaptable and possess the necessary personal qualities to handle complex and tough work environments. Employers are seeking to hire recent graduates who can contribute to the corporate culture and alter the business through fostering creative cooperation (Harvey et al. 1997).

In addition, the traditional educational system negatively impacted the performance of students and teachers in times of COVID-19 (García-González et al. 2020). Thus, by viewing innovation as a pressing issue and an opportunity, it can be enhanced over time for new knowledge contexts. In 200 countries, home to a billion children and youth, 98.6% of all students were affected by the pandemic as of July 2020 (United Nations 2020). Making learning possible and accessible through online learning was the solution (Boustani et al. 2022). Lebanese governments began closing schools and institutions all around the nation to stop the new coronavirus from spreading.

When COVID-19 hit Lebanon, the country was also facing an economic crisis, so majors changed and interests changed for graduate students. The Lebanese context has shown many disparities in the workforce, purchasing power, the sanitary crisis, and economic inflation in the years since 2019. Lebanese graduate students witnessed these changes in terms of teaching methods (online) and in terms of leaving the country to study abroad with financial support and aid. Moreover, many businesses ceased functioning, specifically banks and financial institutions, which made students think about other alternatives or specialization.

In this context and for the purpose of the study, the authors chose top business schools in Lebanon to examine how these institutions adapt to changing market demands and to determine if their curricula should place more emphasis on students' personal growth than the more traditional technical training with the new challenges of learning from home due to the pandemic. These issues arose because of several programs' declining enrolment, which forced some institutions to discontinue a particular curriculum. On the other hand, the researchers investigated the key developments and methodology that enable human resources (HR) recruiters to effectively match candidate profiles with positions or jobs that require constant change, as well as other HR-related issues that enable universities to innovate through their various curricula.

This research intends to answer the following questions: How important is innovation in higher education, as it enables students to improve their practical and soft skills needed in critical times and in online working? What would be the new requirements for human resources qualifications, another matter that innovative and leading universities should focus on specifically in times of crisis?

In conducting this exploratory study, the research aims to make recommendations for steps that business schools should take to carry out their mandate of providing innovative education. Therefore, theoretical background gaps in innovation, employability, and education theories are covered in part two, while the methodology used, and the conceptual framework are covered in part three. Part four emphasis is on the study's results and findings. Finally, in part five the findings are presented and discussed, and part six provides the conclusions, limitations, and future work.

2. Theoretical Background

2.1. Innovation Theory

Over the past few years, the word "innovation" has become increasingly common. Politics, businesses, start-ups, international organizations, and other fields all exhibit it. Innovation management is still a developing "science" despite its popularity. According to Schumpeter (2017), innovation, entrepreneurship, and market power are the driving forces behind economic progress. Market strength resulting from innovation may produce outcomes superior to the invisible hand and price competition. Technical innovation frequently results in transient monopolies that permit exceptional gains that would soon be displaced by rivals and copycats. These brief monopolies were required to provide businesses the motivation to create novel goods and procedures.

A higher rate of economic contact is required due to the complexity of modern economies, which is expanding. The knowledge-based economy of today relies on rapid technological advancement. Innovation now relies on the collaboration of many diverse players rather than on certain individuals.

The COVID-19 pandemic has compelled higher education institutions to embrace virtual platform-based online learning activities, leaving little time to prepare and train staff members to familiarize students with digital technology. While prior research has looked at how students used digital tools in their learning activities, the features of student participation in online learning have received less attention. Salas-Pilco et al. (2022) synthetized student involvement in Latin American higher education institutions during the COVID-19 epidemic from behavioural, cognitive, and affective dimensions, identifying the primary features of student engagement from these three dimensions. Shortage of technological resources, poor internet quality, and a lack of awareness of information and communication technology (ICT) have all been significant impediments, particularly in developing nations (Aung and Khaing 2016).

Despite the fact that interaction and variation were important components of successful online learning, instructors, on the other hand, experienced challenges motivating students, particularly when there was no visual connection. Ultimately, even with innovation, variety, and interaction, some majors of practical and social nature such as sport and physical education do not fully translate to the online setting (Moustakas and Robrade 2022).

The most important theories of innovation management propose a comparison so that the knowledge from one theory may be utilized to fill in the gaps of another. Numerous authors have written about innovation (Henderson and Clark 1990; Abernathy and Utterback 1978; Tushman and Anderson 2018). A great place to start for identifying and categorizing innovations is defined by Henderson and Clark (1990). They provided a fourfold typology by drawing from past work. While modular and architectural innovations have been only briefly discussed in the literature, two of the categories, radical and incremental innovation, have considerable literatures of their own:

- Ettlie et al. (1983) and Tushman and Anderson (2018) discuss incremental innovation. It brings about improvements in basic component quality. This kind of innovation is better characterized as remodelling.
- Contrarily, radical innovation intersects with other characteristics of innovation, including technical discontinuities, and is mentioned in a variety of sources, including Schumpeter (2017). A new meaning is introduced that may lead to a paradigm shift. The use of e-learning technologies in higher education comes with a firm promise that

the learning process will have improved performance. At present, the field of e-learning is at the intersection of commercial, educational, and technological interests, trying to obtain a dominant position in higher education (Dospinescu and Dospinescu 2020).

Finally, effective diffusion of innovation is essential. According to Rogers (2003), diffusion is the process through which an invention is gradually spread among the members of a social system. Further, those who accept any new product or concept may be divided into innovators (2.5%), early adopters (13.5%), early majority (34%), late majority (34%), and laggards (16%). These adoption percentages are based on the bell curve.

2.2. *Employability Theory*

Finding an employability theory might be challenging. It is a multi-dimensional notion, according to Little (2001), and it is important to distinguish between elements that are pertinent to the job and factors that are related to preparation for the job (Knight and Yorke 2002). Morley (2001) adds that employability is "a synergic blend of personal traits, abilities of various sorts, and subject comprehension", Knight and Yorke (2002) assert that employability is not merely about students making deposits in a bank of skills.

For work security, the current economic crisis brought on by the pandemic presents difficulty (Sanchez et al. 2020). Yet, prior study has neglected the effect of employment security on employees' psychological health and well-being during the COVID pandemic and other crises (Pacheco et al. 2020). Employees' hopes for continuing secure and long-term positions inside the company are referred to as their expectations of employment security (Piccoli et al. 2017). Employee stress, worry, and depression can be brought on by employment instability during the pandemic (Pacheco et al. 2020; Wang et al. 2020). Most of the earlier research explored the impact of employment uncertainty on employee wellbeing and identified it as a predictor of unhappiness (Blom et al. 2018).

Given that Dearing (National Committee of Inquiry into Higher Education 1997) placed more stress on students' personal attributes than on their general academic talents, it is possible to see that the concept of employability is more complex.

Knight and Yorke (2002) claim that students' self-theories, personal traits, and views about their own efficacy all have an impact on their employability. They emphasize that the degree to which students believe they can "make a difference" is what is very important. This broadens the focus to cover a larger variety of qualities needed for work success, as well as qualities needed to manage one's professional progression in a way that will keep one's employability.

Employability, according to Nabi (Nabi and Bagley 1998), is about graduates learning a suitable level of abilities and traits and being able to use them to obtain and hold a suitable position. Employability is a notion that originated in the 1990s from the perspective of human resource development along with a rising belief among employees that they cannot rely on their employers for long-term employment. According to Baruch (2001), employability is a promise made to workers that they will have the ability to rapidly find new employment if their current position ceases abruptly. Previously, employability was defined in a variety of ways from both the individual and institutional viewpoints by Harvey (Harvey et al. 1997). Graduates' individual employability is defined as their capacity to exhibit the qualities needed to land jobs.

2.3. *Constructs and Research Hypothesis*

To answer the research questions and to assess the impact of job market employment and majors offered by higher education institutions, the author considers the needs feedback changes and the innovations required for universities to improve their curriculum and offer (and disseminate) these changes among stakeholders (Figure 1).

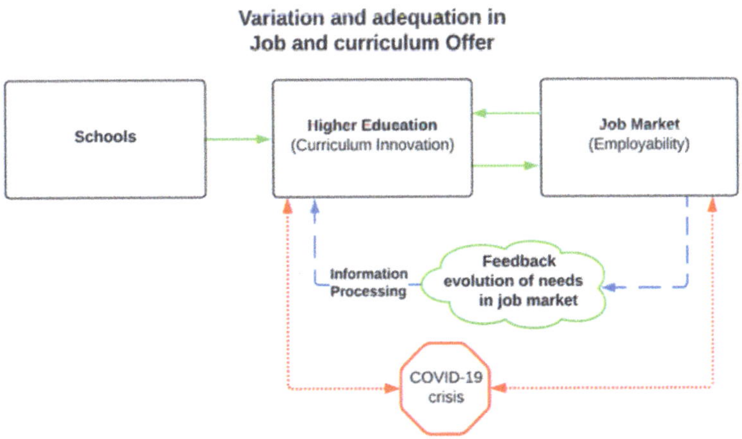

Figure 1. Relation among schools, universities, and job market.

The researcher used the conclusions as a foundation to think about the following research questions: How can universities adapt to the changing demands of the job market? Should business school curriculum and programs emphasize students' personal growth above their technical and conventional education?

Based on the literature analysis, two theories were chosen in this regard:

Hypothesis 1. *Curricula should emphasize technical and conventional learning in stable sectors that are not very innovative.*

Hypothesis 2. *Curricula should emphasize students' personal growth considering the rapidly evolving and innovative workplaces.*

3. Methodology

3.1. Lebanese Context of the Study

In an exploratory manner, our study employs a qualitative approach (Eisenhardt 1989). The authors choose the interpretative approach that enables them to take "a phenomena in its natural surroundings" into consideration. In a case study, the researcher delves deeply into a plan, occasion, activity, process, or person, in accordance with Creswell (2003), that enables the authors to comprehend the whole context of innovation in higher education. Researchers that are attempting to comprehend the social and cultural background most frequently use this method.

The Lebanese context revealed many disparities in the workforce, purchasing power, the sanitary crisis, and economic inflation in the last years, so the years 2019 to today witnessed these changes in terms of teaching methods (online), students leaving the country to study abroad with financial facilities and scholarships from many countries, and businesses that ceased operating, specifically banks and financial institutions.

3.2. Mixed Methodology

A mixed methodology is used, including:

Quantitative focusing on the last 4 years where data (almost 400 students) and statistics related to gender, number, concentration and causes of not enrolling in graduate studies are computed by the author.

Qualitative where the researcher has conducted two types of survey: with university responsible in different business schools and with HR responsible for the employability purpose.

Interviews that were moderated and semi-structured were used to acquire the data (during November and December 2021). One may better understand the process of innovation in education and its effects on employability by using centred semi-structured interviews as a trustworthy data gathering technique (Romelaer 2005).

Three separate sets of questions were employed for this study's purposes:

First, to determine the demands of Bachelor of Arts (BA) students, the authors looked at the quantitative evolution of the student population. Then, as part of this study and with the goal of tracking the innovation and progress of different programs and curricula at business schools, the authors spoke with two deans of Balamand University and of Lebanese University and with a rector of Sagesse University in Lebanon. They made it obvious that they take the lead in introducing innovation into many of their programs and courses. To preserve the participants' intended anonymity, the names of these professors have been coded using P1, P2, P3.

In addition, the authors conducted five in-depth interviews with human resources managers from significant, cutting-edge Lebanese and worldwide businesses. To preserve anonymity, the assigned codes for HR managers are HR1, HR2, HR3, HR4, HR5.

The authors carried out two qualitative investigations in order to properly conduct the study and take into account the significance of innovation in higher education, particularly in the context of crisis and online learning. The first one dealt with various innovations that Lebanese business schools have implemented, while the second research focused on the need for innovation in the labour market.

3.3. *Questionnaire Set and Validation*

The author developed two sets of questions and validated them in the university job setting; nevertheless, this university was not included in the study to avoid bias or any subjective opinion on the students' conditions of employment. The set of questions is based on two fundamental theories: employability and innovation, as well as the impact of COVID-19 on education.

The following inquiries were used in the semi-directive interviews:

Concerning Business School interviews:

1. Do you regularly monitor how the market's requirements for the competencies and abilities of recent graduates are changing, especially during COVID-19?
2. How many times do you consider making changes in your curriculum? Is it a result of proposals from students or market demand or any other disruption such as COVID-19 crisis?
3. What and how do you suggest stakeholders modify the curriculum?
4. What kind of innovation have you implemented over the past three years at your university due to COVID-19?

Human Resources related questions:

1. What are the roles in your company that demand innovation and changes all the time and you have noted during the sanitary crisis?
2. What is the process you use to successfully match a candidate's profile to a position or a job that is evolving or changing constantly?
3. What criteria and procedures do you employ in your company to look for prospective job changes, work enrichment opportunities, or position expansion?
4. Who makes changes to the job proposals, the corporation, or the employees in crisis times?

4. Findings

4.1. *Site Survey*

The site survey is divided into two parts: a first survey was conducted in 2016–2017 and 2018, and a second survey was conducted in 2019 to 2023, revealing a decrease in the number of business students at a well-known Lebanese university, as presented and analysed below. This calls into question the relevance of incorporating innovation into some of this university's master's programs, as well as learning more about the impact of the COVID-19 period on graduate student choices. See Figure 2 below.

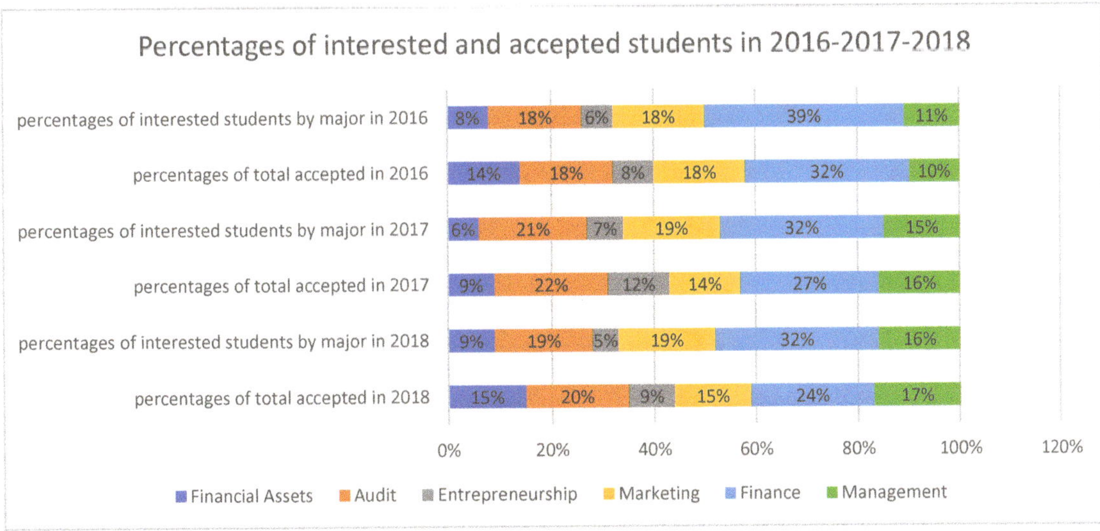

Figure 2. Percentages of accepted students in master's programs (2016–2017–2018).

The authors analysed the percentages and studied the number of students applying for different master's programs in the business school over the total number of students who enrolled in a specific master's program. The findings showed that over the past three years:

- In some majors, such as finance and audit, the percentages are in constant growth either for the demand of these master's programs or in the enrolled for the academic years.
- In other majors, such as management and marketing, the percentages are more or less stable and are not subject to enormous variation.

Furthermore, this business school has a very reasonable number of students (around two hundred receive their bachelor's degree yearly) and has expanded geographically over the Lebanese market. Therefore, for the same business school, in 2022 another qualitative study was conducted on 131 students enrolled in the BBA program but did not express interest in the business school master's program. This survey included several questions concerning their choices of master's program, their preferences for work or for any other university in Lebanon or abroad, in addition to identifying the reason for their intention. The results are shown in Figure 3.

The results of this survey showed that many of these students prefer to access the employment market with a bachelor's degree, whereby 30.53% of the sample have claimed no benefit from a master's degree. Additionally, 5.34% of the sample expressed no interest in pursuing a master's program due to their work schedule that forbids them to attend the courses on time, especially since some courses start at 3 p.m. Their primary objective was a job award then a master's degree which comes later as a secondary requirement.

Moreover, to get the importance of enrolled students post, during, and pre COVID-19, a statistical study related to specialization according to gender was also conducted in 2022–2023 going back from year 2019 until 2023. The above findings constituted a base for the researcher to consider different changes in the interest of students by gender. See Table 1 below.

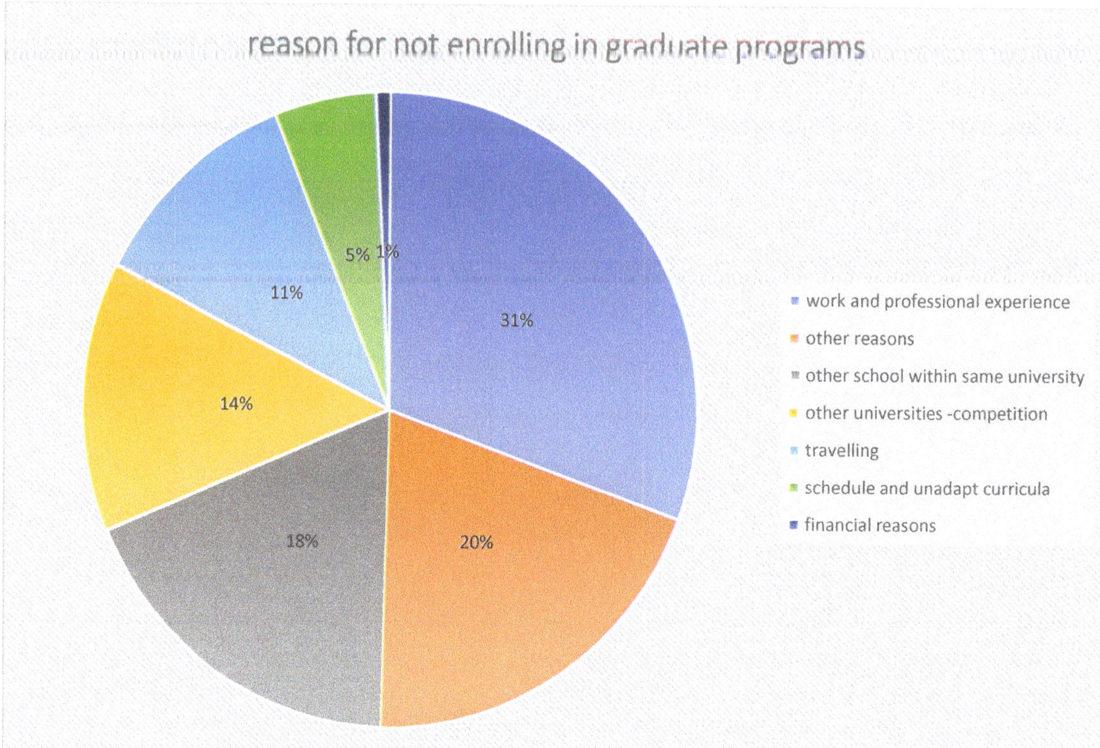

Figure 3. Reasons for disinterest in master's programs.

Table 1. Registration figures by gender in master's programs in a well-known Lebanese business school.

Year	Concentration	Number of Students Enrolled	F	M	% Female	% Male	% From Population	
2022–2023	Accounting and audit	24	14	11	58.33%	45.83%	22.64%	
	Entrepreneurship	9	4	5	44.44%	55.56%	8.49%	43.40%
	Finance	15	6	9	40.00%	60.00%	14.15%	56.60%
	Financial assets	21	8	13	38.10%	61.90%	19.81%	
	Management	24	15	9	62.50%	37.50%	22.64%	
	Marketing	13	8	5	61.54%	38.46%	12.26%	
Total		107	55	52	51.89%	49.06%	100.00%	
2021–2022	Accounting and audit	18	11	7	61.11%	38.89%	26.47%	
	Entrepreneurship	0	0	0	0.00%	0.00%	0.00%	
	Finance	15	9	6	60.00%	40.00%	22.06%	36.76%
	Financial assets	10	2	8	20.00%	80.00%	14.71%	63.24%
	Management	13	8	5	61.54%	38.46%	19.12%	
	Marketing	12	8	4	66.67%	33.33%	17.65%	
Total		68	38	30	55.88%	44.12%	100.00%	
2020–2021	Accounting and audit	20	15	5	75.00%	25.00%	23.81%	
	Entrepreneurship	10	5	5	50.00%	50.00%	11.90%	44.05%
	Finance	21	12	9	57.14%	42.86%	25.00%	55.95%
	Financial assets	6	1	5	16.67%	83.33%	7.14%	
	Management	7	4	3	57.14%	42.86%	8.33%	
	Marketing	20	14	6	70.00%	30.00%	23.81%	

Table 1. *Cont.*

Year	Concentration	Number of Students Enrolled	F	M	% Female	% Male	% From Population	
Total		84	51	33	60.71%	39.29%	100.00%	
2019–2020	Accounting and audit	24	16	8	66.67%	33.33%	23.53%	
	Entrepreneurship	11	4	7	36.36%	63.64%	10.78%	50.98%
	Finance	23	11	12	47.83%	52.17%	22.55%	49.02%
	Financial assets	3	0	3	0.00%	100.00%	2.94%	
	Management	18	9	9	50.00%	50.00%	17.65%	
	Marketing	23	14	9	60.87%	39.13%	22.55%	
Total		102	54	48	52.94%	47.06%	100.00%	

The findings revealed the existence of gender differences in terms of specialization:

- Throughout the four academic years, it is shown that male graduate students prefer finance related majors whereas the predominance of females is evident in the managerial and marketing concentrations.
- As for the accounting and auditing fields, a large majority of females are enrolled in this specialization.
- Entrepreneurship specialization suffered during COVID-19 crisis as noted in the non-existing/null percentage of enrolled graduate students in 2021–2022.

To have the trend of these specializations by two different majors, the authors combined accounting and audit with finance and financial assets as a first major and the remaining specialization, marketing, management and entrepreneurship in another major; the results showed that after and during COVID-19 the percentages of enrolled students in these fields is the highest at 55.95% in 2021, 63.24% in 2022 and 56.6% in 2023. See Figure 4 below.

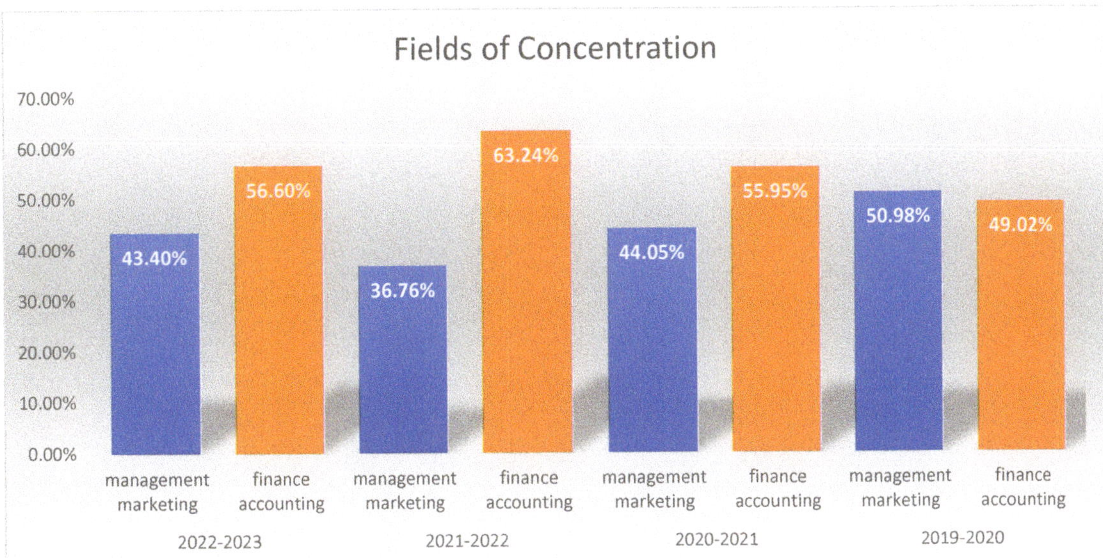

Figure 4. Fields of concentration in master's programs.

4.2. Results from the HR Interviews

The human resources directors of five different organizations were coded by the authors as follows (see Table 2 below):

HR 1: Human resources director for a major organization (distribution, dining . . .)
HR 2: Human resources manager for a holding (of services, goods, or automobiles)
HR 3: Director of human resources and talent for a big marketing and communications company
HR 4: A multinational holding's human resources manager (service, distribution, industry).
HR 5: Human resources director for a major organization (retail).

Table 2. Key answers from the HR questionnaire.

	HR1	HR2	HR3	HR4	HR5
Q 1	Old School People have been in the same place for more than ten years, unable to change and advance with the times.	Following up on emerging trends and strategies in social media and marketing, as well as innovation in high technology departments like the IT and marketing ones.	Any positions within a department that have direct client or market interaction must constantly adapt to changes in the industry.	Due to outside circumstances like escalating competition, shifting market demands, shifting consumer behaviour, or because of corporate expansion. observe consumer behaviour, trends, and preferences.	All departmental jobs that directly interface with customers or the market must continuously adapt to changes in the sector.
Q 2	-Selection by interview -Personal and technical evaluation, as determined by performance review tools -An assessment of one's capacity to adapt to change	upgrading the skill system through repeated seminars, on-the-job training, and outsourced trainings conducted by professionals and consultants.	Compare the profile to the standards established by the company's culture, which states that "we recruit for attitude and train for skills." Our Talents are trained, coached, and developed to be proactive about change.	There are other methodologies, including staying current on trends and innovations through being close to the market and going to conferences and seminars.	With frequent seminars, on-the-job training, and outside trainings led by consultants and specialists, the skill system is upgraded.
Q 3	-The company's potential for development plus the business expansion signal Modernization of real-world commerce -Continuous research into and evaluation of the opposition. -Adopting new technologies, managing change, and evolving markets	When a new product was introduced, the concerned workers were given access to training and awareness-raising opportunities from upper-level employees of our organization and an outside company.	To be prepared for any impending changes in responsibilities, both locally and worldwide, trends progression and market analyses.	a number of methods, including conducting job analyses, shadowing workers to comprehend the task in issue and how the change will be implemented, creating updated job descriptions, etc.	should be ready for any upcoming changes in roles, both locally and globally, as well as for trends' advancement and market evaluations.
Q 4	Both: From employee input and, when necessary to adjust to changes, from management proposals.	In light of the changes affecting the business, the organization suggests adjustments to the job.	Most of the time, the Agency works in tandem with the Talents.	On an organizational level as well as an individual personnel level, this occurs.	From suggestions made by employees and, as needed to adapt to changes, from management recommendations.

Although such research looks at the association between the requirements of labour market through the opinion of IIR and innovation in curriculum, it is necessary because several academics have noted that the studies are few (Waheed et al. 2019). The literature states that innovation activities need to be a focus in businesses by employing new human resources management techniques that can boost employees' involvement in new knowledge. Innovation within the curriculum can grow by implementing new courses and new skills that showed up after the pandemic and due to the digitalized process and methods. Additionally, innovative work is expected to produce innovative results (Crossan and Apaydin 2010), which presents an opportunity for ability enhancing skills and practices as well as providing challenges and motivation (Bos-Nehles et al. 2017).

Transforming learning environments into ones that are more effective and innovative, including e-training, having a hybrid system of alternating between online courses and on campus ones, moreover, focusing on new technologies in the curriculum is required in several jobs. The findings of our research suggest that Lebanese government should develop a strategy to provide more adequate regulations for the internet. To support an electronic and resilient educational system, backup measures are also required (Alsoud and Harasis 2021).

However, in changing scenarios from the past to the present, government organizations are attempting to promote their strategists to produce innovations while dealing with rapidly changing and unpredictable circumstances by focusing on distinctive employees and enhancing operating systems, as organizations need technological processes and practices to produce innovations (Ebersberger and Kuckertz 2021).

4.3. Results from the Academics/Faculties

The academics (professors/deans and rector) in three distinct business schools were classified by the authors as follows (see Table 3 below):

P1: Business school dean
P2: Business school dean
P3: Rector/president of a university

Table 3. Key answers from the academics questionnaire.

Questions	Answers
Q1 P1	We do. But not sufficient. However, we frequently receive alerts from our professional and social connections about the flaws in our programs and courses.
Q1 P2	I suppose that I, myself, have a longer-term perspective on changes; that is, when I arrange my courses, I pay more attention to long-term trends than to market demands. I frequently use research conducted by respectable professional companies like McKinsey, Deloitte, etc. to identify these patterns.
Q1 P3	Of course I (we) keep an eye on the shifting market demands, but this is more of a periodic process.
Q2 P1	Both in actuality, as well as our understanding of global evolution and tendencies We sometimes check over the main colleges' websites.
Q2 P2	As these patterns alter and as we see demands during classroom lectures, we adjust our curriculum accordingly.
Q2 P3	In general, we update our curriculum every five years. Each speciality has a committee that is suggested by the dean. The required improvements are proposed by the committee. Not the recommendations of the pupils, but rather changes in the market.
Q3 P1	The BBA curriculum have changed, and top professors have been met to discuss the adjustments. We sent the updates to the other teachers through email. Together with the External Consultative Council, we discussed the revisions.
Q3 P2	A curriculum can be changed without a drawn-out procedure.
Q3 P3	Adding or removing courses is one way to make modifications, as is recommending (adding or removing) new chapters for a current course. To the Faculty Council, then to the University Council, as a suggestion from the scientific committee.

Table 3. *Cont.*

Questions	Answers
Q4 P1	-New elective courses; -New BBA curriculum; -EDBA -Began the accreditation procedure -New teaching regulations established by the university were introduced at the FGM level. Systematic research and worldwide publishing were also encouraged.
Q4 P2	Employing flipped classrooms, where students provide lectures on certain subjects, and heavily utilizing case studies across the courses
Q4 P3	Expanding the range of multimedia used in literacy lessons.

The authors found that the country's economic destiny is strongly influenced by higher education. Several Lebanon-born students left Lebanese universities and joined other international universities abroad. Therefore, the demand for foreign higher education is anticipated to increase because of the current economic crisis Lebanon is facing. The impact of the pandemic on the rate of employment is the biggest worry and graduates are incapable of finding an appropriate job despite the high level of expertise and the quality of learning and the adequacy of the curriculum—the subjects of this research.

Since the outdated chalk-talk model has been replaced with the modern technology, teaching and learning are made feasible by e-learning systems (Mallah Boustani and Sayegh 2021). It will promote employability, happiness, health, and productivity through the development of new skills to assure the general progress in Lebanon. Employers evaluate applicants based on their educational credentials, such as grade point averages and degree classifications, as demonstrated by Piopiunik (Piopiunik et al. 2020).

Moreover, the placement of recent graduates in the labour market is thus also being impacted, which is causing higher job-separation rates and slower wage growth (Fredriksson and Ihlen 2018).

E-learning resources have been essential in facilitating student learning while schools and universities have been closed due to the pandemic (Subedi et al. 2020).

Therefore, Hypotheses 1 and 2 are supported by the data collected from the interviews and the authors found that: many majors, such as finance and management which are in the traditional learning, accept Hypothesis 1, yet the influence of changes is evident in these fields less than in those of other business majors. Additionally, Hypothesis 2 is considered viable for industries such as marketing, distribution, and entrepreneurship that need newer skills and more ICT expertise that COVID-19 lockdown led all industries to use, therefore digitalization and e-training are frequently the topic of innovation. The percentage of students has been decreasing as a result, according to the responsible participants in the different institutions interviewed, which has led to the radical innovation of these majors as well as a critical necessity for ongoing education to keep up with the evolving and changing demands of the labour market.

5. Discussion

According to the literature, firms should prioritize innovation activities by implementing new human resource management approaches that can increase employees' involvement in new knowledge. Curriculum innovation can increase by adopting new courses and skills that emerged after the pandemic and as a result of digitalized processes and procedures. Furthermore, innovative work is expected to provide novel results, which provides an opportunity for capacity enhancement skills and practices required for new employments, as well as difficulties and motivation, which align with the work of (Bos-Nehles et al. 2017).

The COVID-19 challenges created opportunities and innovation, while transforming learning settings into more effective and innovative ones, including e-training, having a hybrid system of alternating between online and on-campus courses, and focusing on new technologies in the curriculum. All are necessary in a variety of positions. The Lebanese government should adopt a strategy to offer more adequate internet laws and regulations adapted to novel methods in learning. However, in changing scenarios from the past to the present, government organizations are attempting to promote their strategists to produce innovations while dealing with rapidly changing and unpredictable circumstances by focusing on unique employees and improving operating systems, as organizations require technological processes and practices to produce innovations. This agrees with the findings of (Ebersberger and Kuckertz 2021).

Moreover, E-learning technologies have made teaching and learning possible specifically during lockdowns as was found by (Mallah Boustani and Sayegh 2021). It will promote employment, happiness, health, and productivity through the development of new skills, ensuring Lebanon's overall progress. As illustrated by Piopiunik (Piopiunik et al. 2020), employers evaluate applicants based on their educational credentials, such as grade point averages and degree classifications. Furthermore, the placement of fresh graduates in the labour market is disrupted, resulting in increased job separation rates and slower wage growth (Fredriksson and Ihlen 2018). Our findings align with (Subedi et al. 2020) in the fact that during Lebanese crisis and pandemic lockdown, while universities were closed owing to the pandemic, e-learning services were critical in aiding student learning.

Online learning does not have a single pedagogy that works for everyone. There are numerous topics with various requirements and new updates and requirements in curriculum. Various methods of online learning are required for various courses and age groups (Doucet et al. 2020). Online learning also gives extra opportunities for students with physical disabilities and more freedom to interact in the virtual environment while learning, needing less movement (Basilaia and Kvavadze 2020). However, the level of academic performance of the students is expected to decline for the classes held for both the year-end assessment and internal examination (Sintema 2020).

The results of this research underscore the necessity to rebuild efforts to focus on the SDGs, especially given the changing higher education scene during COVID-19. While there are still many students who face poor on-line higher education conditions, this provides a key foundation for speeding our understanding of accomplishing SDGs in higher education during and after the pandemic, as demonstrated by the findings of (Crawford and Cifuentes-Faura 2022).

Deng et al. (2022) investigated the impact and strength of the COVID-19 event, which was found to be negatively associated with perceived external employability and, as a result, lowered employee turnover intention. Furthermore, Deng et al. observed that organizational identification not only dampened the positive effect of perceived external employability on turnover intention, but also amplified the negative impact of perceived organizational growth on turnover intention. Moreover, Zhou et al. (2022) noted that employees' career commitment mediated its effect on their work engagement in times of COVID-19. These findings agreed with this research result when, in times of crisis, curriculum innovation and its positive implication on employability is a necessity, leading students to change in their specialization and career paths.

6. Conclusions and Limitations

Based on the findings, policymakers should enhance the educational sector to support e-learning, which improves the learning environment by encouraging innovation, creativity, and efficiency and adopting new changes to the curricula for universities. Government organizations should demand more innovation, particularly in developing nations where having the skills and capacities to spark original solutions is valuable human capital (Shahzad et al. 2016).

According to the findings of the current study, human resources are a crucial instrument for the education sector. As a result, the Ministry of Education should foster innovation across all ministries to promote HRM strategies.

The possibility of evaluating the feasibility of innovations developed by other institutions without necessarily needing to apply them themselves exists in varied higher education systems. By allowing low-risk testing, creative behaviour opens the potential of studying its impacts without requiring that all institutions adopt it at once. This fosters better levels of customer focus, social mobility, efficiency, flexibility, innovativeness, and stability (both with reference to student and labour market demands).

Universities must also equip and retrain professionals for a world where the necessity for regular labour is steadily declining as robots, AI and ICT gradually take the place of employees. The need for more complex design thinking is still growing. Therefore, universities may and should play a significant role in lifelong learning in addition to their initial teaching and training functions at the tertiary level, particularly in several disciplines that the current research highlighted as having a high rate of innovation and change.

The online teaching approach has taken the place of the conventional teaching method. Students have the option to discover another perspective thanks to online instruction. The new teaching technique comes with several problems. Education institutions are working to make up for the lost learning while looking for solutions to the problems brought on by the lockdown. Universities require resources to make up for the learning lost when they reopen. New policies should be developed to aid recent graduates in their entry into the labour market in order to prevent the lengthier unemployment duration.

The creator of innovative educational tools may choose to concentrate on customization of the needs of workforce to address the difficulty of accessibility for all students from different economic backgrounds. Considering the current situation, educational systems all over the world need to invest in the professional development of teachers, particularly in ICT and effective pedagogy. The COVID-19 pandemic has shown us that teachers and students/learners need to be trained in how to use a variety of online educational resources. Teachers and students should be encouraged to continue using such online tools to improve teaching and learning after the COVID-19 pandemic when regular classes begin.

The limitation of the research is that it was conducted within a particular services industry (the educational business in three universities) in one country, and the small number of corporations in this study are the study's major limitations. In view of the students enrolment in the same university and country, the results produced are limited, although useful. As a result, additional research might be conducted in the future to investigate the causes of this obvious difference in perception among students from other countries, as well as within occupational skills and requirements.

Future researchers might try to replicate the findings in different contexts during the COVID-19 crisis to conduct a comparative analysis and maximize the value of the research findings. They could also conduct longitudinal and cross-sectional data to test the impact of evolutionary perspectives on innovation performance on curriculum since the current study concentrated on a qualitative approach. The invention can be modified over time to consider fresh information contexts and test the relationship in unanticipated situations such as the current economic crisis in Lebanon.

Funding: This research received no external funding.

Institutional Review Board Statement: The study was conducted in accordance with the Declaration of Helsinki, and approved by the Institutional Review Board USJ-2020-206.

Informed Consent Statement: Informed consent was obtained from all subjects involved in the study.

Data Availability Statement: Data are available from the corresponding author upon request.

Conflicts of Interest: The author declares no conflict of interest.

References

Abernathy, William J., and James M. Utterback. 1978. Patterns of industrial innovation. *Technology Review* 80: 40–47.

Alsoud, Anas Ratib, and Ahmad Ali Harasis. 2021. The impact of COVID-19 pandemic on student's e-learning experience in Jordan. *Journal of Theoretical and Applied Electronic Commerce Research* 16: 1404–14. [CrossRef]

Aung, Than Nwe, and Soe Soe Khaing. 2016. Challenges of implementing e-learning in developing countries: A review. In *Genetic and Evolutionary Computing. Proceedings of the Ninth International Conference on Genetic and Evolutionary Computing, Yangon, Myanmar, 26–28 August 2015*. Berlin/Heidelberg: Springer International Publishing, Volume II 9, pp. 405–11.

Baruch, Yehuda. 2001. Employability: A substitute for loyalty? *Human Resource Development International* 4: 543–66. [CrossRef]

Basilaia, Giorgi, and David Kvavadze. 2020. Transition to online education in schools during a SARS-CoV-2 coronavirus (COVID-19) pandemic in Georgia. *Pedagogical Research* 5: em0060. [CrossRef]

Blom, Victoria, Anne Richter, Lennart Hallsten, and Pia Svedberg. 2018. The associations between job insecurity, depressive symptoms and burnout: The role of performance-based self-esteem. *Economic and Industrial Democracy* 39: 48–63. [CrossRef]

Bos-Nehles, Anna, Maarten Renkema, and Maike Janssen. 2017. HRM and innovative work behaviour: A systematic literature review. *Personnel Review* 46: 1228–53. [CrossRef]

Boustani, Nada Mallah, May Merhej Sayegh, and Zaher Boustany. 2022. Attitude towards Online Shopping during Pandemics: Do Gender, Social Factors and Platform Quality Matter? *Journal of Risk and Financial Management* 15: 474. [CrossRef]

Carnevale, Joel B., and Isabella Hatak. 2020. Employee adjustment and well-being in the era of COVID-19: Implications for human resource management. *Journal of Business Research* 116: 183–87. [CrossRef]

Crawford, Joseph, and Javier Cifuentes-Faura. 2022. Sustainability in higher education during the COVID-19 pandemic: A systematic review. *Sustainability* 14: 1879. [CrossRef]

Creswell, John. 2003. *Research Design: Qualitative, Quantitative, and Mixed Methods Approaches*. Los Angeles: Sage.

Crossan, Mary M., and Marina Apaydin. 2010. A multi-dimensional framework of organizational innovation: A systematic review of the literature. *Journal of Management Studies* 47: 1154–91. [CrossRef]

Deng, Hui, Wenbing Wu, Yihua Zhang, Xiaoyan Zhang, and Jing Ni. 2022. The Paradoxical Effects of COVID-19 Event Strength on Employee Turnover Intention. *International Journal of Environmental Research and Public Health* 19: 8434. [CrossRef]

Dospinescu, Octavian, and Nicoleta Dospinescu. 2020. Perception over e-learning tools in higher education: Comparative study Romania and Moldova. Paper presented at the IE 2020 International Conference, Madrid, Spain, July 20–23; pp. 59–64.

Doucet, Armand, Deborah Netolicky, Koen Timmers, and Francis Jim Tuscano. 2020. *Thinking about Pedagogy in an Unfolding Pandemic: An Independent Report on Approaches to Distance Learning during COVID19 School Closures*. Brussels: Education International.

Duraku, Zamira Hyseni, and Linda Hoxha. 2020. The impact of COVID-19 on higher education: A study of interaction among Kosovar students' mental health, attitudes toward online learning, study skills and changes in students' life. In *The Impact of the COVID-19 Pandemic on Education and Wellbeing*. Edited by Zamira Hyseni Duraku. Pristina: University of Prishtina "Hasan Prishtina" Faculty of Philosophy, Department of Psychology. ISBN 978-9951-00-293-6.

Ebersberger, Bernd, and Andreas Kuckertz. 2021. Hop to it! The impact of organization type on innovation response time to the COVID-19 crisis. *Journal of Business Research* 124: 126–35. [CrossRef]

Eisenhardt, Kathleen M. 1989. Building theories from case study research. *Academy of Management Review* 14: 532–50. [CrossRef]

Ettlie, John E., William P. Bridges, and Robert D. O'Keefe. 1983. Organizational strategy and structural differencesfor radical vs incremental innovation. *Management Science* 30: 682–95. [CrossRef]

Finsterwalder, Jörg, and Volker G. Kuppelwieser. 2020. Equilibrating resources and challenges during crises: A framework for service ecosystem well-being. *Journal of Service Management* 31: 1107–29. [CrossRef]

Fredriksson, Magnus, and Øyvind Ihlen. 2018. Introduction: Public Relations and Social Theory 1. In *Public Relations and Social Theory* London: Routledge, pp. 1–16.

García-González, Marialuz Arántzazu, Fermín Torrano, and Guillermo García-González. 2020. Analysis of stress factors for female professors at online universities. *International Journal of Environmental Research and Public Health* 17: 2958. [CrossRef]

Harvey, Lee, Sue Moon, Vicki Geall, and Ray Bower. 1997. *Graduates' Work: Organisational Change and Students' Attributes*. Birmingham: Centre for Research into Quality.

Henderson, Rebecca M., and Kim B. Clark. 1990. Architectural innovation: The reconfiguration of existing product technologies and the failure of established firms. *Administrative Science Quarterly* 35: 9–30. [CrossRef]

Knight, Peter T., and Mantz Yorke. 2002. Employability through the curriculum. *Tertiary Education and Management* 8: 261–76. [CrossRef]

Kohntopp, Thomas, and Jack McCann. 2020. Leadership in virtual organizations: Influence on workplace engagement. In *The Palgrave Handbook of Workplace Well-Being*. Berlin/Heidelberg: Springer, pp. 1–26.

Little, Brenda. 2001. Reading between the lines of graduate employment. *Quality in Higher Education* 7: 121–29. [CrossRef]

Mallah Boustani, Nada, and May Merhej Sayegh. 2021. E-learning: Factors affecting students online learning during COVID-19 quarantine in a developing country. In *Information and Knowledge Systems. Digital Technologies, Artificial Intelligence and Decision Making: 5th International Conference, ICIKS 2021, Virtual Event, 22–23 June 2021*. Cham: Springer International Publishing, pp. 17–28. [CrossRef]

Morley, Louise. 2001. Producing new workers: Quality, equality and employability in higher education. *Quality in Higher Education* 7: 131–38. [CrossRef]

Moustakas, Louis, and Denise Robrade. 2022. The challenges and realities of e-learning during COVID-19: The case of university sport and physical education. *Challenges* 13: 9. [CrossRef]

Nabi, Ghulam R., and David Bagley. 1998. Graduates' perceptions of transferable personal skills and future career preparation in the UK. *Career Development International* 3: 31–39. [CrossRef]

National Committee of Inquiry into Higher Education (Great Britain). 1997. *Higher Education in the Learning Society: Report of the National Committee*. Wales: National Committee of Inquiry into Higher Education.
Onyema, Edeh Michael, Nwafor Chika Eucheria, Faith Ayobamidele Obafemi, Shuvro Sen, Fyneface Grace Atonye, Aabha Sharma, and Alhuseen Omar Alsayed. 2020. Impact of Coronavirus pandemic on education. *Journal of Education and Practice* 11: 108–21.
Pacheco, Tyler, Simon Coulombe, Christine Khalil, Sophie Meunier, Marina Doucerain, Emilie Auger, and Emily Cox. 2020. Job security and the promotion of workers' wellbeing in the midst of the COVID-19 pandemic: A study with Canadian workers one to two weeks after the initiation of social distancing measures. *International Journal of Wellbeing* 10: 58–76. [CrossRef]
Piccoli, Beatrice, Antonino Callea, Flavio Urbini, Antonio Chirumbolo, Emanuela Ingusci, and Hans De Witte. 2017. Job insecurity and performance: The mediating role of organizational identification. *Personnel Review* 46: 1508–22. [CrossRef]
Piopiunik, Marc, Guido Schwerdt, Lisa Simon, and Ludger Woessmann. 2020. Skills, signals, and employability: An experimental investigation. *European Economic Review* 123: 103374. [CrossRef]
Rogers, Everett. M. 2003. *Diffusion of Innovations*, 5th ed. New York: Free Press.
Romelaer, Pierre. 2005. L'entretien de recherche. In *Management des Ressources Humaines: Méthodes de Recherche en Sciences Humaines et Sociales*. Edited by P. Roussel and F. Wacheux. Bruxelles: De Boeck, Chapter 4. pp. 101–34.
Salas-Pilco, Sdenka Zobeida, Yuqin Yang, and Zhe Zhang. 2022. Student engagement in online learning in Latin American higher education during the COVID-19 pandemic: A systematic review. *British Journal of Educational Technology* 53: 593–619. [CrossRef]
Sanchez, Daniel Garrote, Nicolas Gomez Parra, Caglar Ozden, and Bob Rijkers. 2020. *Which Jobs Are Most Vulnerable to COVID-19? What an Analysis of the European Union Reveals. What an Analysis of the European Union Reveals*. World Bank Research and Policy Briefs. Washington: World Bank Group.
Schumpeter, Joseph A. 2017. *The Theory of Economic Development: An Inquiry into Profits, Capita I, Credit, Interest, and the Business Cycle*. London: Routledge.
Shahzad, Khuram, Sami Ullah Bajwa, Ronda B. Ansted, and Dawood Mamoon. 2016. Evaluating human resource management capacity for effective implementation of advanced metering infrastructure by electricity distribution companies in Pakistan. *Utilities Policy* 41: 107–17. [CrossRef]
Sintema, Edgar John. 2020. Effect of COVID-19 on the performance of grade 12 students: Implications for STEM education. *Eurasia Journal of Mathematics, Science and Technology Education* 16: em1851. [CrossRef]
Subedi, Suraksha, Suvash Nayaju, Sweta Subedi, Sanjeev Kumar Shah, and Jennifer Mathias Shah. 2020. Impact of E-learning during COVID-19 pandemic among nursing students and teachers of Nepal. *International Journal of Science and Healthcare Research* 5: 68–76.
Suneson, Grant. 2020. Industries hit hardest by coronavirus in the US include retail, transportation, and travel. *USA Today*. March 20. Available online: https://eu.usatoday.com/story/money/2020/03/20/us-industries-being-devastated-by-the-coronavirus-travel-hotels-food/111431804/ (accessed on 20 September 2022).
Tushman, Michael L., and Philip Anderson. 2018. Technological discontinuities and organizational environments. In *Organizational Innovation*. London: Routledge, pp. 345–72.
Tuzovic, Sven, and Sertan Kabadayi. 2021. The influence of social distancing on employee well-being: A conceptual framework and research agenda. *Journal of Service Management* 32: 145–60. [CrossRef]
UNESCO. 2020. *Global Education Meeting Extraordinary Session on Education Post-COVID-19 Background Document*. Paris: UNESCO.
United Nations. 2020. *Policy Brief: Education during COVID-19 and beyond*. New York: United Nations. Available online: https://www.un.org/development/desa/dspd/wp-content/uploads/sites/22/2020/08/sg_policy_brief_covid-19_and_education_august_2020.pdf (accessed on 20 September 2022).
Waheed, Abdul, Xiaoming Miao, Salma Waheed, Naveed Ahmad, and Abdul Majeed. 2019. How new HRM practices, organizational innovation, and innovative climate affect the innovation performance in the IT industry: A moderated-mediation analysis. *Sustainability* 11: 621. [CrossRef]
Wang, Wen, Kim Mather, and Roger Seifert. 2020. Job insecurity, employee anxiety, and commitment: The moderating role of collective trust in management. In *Multilevel Trust in Organizations*. London: Routledge, pp. 84–101.
Wiradendi Wolor, Christian, Solikhah Solikhah, Nadya Fadillah Fidhyallah, and Deniar Puji Lestari. 2020. Effectiveness of e-training, e-leadership, and work life balance on employee performance during COVID-19. *Journal of Asian Finance, Economics and Business* 7: 443–50. [CrossRef]
Zhou, Wenxia, Zhen Pan, Qiuping Jin, and Yue Feng. 2022. Impact of self-perceived employability on sustainable career development in times of COVID-19: Two mediating paths. *Sustainability* 14: 3753. [CrossRef]

administrative sciences

Article

An Enneagram Approach to Strategy

Issam Tlemsani [1], Mohamed Ashmel Mohamed Hashim [2], Robin Matthews [3], Vera Ndrecaj [2,*] and Rachel Mason-Jones [2]

1 The Centre of International Business, Gateshead NE8 1AN, UK; i.tlemsani@tcib.org.uk
2 Cardiff School of Management, Cardiff Metropolitan University, Cardiff CF5 2YB, UK; mmohamedhashim@cardiffmet.ac.uk (M.A.M.H.); rkmason-jones@cardiffmet.ac.uk (R.M.-J.)
3 Department of Strategy, Marketing and Entrepreneurship, Kingston University London, Kingston upon Thames KT2 7LB, UK; robindcmatthews@gmail.com
* Correspondence: vndrecaj@cardiffmet.ac.uk

Abstract: This conceptual paper examines how Enneagram as a personality typing system could be adopted as a meta-model across multiple domains. The enneagram strategy is a structural typology that many organisations and individuals use to effectively understand their business profiles and interpersonal patterns, despite scholars showing concerns about its robustness and dynamism. This paper extends the argument as to how the mechanics of the Enneagram are intrinsically interrelated with (a) organisational systems and networks, (b) organisational grammar, (c) supply chains, and (d) creativity. It also emphasizes the importance/close association of the Enneagram tool with interdependence, business model, networks, statics, and system dynamics. The paper uses a combination of research methods (theoretical, exploratory, and descriptive) to evaluate the adaptability of the enneagram. The findings demonstrate that enneagram as a strategy could be applied to various business phenomena but requires more cross-domain empirical research. The enneagram can also be used to develop meta-models/toy models that can relate to the organisational outcomes by integrating tangible and intangible process assets. This paper reliably lays a set of principles/foundations to launch the enneagram approach to a complex, broad, systematic, and creative scope.

Keywords: enneagram; creativity; meta-model; networks; systems; grammar; organisation

Citation: Tlemsani, Issam, Mohamed Ashmel Mohamed Hashim, Robin Matthews, Vera Ndrecaj, and Rachel Mason-Jones. 2023. An Enneagram Approach to Strategy. *Administrative Sciences* 13: 119. https://doi.org/10.3390/admsci13050119

Received: 1 March 2023
Revised: 4 April 2023
Accepted: 6 April 2023
Published: 28 April 2023

1. Introduction

The Enneagram is a personality topology tool or system that formulates nine distinct but closely interconnected personality types. It fosters the idea that people have one powerful or dominant personality type which guides and influences them to interact with the external world, respond to conflict situations, and manage stress. The diagram below (Figure 1) describes the typical enneagram systems made up of nine different types with their own pattern of behaviour.

We present the enneagram of strategy as a meta-model. Then, we undertake some fine graining which we hope will reveal the components of the meta-model bearing in mind two things; first, that the meta-version is more than the sum of its components and second, the enneagram is a mandala (Cusack 2020; Kam and Fluit 2021).

The Enneagram methodology is intended to point to a route towards creative imagination. First, there is the situation now (the system state). The business issue or problem is situated in a particular setting or environment of internal and external dynamics. Since the problem has unique as well as general characteristics, an element of creativity is required. Often managers are required to reconcile opposites; growth versus short-term returns; exploitation of existing assets (to recover sunk costs) and exploration (investment in the discovery of new markets, products, and ways of doing things); operating effectively and innovating; implementing and adapting. Such activities contradict present dilemmas and paradoxes. They require creativity as well as analysis. They require a balance between all the Jungian functions (Kam 2018; Akhvlediani et al. 2020).

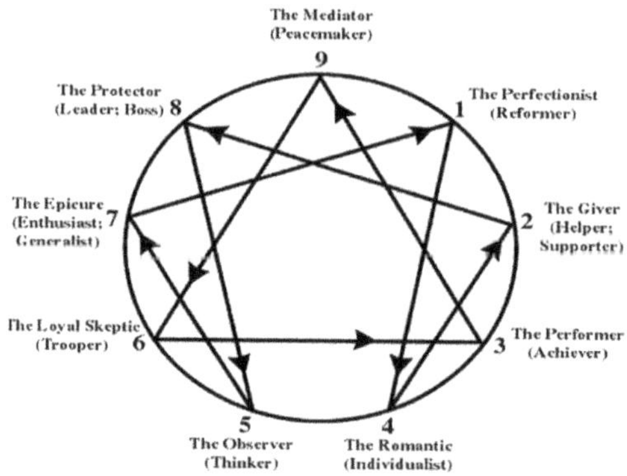

Figure 1. The standard Enneagram. Based on (Navabifar et al. 2020; Dooley and Van de Ven 1999).

First, we must remember that the process is continuous. Things never turn out as expected. The internal and external dynamics change. Therefore, there is always a need to adapt. Second, organisations (or individuals) do not pursue single strategies: at any one time, many strategies must be coordinated. This leads to a third observation (Matthews 1996): Strategy and decisions do not take place at a single point in an organisation. They are distributed, by many decisions and decision-makers (Navabifar et al. 2020; Schwarz and Zarrabi 2017; Dooley and Van de Ven 1999). The research questions are: (1) How can the enneagram approach be adopted systematically to strategic management disciplines? (2) How can its close association be sized using realistic measures?

The Enneagram strategy developed by the authors is an application of the Enneagram to management decision-making. The Enneagram has a long history. In the 20th century, behavioural psychologists adopted the Enneagram as a mainstream tool to investigate personality typologies and personal growth therapy. Many accounts of the personality Enneagram exist in clinical studies, specialist literature, and clinical practice. This research uses the Enneagram as an organisational mechanism to shape organisational conditioning in a systematic approach to reconciling organisational priorities, which are generally complex, specific, and time-bound.

We claim that the mechanics of the nine-pointed Enneagram could be used to discover solutions to many distinct organisational problems as long as the organisation knows how to use both (a) analytical and (b) creative skills through a standard routine. We extend the utility of the Enneagram methodology by integrating dynamic variables such as interdependence, business models, networks (statics and dynamics), and organisational grammar.

2. Literature Review

The personality topology system has undergone significant development over decades by various scholars, strategists, and practitioners. The core theoretical foundation is progressively developed based on the combination of psychological, spiritual, sociological, and philosophical perspectives. It also draws on various schools of thought and traditions such as Jungian psychology and the Sufi tradition. The strategic enneagram approach describes nine distinct but closely associated personality types. Each type has its own set of traits, including motivation, fear, and behavioural pattern. Thus, the enneagram is often used as a tool for building self-awareness, personality development, personal growth and understanding, and building relationships (Cusack 2020; Kam and Fluit 2021).

The existing research gap in the literature is the effectiveness of the enneagram that cannot be validated without sufficient empirical evidence across multiple domains. Thus, there is a paucity of literature and much scepticism (Riso and Hudson 1999; Rohr and Ebert 2001). The paucity of cross-cultural research can lead to issues with its contextual application, generalization, and customization. Unless the application of the enneagram is extended across hybrid systems or domains, it may lead to inconsistent awareness, education, and training of scholars and practitioners. In conclusion, the enneagram strategy can lead to various incompatibilities with the Enneagram's perceived validity, its cross-cultural applicability, and training unless supported by conceptual and implementable models (Kam 2018, 2022). This paper contributes by filling the existing gaps considerably, exhibiting a unique approach to the enneagram application.

The literature validates that there is no unifying theory for the application of the Enneagram. However, various individuals and groups have progressively developed closely associated theories to the enneagram (i.e., the wisdom of enneagram which examines psychological and spiritual growth for the nine personality types); Riso and Hudson (1999), and Rohr and Ebert (2001) discussed the introduction of the enneagram in Egypt by the Desert Fathers and its revival by a Franciscan in the 14th century.

Currently, the application and development of enneagram ideologies are related to three distinct aspects of enneagram theory, namely: personality development, its structure, and emphasis on personality growth (Hook et al. 2020). In this paper, we focus on how the enneagram theory and its structure can be applied to complex phenomena in hybrid domains, including but not limited to networks, systems, creative imagination, and supply chains. We also highlight areas for potential development.

2.1. *The Enneagram and Organisational Networks*

In this section, (i) we distinguish statics and dynamics; statics refers to the state of a system (organisation or firm) at the moment in time, while dynamics describe how the system (organisation or firm) behaves over time; (ii) we describe the role of organisational grammar in a network; and (iii) we present the supply chain as an archetypal network. We distinguish between the system states of an organisation and its trajectory over time. Clarifying the distinction before we get into more technical details with an example is a good idea. Here, we illustrate system states with concrete examples. They are actual situations. Hence, there is a need to preserve anonymity (Kartikeyan 2020; Blose et al. 2023).

The state of A's company (the system state) is that they produce a variety of security services (some high-value services, corporate and personal protection, and some lower value-added, such as alarms and security guards). The system was initially static at the beginning of our executive programme, but the problem A faced was to reorganize and restructure so that; (a) he could hand over day-to-day operational problems to other managers and (b) devote more time to longer-term strategic issues (the trajectory over time).

Another executive, B, manages part of a holding company that produces and sells low-value medical supplies. The state is this: Customers require quick responses to their orders. Production takes time, so holding stocks of goods (working capital) is necessary. There are many competitors. However, corporate headquarters, as is often the case with holding companies, are unwilling to tie up much cash in working capital. In this case, corporate debts must be refinanced at a relatively high-interest cost. Thus, there is a dilemma; orders are delayed, and customers are lost. Therefore, B cannot meet sales targets. However, if he holds higher stocks, debt levels cannot be reduced. He is pessimistic about future system states.

The Enneagram and Organisational Dynamics

As we noted, these are temporary states, subject to change because of inner and outer dynamics. Grammar is also subject to change; if payoffs are unsatisfactory for one reason or another, this will bring about change.

As shown in Figure 2, the state of an organisation in the present time is represented as an intersection of inner and outer dynamics with payoffs, all expressed in the context of grammar. Inner dynamics (ID) are an organisation's assets, capabilities, and competencies.

Outer dynamics (OD) include forces of competition and cooperation, and interacting macro forces; economic, environmental, governmental, legal, technological, and so on (Tlemsani 2010, 2020; Tlemsani and Matthews 2010; Tlemsani et al. 2022).

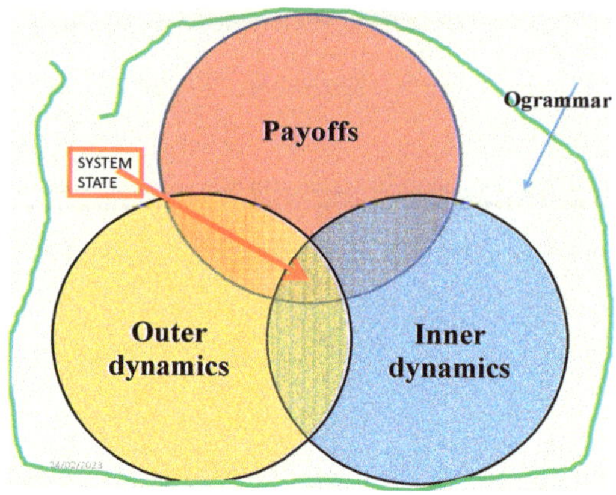

Figure 2. The meta-model and the system state. Based on (Tlemsani et al. 2022).

The system state is not an equilibrium state. It shows where an organisation happens to be at a moment in time. It is an intersection of dynamics, payoffs, and grammar. Note that the system state is not a point but an area. We should think of the system state indicated in the diagram as a vector that includes the relevant aspects of inner and outer dynamics, grammar, and payoffs that exist at a point in time.

2.2. The Enneagram as Meta-Model and Organisational Grammar

The enneagram model (and its sub-model, the meta-model) typifies the first type but adds another dimension. The purpose is to tune into the creative imagination of individuals or groups. In that sense, the Enneagram model is a Mandala. Mandalas are used in some Buddhist traditions as objects of contemplation. The nine-point enneagram (Figure 3c) Mandalas are symbols used in Hinduism and Buddhism to focus attention, an aspect of mindfulness, and develop creative or active imagination. Geometrically, enneagrams are a class of nine-point figures.

The strategic enneagram referred to here is made up of an inner triangle (Figure 3a) and an irregular hexagon, a six-point figure (Figure 3b). It originates in Sufi psychological and mystical teaching, the Pythagorean number system, and traditional religions. In the 20th century, the enneagram was developed by Gurdjieff, Ouspensky, and John Bennett. The Strategic Enneagram is symmetric. It is based on recurring decimals; 1/3 (0.333 ...) 2/3 (0.666 ...) and 3/3 (0.999 ...) expressed in the inner triangle, representing the meta-model and hexagram based on 1/7 (0.1428571 ...), 2/7 (0.285714 ...), and so on.

As a mandala, the Strategic Enneagram, illustrated in Figure 3, is used to focus attention, linking it to mindfulness and developing creative or active imagination. Mindfulness has been recently imported into management thinking from Buddhism as a technique for reflection and managing stress.

The Enneagram methodology is a framework for analysing strategic problems and designing creative strategies, combining intellect (analysis and logic) and imagination (creativity and intuition). Application of the Enneagram mandala to management to business and strategy reflects the proposition that spiritual and mystical aspects of life are not separate from material and practical aspects.

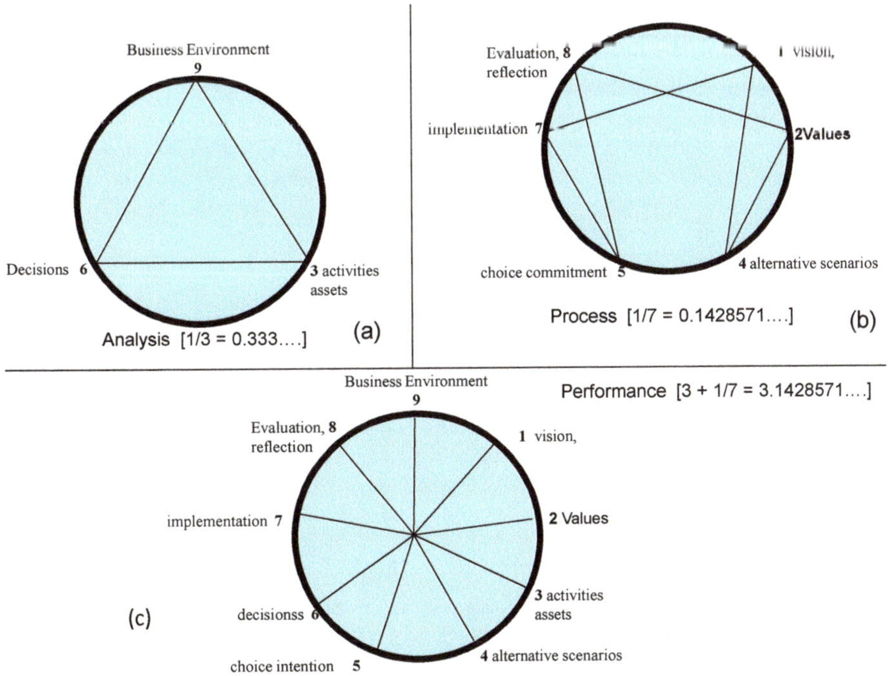

Figure 3. Enneagram as Meta-Model. Based on Kam and Fluit (2021) and Kam (2022).

Creativity probably cannot be taught. However, it can be encouraged. The Strategic Enneagram as a mandala is a way of evoking creativity in individuals and groups. It incorporates strategy, mindfulness, and creativity (Kam and Fluit 2021; Kam 2022). The phrase strategic process is a series of states of a system over time. Thus, we begin by considering the process which we think of as being to some extent deliberate and second the state itself, represented by the inner triangle in Figure 3a. In Figure 3, both the process and system state are embedded in grammar. The process is traced out by the hexagram 1, 4, 2, 8, 5, … It has a cognitive aspect (1, 4, 2), an implementation aspect (8, 5, 7), and, since the situation is dynamic and adaptive, a learning process (7, 1) relating what is implemented to what is intended and what values were intended and what was achieved (2, 8). The enneagram is symmetric around risk which arises when the purely cognitive is implemented. Purely cognitively, anything is possible.

Networks and Grammar

Grammar determines (a) the nodes (the aspects of the world we choose to focus on) and (b) connections (how the nodes are linked). Nodes correspond to the parts of speech (nouns, verbs, adjectives, prepositions and so on) in ordinary grammar. Linkages correspond to the syntax (grammatical rules such as declension and conjugation) that connects parts of speech. Figure 3 can be seen as a picture of five different grammars; a and f are connected networks, b, c, and e are only partly connected, and c is a star-shape network emanating from the central node A. Linkages can be thought of in the static sense as synergies or in the dynamic sense of feedbacks (Feltsan 2019). Incorporated below are some of the characteristics of grammar (Akande 2009; Kim and Kim 2010).

- If we think of strategy as the interaction of interdependent networks (inner and outer dynamics and payoffs), Grammar determines the nature of the interactions; which parts (or nodes) are linked and how they are linked.

- Grammar is a complex form of conditioning. It includes rules, laws, regulations, cultures, ways of thinking about problems, and so on. It includes formal and informal routines (R), the architectures that bind routines together (A), influences of national and corporate culture (C), and mindsets (MS); the acronym MARCS is a useful way of describing the influence of grammar.
- Grammar is a complex adaptive system (CAS); its elements (nodes) interact with one another, conflicting with, reinforcing, or dampening one another whilst still retaining an internal cohesion.

2.3. Interdependence and Enneagram Methodology

The Enneagram methodology, which is outlined here, spans both requirements of management in a new era. Most of the analytic techniques taught in business schools can be shown to be sub-models of the Enneagram methodology, which is a general model. In addition, the Enneagram methodology provides a technique for encouraging creative imagination. To be credible to businesspeople, we need to focus on practical aspects. Can the methodology be used to increase sales/profits? Can it expand the capabilities of the individual and team? Can it help managers achieve the task as defined above? We maintain that the answer is yes and attempt to show why this is so in this paper. The previous section outlines some theories. The remainder of the paper focuses on practical issues (Cusack 2020; Petsche 2016). According to (Edwards 1992; Moore 1992; Schwarz and Zarrabi 2017) the rationale for the Enneagram methodology is as follows.

1. Thinking, feeling, analysing, and responding are conditioned or programmed.
2. Conditioning or programming is achieved as a result of a mechanism that we call organisational grammar (grammar).
3. Organisational grammar has positive and negative aspects: both are necessary.
4. It is functional (positive), enabling us to make sense of the world. It introduces a degree of stability and predictability into the world.
5. However, it is a form of conditioning/programming (negative). It limits creative imagination.
6. There are many alternative organisational grammars, i.e., new ways of thinking, feeling, analysing, and responding.
7. Christian mystics, the Kabala, Buddhists, and Sufis have made the point about conditioning in a variety of ways for generations.
8. Some computer scientists make the same point in a different way: that the possibility exists in the near future of creating spiritual machines capable of creative imagination.
9. Some scientists see the transition to alternative organisation grammars as happening around 2040/2050. They speak of the Singularity.

2.4. The Enneagram as a Network—Statics and Dynamics

The distinction between statics and dynamics is convenient because it enables us to distinguish where an organisation is now (the system state) and where it may be in the future (its path or trajectory over time). Although the distinction between statics and dynamics is useful, it is artificial. The present moment is never static: it moves continually into the future. A business is subject to dynamic (constantly changing) pressures at any time.

Nodes in Figure 4 could be interpreted in many ways; as elements of outer dynamics (A = political, B = economic, C = technological ... D = creativity and E = innovations). Alternatively, they could refer to inner dynamics (marketing, sales, operations, logistics and so on, in the value chain); or they may represent the tangible or intangible assets of a firm; or they might be interpreted as payoffs to different stakeholder groups (A, B, C ... stand for profits, returns to shareholders, creditors, employees, the community, as taxes, or to customers as quality products or services, the environment); or they may be different aspects of organisational grammar.

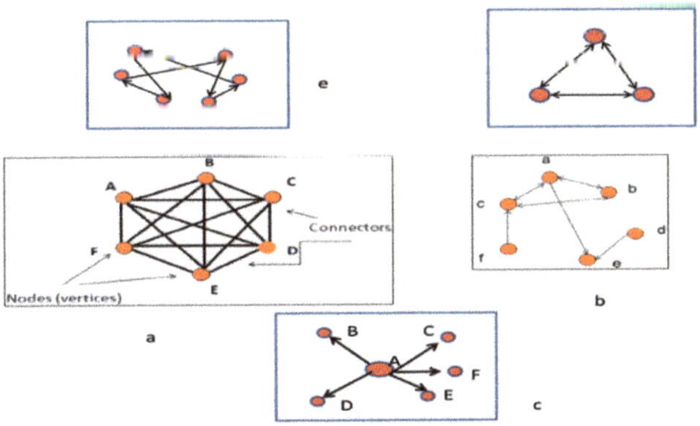

Figure 4. The variety of networks. Generally, networks consist of nodes (vertices) and connections (edges), The (**a–e**) demonstrate the unique connection and complexities in networks. Based on (Gulati et al. 2011).

The Enneagram is a connected network.

- In Figure 1 the system state corresponds to the numbers 3, 6, and 9. The trajectory corresponds to the other numbers 1, 2, 4, 5, 7, and 8.
- Note an important aspect of the trajectory (process) Enneagram: It is a connected network that is ordered 142 (cognitive), 857 (action), 142, 857, and so on since decisions or strategies and continuous processes through time.
- Mathematically, the Enneagram is based on the idea of sevenths (1/7, 2/7, 3/7, and so on). Translating sevenths into fractions, 1/7 becomes 0.142857142 … an infinite series. The fraction 2/7 is 0.285714285 … , and so on. This mathematical aspect of the Enneagram is why it operates like a mandala. In some traditions, a mandala approaches creativity (Khavul et al. 2010).

System States and Trajectory; The Enneagram Mandala in Time

The system state describes where an organisation is now. It is like a snapshot of a moment in time. A complete description of a system state would specify all the inner and outer dynamics, the payoffs generated to stakeholders and the state of organisational grammar. The trajectory is the path over time as inner and outer dynamics, grammar, and payoffs change. The trajectory describes the succession of system states that occur over time. If St[0] describes the system state now t(0), the succession of system states is St[0], St[1], St[2], … , St[n].

Economists like to talk about equilibrium. However, it is mistaken to think that a system state is an equilibrium. There are too many variables and too many decision-makers to consider, the elements of inner and outer dynamics, payoffs, and so on. The system state is simply the state an organisation happens to be in now.

We could think of equilibrium in connection with the system state as a Nash equilibrium; Nash equilibrium is a situation in which no one has the incentive to change their strategy so long as no one else does. Equilibrium as a trajectory could be thought of as an evolutionarily stable strategy; a set of agreed strategies that could not be upset or disturbed if a small number of decision-makers diverted from the agreed strategy.

2.5. The Enneagram and Supply Chain Network

The nodes are suppliers (raw materials and energy, equipment, labour, management and so on), the firm's value chain in question (showing how it adds value), distributors, retailers, and final customers. As demonstrated in Figure 5, connections summarize transforming inputs from suppliers into outputs distributed to final customers, distributors, and retailers (Elsaleiby 2019).

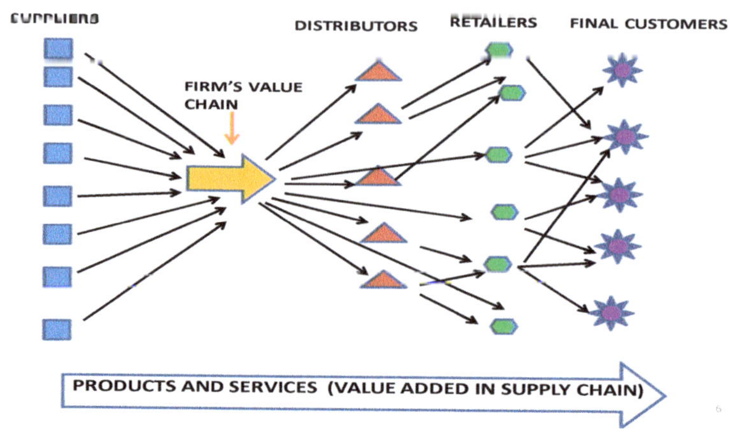

Figure 5. Businesses as networks. Based on (Elsaleiby 2019).

In I's company, the state is such that outer dynamics in the company are seasonal: meaning that there is an uneven pattern of revenues over the year. In most companies inner and outer dynamics as well as organisational grammar are problems, meaning that payoffs (profits and margins) are being squeezed. The recession (outer dynamic) is a problem, and such problems are likely to become more and more intense. As the world macro situation worsens, competition will intensify; nations will try to export their way out of recession. Of course, every country cannot do so (exports = imports in total). Often regulation brings new problems. Beer companies, for example, have regulations to limit the consumption of alcohol. Then, there is the recession (outer dynamic) and grammar in the form of low management skills of retailers/partners; all these things squeeze sales and profits (and other payoffs). Consider the firm as a transformer of inputs into outputs via the supply chain. If the arrows pointed left to right instead, this would indicate cash flows from final customers to profit margins and costs (rents, wages, and capital costs) in the supply chain. Alternatively, the left–right arrows could indicate financial accounting relationships in the supply chain. Figure 5 presents the connectors as two-way messages: in one direction, supplies are transformed into final outputs of consumer goods and services via distributors and retailers; in the other direction, customer cash is absorbed into profit margins and costs of production at various stages in the supply chain (Figure 6).

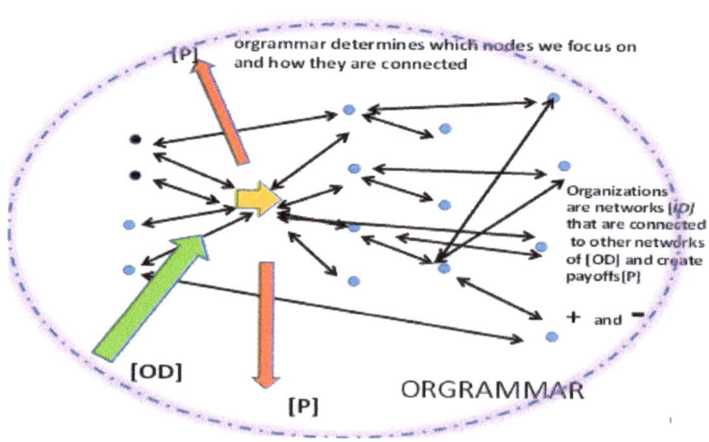

Figure 6. The Meta-Model as a network. Based on (Kam and Fluit 2021; Kam 2022; Schwarz and Zarrabi 2017).

The problem of capitalism has never really been one of supply. It is intrinsically dynamic. The problem has always been having demand keep pace with supply. The answer is sometimes yes and sometimes no. When it does, we have economic growth. When it does not, we have stagnation and recession. We trace the roots of the current Great Recession fundamentally to problems of deficient demand. Marketers understand the importance of demand. That is why such a high proportion of a firm's expenditures are on marketing and promotion (and built-in obsolescence so that products wear out and must be renewed). The great economist Keynes recognised the primary importance of demand (he called it Effective Demand). He recognised that deficiency of demand brought instability and required that to void recessions and depressions governments would have to fill the gap left by insufficient investment demand by corporations and consumer demand by households by creating demand through government expenditures. However, Keynes has been forgotten. We could argue that the Keynesian revolution never happened. Sooner or later, governments will have to recognise that government expenditure (government demand) is the precondition for the survival of their economic and social systems (Figure 7).

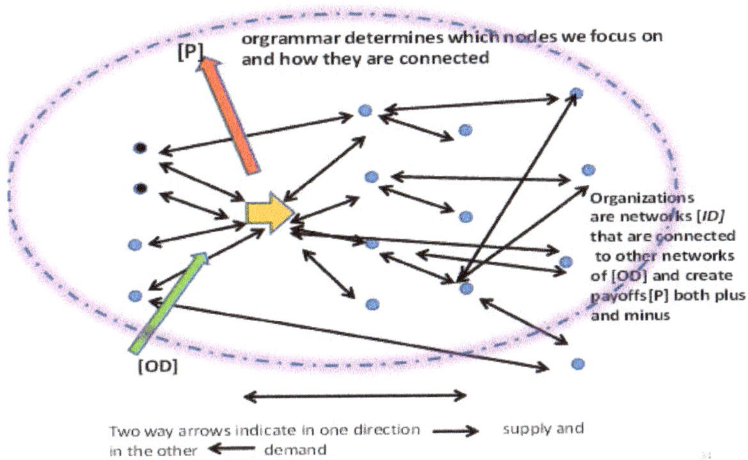

Figure 7. Bringing demand into the picture. Based on (Tlemsani et al. 2022; Tlemsani and Matthews 2010).

2.6. Meta-Model

The meta-model is a way of describing the current state of an organisation. The current state never lasts for long. It is always subject to change. Hence the following categories are identified in the meta-model shown in Figure 8.

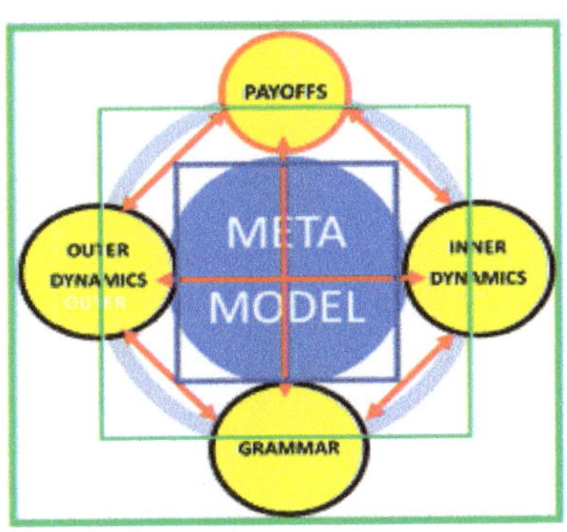

Figure 8. Meta-Model of Enneagram. Based on (Tlemsani et al. 2022; Tlemsani 2020).

An organisation cannot happen without formal and informal rules, including laws, traditions, regulations, systems and structures, cultural and historical influences (including religions), and the mindsets and ways of thinking and doing common to individuals and groups in an era, nation, region, society, and family. The examples in the previous sentence are organizing principles. In computer programming, they are described as Standard Operating Procedures. They make organisations work. All these organizing principles are represented by artefacts; architecture, the layout of cities, works of art, the creative arts (music, literature, cinema, theatre, sculpture, new media), consumer goods and services, and the technology of an era; generally, what we see and experience around us that changes like all things over time. We refer to these organizing principles as organisational grammar, grammar, for short. Grammar summarizes the core organisational principles that underlie outer and inner dynamics and payoffs.

Outer dynamics are outside forces; for example, competition, new and often disruptive technologies, political, economic, ecological, etc. Organisations live in a global capitalist environment (sometimes described as neo-liberalism), which evolves and changes. Grammar, and all the informal and formal rules, including culture, referred to in the previous paragraph, is to some extent homogeneous (global similarity), spread by information networks, media, advertising, the internet, and social networks created by science, technology, and the arts.

Inner dynamics include tangible and intangible assets and the organisation's dynamic capabilities (competencies). Tangible assets include physical capital, human beings, natural resources, access to finance (debt and equity), access to information and data, and tacit and explicit knowledge contained in an organisation. Intangible assets include brand, reputation, and corporate image. Included in intangible assets are the elements of corporate culture; mindsets, ways of thinking and doing by individuals and groups, and their assumptions and traditions (Mohamed Hashim et al. 2022a).

Payoffs to stakeholders include financial returns, measured by profit, EBIT, CAGR, sales, market share, and many financial ratios. Capitalism focuses on returns to shareholders; market capitalization, returns to equity, and debt liabilities, for example, and teachers often parrot the importance of shareholder value. The enneagram approach to strategy is viewed as a meta-model that provides a home and a place for these toy models. The meta-model is a sub-model of the full Enneagram model. It focuses on elements of the system state of an organisation, inner and outer dynamics, payoffs, and grammar. Figure 8 illustrates the basic elements of a system state as a network of relationships between inner (ID) and outer dynamics (OD), payoffs (P), and organisational grammar (G), the meta-model.

The term *meta* is used because (as we see below) many economic and management models relate to one or other of the categories of the meta-model. This picture is rather misleading in that it treats the four elements of the meta-model separately. No diagram can give a complete picture, but the purpose of Figure 9 is to relate the meta-model to some of the standard economic/strategic models (Koutsopoulos et al. 2020; Boritz and White 2016).

Figure 9 is an archetypal picture of the meta-model of a company. Some companies' suppliers are equipment manufacturers. In others, for example, consultancy companies, the key suppliers are people with the right skills, and the company sells directly to the final customer: a Business-to-Business relationship. In security companies, the relationship is both Business-to-Business and Business-to-Customer. Some companies are retailers, distributors, and sellers to final customers.

Figure 9 must be adapted to suit the situation. The company and its value chain are at the centre of the picture. The main activity in the value chain may be bottling and distribution. In a retail company, the main activities are sales and marketing and, most importantly, working capital management. In a furniture design company, the main activities may be designing tailor-made products and layouts. Very often, add-on services offer the highest margins, e.g., retailers who offer an extended warranty, phones, and IT companies that offer apps (Mohamed Hashim et al. 2022b).

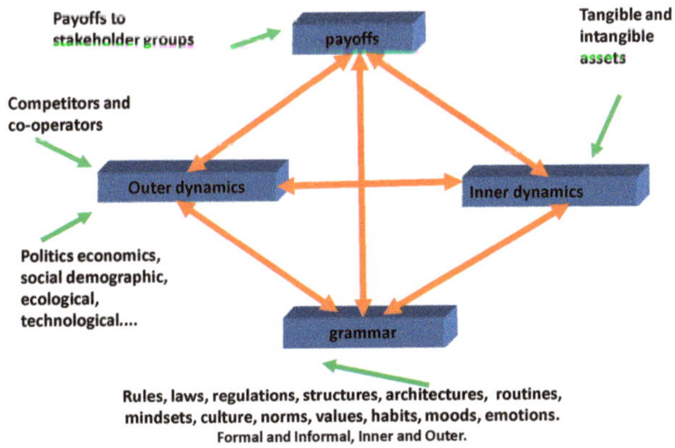

Figure 9. Meta-Model (simplified version). Source: (Koutsopoulos et al. 2020; Boritz and White 2016).

3. Methodology

This paper uses a combination of theoretical, exploratory, and descriptive research methods to evaluate the enneagram's adaptability. The methodology adopted two distinct parts. The theoretical part is signified by providing a critical review of the literature and the meta-models. The theoretical review synthesizes the adoption of an enneagram in the organisation using analytic and creative imagination. It stems from the idea that the solution to business problems requires creativity as well as analysis. Thus, the methodology deployed in this paper is an ontological analysis. We captured, described, and examined different meta-models, approaches, real-world examples, and creative imaginations using Enneagram. Therefore, the Methodology Section exemplifies how Enneagram as a mechanism can help various organisational reconciliations.

Let us denote the real world of everyday experience as [R] and the spiritual world as [Ω]. [R] and [Ω] are distinguished by having different organisational grammars (grammars). The purpose of grammar in [R] is (a) to enable us to make sense of the (real) world and (b) to introduce some order and stability into it. Other spheres of Being, which we summarize as [Ω], have different grammars.

- One proposition of our analysis is that creativity in [R] involves being able to see the world through a different grammar. Put another way, spiritual techniques in Buddhism, Zen, Sufism, and so forth enable the individual to access (perhaps briefly) an alternative grammar.
- A second proposition is that issues in business, such as innovating and developing new products, markets, and technologies, are not so different from the issues facing the painter or the sculptor in that they involve creating something new.

Thus, by applying a well-aligned methodology (theoretical, exploratory, and descriptive), we developed, estimated, and regularized a cross-sectional enneagram approach to strategy. It is important to set out the methodology as a process over time, as described in Table 1.

The need for urgent creative analysis and the importance of setting out a dependable methodology (captured in Table 1) can be expressed in a mandala, such as the diagram below (Figure 10). The mandala captures and summarises the holistic approach of the Enneagram methodology. The enneagram methodology applied to reconcile organisational activities is shown in Table 1.

Figure 10 represents the nine pointers of reconciliations adopted by organisations to shape and react themselves in different conditions. The Enneagram methodology is explained further in the next section. We note now that it draws on many disciplines and intellectual concepts, such as networks from physics as well as disciples such as Buddhism, Zen, and Sufism which have much in common and are concerned practically with developing the creative capabilities that most people possess (Moore 1992; Kern et al. 2011). Thus, we attempted to explore how the Enneagram methodology as an organisational tool could be used to achieve organisational conditioning/determines priorities.

Table 1. The methodology as a process. Source. Based on (Navabifar et al. 2020; Schwarz and Zarrabi 2017; Dooley and Van de Ven 1999).

Intent	Intent describes the projection present in mind for new goals and new possibilities.
Deconstruct the problem	Be specific about the issue: what stories/experiences illustrate it? Discover other approaches.
Energy	Explicitly recognizing the contradiction, ambiguities, and dilemmas in a problem can create an energy surplus to generate insights.
The way forward	Find solutions and implement them, recognizing that continuous adaptation is necessary as new issues emerge.
Decision	The transition from thought to action: Onset of risk and uncertainty.
Implementation	Making things happen. The difficult part is realising that you are part of the problem.
Adaptation	Decisions and strategies are continuing processes. Adaptation is always necessary.

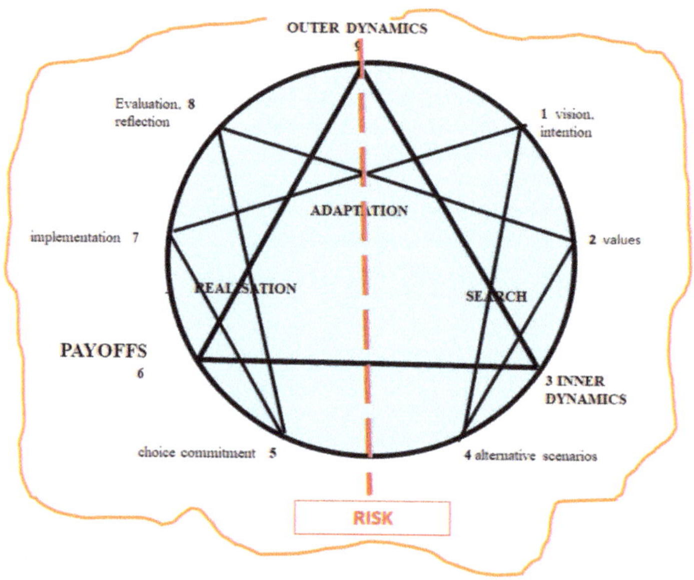

Figure 10. The enneagram methodology applies to reconcile organisational activities. Based on (Navabifar et al. 2020; Schwarz and Zarrabi 2017; Dooley and Van de Ven 1999).

4. Results

The enneagram approach to strategy is widely used in the personal and spiritual growth, personality development, and psychological development domains (Hook et al. 2020; Riso and Hudson 1999). That means the enneagram is viewed as a powerful tool for self-discovery, a personality model that describes nine distinct types of people, each with its own traits, motivations, and behaviours. However, its applicability beyond these domains continues to be a grey or underdeveloped area.

By examining the results derived from the existing literature, we have assessed how an enneagram approach can be used in business strategy, organisational decision processes, and interventions. We relate the application specifically to (i) organisational network, (ii) organisational dynamics, (iii) organisational grammar, and (iv) network and grammar. Our evidence derived across various domain indicate the enneagram could be applied/adopted as a robust tool in organisational interventions at various levels and in multiple dimensions, i.e., at an individual level, organisation level, and society level.

The results of interdependency and the enneagram methodology indicate that the combination can be applied to trigger creative imagination in business management. Thus, the application of the enneagram can be widened to static and dynamic system states. Further, this paper also explored the effectiveness and outcomes of the enneagram-based interventions in supply chain and organisation grammar. It showed a road map of how the application of the enneagram can be transformed into meta-models. Thus, these results justify the effectiveness of enneagram-based intervention, which can be useful in improving payoffs, enhancing communication, examining the strengths and weaknesses of organisational networks, and gaining productivity.

Another way the Enneagram approach can be applied to strategy is by identifying potential conflicts and finding ways to address them. Each Enneagram type has its own set of strengths and weaknesses and understanding these can help teams/organisations work more effectively together.

The Enneagram approach to strategy has several key findings that can help organisations, individuals, and teams improve their strategic planning and decision-making processes. Some of these findings include the following:

1. Each Enneagram type has unique strengths and weaknesses that can be leveraged in the strategic planning process. By understanding these traits, individuals and teams can optimize their performance and make more effective decisions.
2. Leveraging Enneagram types of problem-solving and decision-making in different ways. For example, some types may be more analytical and detail-oriented, while others may be more creative and innovative. Understanding these differences can help teams make more informed decisions considering different perspectives.
3. Improved leadership: Leaders who understand the Enneagram approach can more effectively motivate and manage their teams. By understanding the strengths and weaknesses of each Enneagram type, leaders can assign tasks and responsibilities that play to each team member's strengths.

Overall, the Enneagram approach to strategy can provide individuals and teams with valuable insights and tools for improving their strategic planning and decision-making processes. By leveraging each team member's unique strengths and perspectives and fostering a culture of self-awareness and growth, teams can create a more successful and fulfilling work environment, which can contribute to a more successful and effective strategy. The theoretical implication of the enneagram approach is that it confronts the traditional personality view that visualises the individual as fixed and static. The enneagram sees that personality development is progressive and dynamic, and it can be shaped effectively by fostering self-awareness and experiences. It also stresses the importance of identifying and addressing the motivation and fears that may steer individuals. Another implication of the enneagram for practical use is that it can be utilized as an effective tool for personal development, growth, and fostering relationships. When the topological tool is applied, individuals can gain deeper insights about themselves and others, and improve their compatibility in building relationships.

5. Discussion

This section develops a constructive discussion derived from the existing literature, its philosophical application, and potential usage in the strategic management discipline. Further, it provides a set of principles via graphical representation to discuss the future state.

5.1. From the Meta Model to the Full Enneagram Model

Figure 11 transforms a system state at any time *t* into three possible trajectories over time: three possible alternative future scenarios. The diagram is a simplification. First, there are many possible scenarios; many of them cannot even be anticipated. The future is a source of risk and uncertainty. Second, as noted below in Figure 11, the system state should be represented by a vector in the space of many dimensions rather than a point in three-dimensional space. However, the diagram does provide a bridge between system states and trajectories over time (Desmarais et al. 2020; Mohamed Hashim et al. 2021; Hellany 2006).

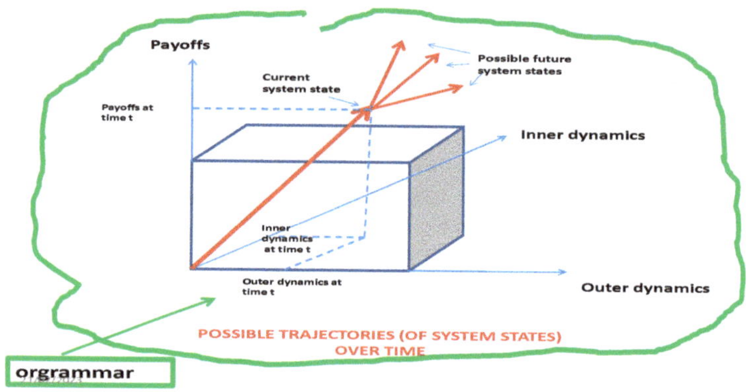

Figure 11. The Meta-Model and the Future. Based on: Desmarais et al. (2020).

5.2. The Enneagram Mandala and Management Strategy

Creative imagination (new ways of thinking, analysing, feeling, and responding to situations) requires stepping outside conditioned patterns and entering an alternative grammar. Hence, both Figure 12 and Table 2 are set in the context of organisational grammar.

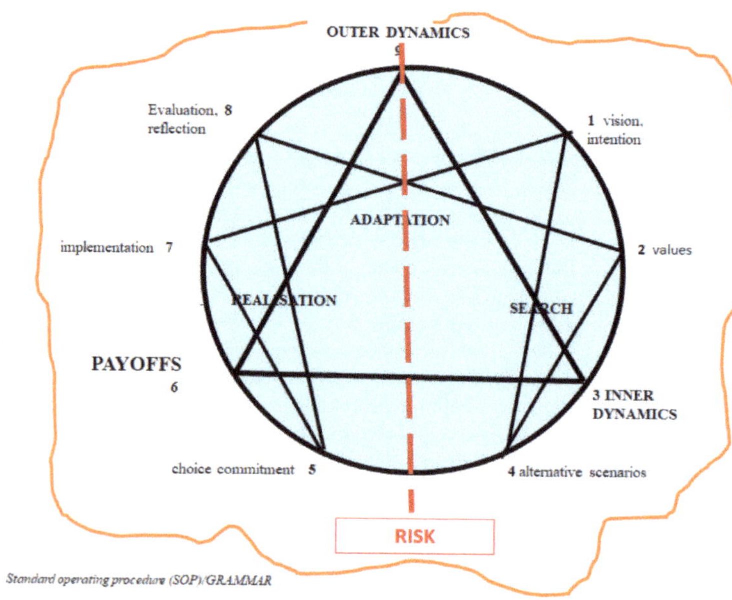

Figure 12. The Enneagram of Strategy. Based on (Navabifar et al. 2020; Schwarz and Zarrabi 2017; Dooley and Van de Ven 1999).

Table 2. A linear version of the Enneagram. Based on Authors Proposal (2023).

9	The meta-model: see Figure 11
1	Intention: What you intend to achieve for yourself and for your company.
2	Energy: Generate energy. That is the purpose of the Enneagram mandala.
3	Business problem: Which problems/issues you intend to focus on.
4	Stories: Illustrate the problems/issues with concrete examples or stories. Try to see them from many points of view. This is deconstruction.
5	Commitment: Decisions are about being committed to achieving something for yourself and for your company.
6	Creative imagination: *New ways of thinking, analysing, feeling, and responding to situations.*
7	You are part of the problem: The most difficult step. Reflection on your own management is required. Management is looking *after the resources (assets) that they are entrusted with.*
8	Trajectory. What to do about inner dynamics and organisational grammar? Implementing negotiation and adaptation as dynamics of the system state change (summarized in the meta-model).

Thus far, we have made several propositions. Now we add two more. The first is that people normally do not make decisions in the way we like to imagine. The second is that many who are called leaders do not lead; they are dependent upon the society that gives them power rather than true leaders or creators.

The apparent contradictions, i.e., decisions that are not really decisions and leaders who do not lead, hinge on grammar. Grammar is conditioning or programming. It enables us to make sense of things (according to the grammar we adopt) and introduces stability and organisation. Here, we focus on what we called the real world [R] in Section 3. Many decisions and strategies are conducted in a way that is completely programmed by conventional grammar. Sometimes this is good enough, but when managers are faced with problems that are extreme, unique, or require creativity, as often happens, then it is not enough. They must create a different grammar.

Similarly, people in a position of leadership are often conformists with respect to conventional grammar. In being conformists to conventional grammar, they are inauthentic. Such people, CEOs, political leaders, and so on, think they are leading and think their role is critical when, in fact, they are merely being swept along by events. Tolstoy portrays Napoleon like this. Napoleon thinks he is controlling events, but he is being swept along by many seemingly trivial circumstances, the actions and moods of individual soldiers, small accidents, the weather, randomness, or luck. Leaders have a habit of attributing their successes to themselves and failures to others or to bad luck. The components of Figures 11 and 12 are explained in a linear version in Tables 2 and 3.

Table 3. The Enneagram as a personal mandala. Based on Authors Proposal (2023).

9	The Enneagram of strategy: See Figure 12.
3	Business problem: Which problems/issues you intend to focus on.
6	Creative imagination: *New ways of thinking, analysing, feeling, and responding to situations.*
1	Intention: What you intend to achieve for yourself and for your company.
4	Stories: Illustrate the problems/issues with concrete examples or stories. Try to see them from many points of view.
2	Energy: Generate energy. That is the purpose of the Enneagram mandala.
8	Trajectory. What to do about inner dynamics and organisational grammar? Implementing negotiation and adaptation as inner and outer dynamics (and outer grammar) states change (summarized in the meta-model).

Table 3. Cont.

6	Commitment: Decisions are about being committed to achieving something for yourself and for your company.
7	You are part of the problem: The most difficult step. Reflection on your own management is required. Management is *looking after the resources (assets) that they are entrusted with.*

6. Conclusions

This paper extends the meta-model to the Enneagram model and introduces changes in system states over time and the role of decision-making. The Enneagram is a mandala that encourages imagination and creativity as well as analytic thinking. Early sections repeat some things from an earlier paper which dealt mostly with the situation now when the underlying dynamics are imagined to be frozen. They never are frozen, but the simplification is useful. The distinction between a moment in time and transitions over time is convenient but artificial. We must always think of time as an interval. There is no timeless moment, only intervals, long or brief.

Previous research findings have indicated that in a few cases it was not possible to generate results which support the nine factors or points of the enneagram (Hook et al. 2020). There is a paucity of the literature in terms of identifying a basis to cluster the nine standard types of enneagrams. Further, how these nine types are applied to various domains and their secondary facets (the logical connection between the wings and movements) continue to be a grey/underdeveloped area.

Strategic management appreciates the growing popularity of the enneagram specifically in the niche disciplines. We emphasized this phenomenon by integrating/scientifically linking the enneagram typology with the organisational network, grammar, supply chain, and creativity. It establishes the vital connection of enneagram to closely associated but extended/hybrid business disciplines. We attempted to provide a balanced scientific view interconnected with distinct enneagram practices; thus, it offers an opportunity to learn from a new hybrid disciplinary tradition. The applicability of the enneagram is limitedly established with rigorous research in the organisational literature; still, the enneagram is under-researched, neglected, and active mostly in unindexed journals. This reveals its limited reputation among scholars and practitioners. As stated, accumulated conceptual and scientific evidence indicates that the reliable application of enneagram can be extended to various organisational disciplines such as networks, grammar, supply chain, and creativity. The application of an enneagram is relatively influenced by contextual setting, or the pattern of connections.

In a certain discipline (i.e., supply chain) the methodology to assess the pattern of connection/the architecture required to conclusively examine the accumulated impact requires a new/innovative approach. Recent research used cluster analysis to determine the pattern underlying the nine interconnected archetypes. Rigorous, diverse, and particularly mixed method research is required to (a) shape the operationalization of the enneagram and (b) align enneagram theory with other disciplines, endorsing its extendibility to hybrid disciplines. We conclude that the Enneagram approach to strategy may serve as a practical and effective tool that can be utilized to shape the development of a new body of knowledge in strategy, organisational grammar and networks, organisational systems, and creative disciplines.

Author Contributions: Conceptualization, M.A.M.H., I.T. and R.M.; methodology, V.N.; software, M.A.M.H. and I.T.; validation, R.M.-J., V.N. and R.M.; formal analysis, M.A.M.H. and I.T.; investigation, M.A.M.H., I.T., R.M.-J., R.M. and V.N.; data curation M.A.M.H., I.T., R.M.-J., R.M. and V.N.; writing—original draft preparation, M.A.M.H., I.T. and R.M.; writing—review and editing, M.A.M.H., I.T., R.M.-J. and V.N.; visualization, M.A.M.H. and I.T. All authors have read and agreed to the published version of the manuscript.

Funding: This research received no external funding.

Institutional Review Board Statement: Not applicable.

Informed Consent Statement: Not applicable.

Data Availability Statement: Not applicable.

Conflicts of Interest: The authors declare no conflict of interest.

References

Akande, Adebowale. 2009. Comparing Social Behaviour across Culture and Nations: The 'What' and 'Why' Questions. *Social Indicators Research* 92: 591–608. [CrossRef]

Akhvlediani, Maia, Zurab Mushkudiani, and Sophio Nikabadze. 2020. Whether we need emotions in business or not. Access to science, business, innovation in digital economy. *ACCESS Press* 1: 71–78. [CrossRef]

Blose, Taylor M., A. Corey Yeates, Mousumi Som, A. Murray Kelly, Matt Vassar, and Jeffrey Stroup. 2023. The Enneagram and its application in medical education. *Proceedings—Baylor University Medical Center* 36: 54–58. [CrossRef]

Boritz, J. Efrim, and Giselle White. 2016. Business Models and Business Model Descriptions: A Research Synthesis. *SSRN Electronic Journal*. [CrossRef]

Cusack, Carole M. 2020. The Enneagram. *Aries* 20: 31–54. [CrossRef]

Desmarais, Michèle, Ondina Galiano, Valerie Gazemar, and Julie Fréchette. 2020. The enneagram model for nursing competencies development-An exploratory qualitative study. *Journal of Nursing Education and Practice* 10: 1. [CrossRef]

Dooley, Kevin J., and Andrew H. Van de Ven. 1999. Explaining Complex Organisational Dynamics. *Organisation Science, (3, Special Issue: Application of Complexity Theory to Organisation Science)* 10: 358–72.

Edwards, Anthony C. 1992. Competitiveness and apartheid in the new age: The Enneagram schools. *Religion Today* 7: 11–15. [CrossRef]

Elsaleiby, Aber. 2019. An integrated strategic-operational supply chain network design: A case of consumer goods supply chain. *International Journal of Supply Chain and Operations Resilience* 3: 275. [CrossRef]

Feltsan, Inna. 2019. Social Networks as a Tool for Development Foreing Grammar Competence of University Students. *Adult Education: Theory, Experience, Prospect* 2: 131–37.

Gulati, Ranjay, Dovev Lavie, and Ravindranath (Ravi) Madhavan. 2011. How do networks matter? The performance effects of inter-organisational networks. *Research in Organisational Behavior* 31: 207–24. [CrossRef]

Hellany, Ali. 2006. Enneagram: Exploring learning styles of engineering students. Paper present at 2006 IEEE GCC Conference (GCC), Manama, Bahrain, March 20–22; pp. 1–6. [CrossRef]

Hook, Joshua N., Todd W. Hall, Don E. Davis, Daryl R. Van Tongeren, and Mackenzie Conner. 2020. The Enneagram: A systematic review of the literature and directions for future research. *Journal of Clinical Psychology* 77: 865–83. [CrossRef]

Kam, Christopher. 2018. Integrating Divine Attachment Theory and the Enneagram to Help Clients of Abuse Heal in Their Images of Self, Others, and God. *Pastoral Psychology* 67: 341–56. [CrossRef]

Kam, Christopher, and Danielle Vriend Fluit. 2021. Combining the Enneagram and Narrative Therapy for Adult ego Development. *Current Psychology* 42: 406–15. [CrossRef]

Kam, Christopher. 2022. Enhancing Enneagram Therapy with Contemporary Research on the Conscious and Unconscious Mind. *Integrative Psychological & Behavioral Science, preprint*. [CrossRef]

Kartikeyan, V. 2020. Embrace the Organisation Unconscious: The Next Steps for Change Leaders and Organisation Development Practitioners. *NHRD Network Journal* 13: 340–51. [CrossRef]

Kern, Erib Kern, Kristi Perryman, and Anthony Suarez. 2011. Using the Enneagram to Facilitate Resolution of Supervisory Conflict. *Journal of Counselor Practice* 8: 136–54. [CrossRef]

Khavul, Susanna, Mark Peterson, Drake Mullens, and Abdul A. Rasheed. 2010. Going Global with Innovations from Emerging Economies: Investment in Customer Support Capabilities Pays Off. *Journal of International Marketing* 18: 22–42. [CrossRef]

Kim, Yungwook, and Soo-Yeon Kim. 2010. The Influence of Cultural Values on Perceptions of Corporate Social Responsibility: Application of Hofstede's Dimensions to Korean Public Relations Practitioners. *Journal of Business Ethics* 91: 485–500. [CrossRef]

Koutsopoulos, Georgios, Martin Henkel, and Janis Stirna. 2020. An analysis of capability meta-models for expressing dynamic business transformation. *Software and Systems Modeling* 20: 147–74. [CrossRef]

Matthews, Robert. 1996. *Corporate Structure as a Cooperative Game: Theory and Applications*. Kingston-upon-Thames: Kingston University.

Mohamed Hashim, Mohamed Ashmel, Issam Tlemsani, and Robin Duncan Matthews. 2022a. A Sustainable University: Digital Transformation and Beyond. *Education and Information Technologies* 27: 8961–96. [CrossRef]

Mohamed Hashim, Mohamed Ashmel, Issam Tlemsani, and Robin Matthews. 2021. Higher education strategy in digital transformation. *Education and Information Technologies* 27: 3171–95. [CrossRef] [PubMed]

Mohamed Hashim, Mohamed Ashmel, Issam Tlemsani, Robin Matthews, Rachel Mason-Jones, and Vera Ndrecaj. 2022b. Emergent Strategy in Higher Education: Postmodern Digital and the Future? *Administrative Sciences* 12: 196. [CrossRef]

Moore, James. 1992. New lamps for old: The Enneagram Débâcle. *Religion Today* 8: 8–12. [CrossRef]

Navabifar, Farnaz, Hamid Atashpour, and Mohsen Golparvar. 2020. The Effect of a Premarital Educational Program Based on 9- Type Personalities Types (Enneagram) on Emotional Expressiveness of Couples Preparing for Marriage. *Majallah-i ravānpizishkī va ravānshināsī-i bālīnī-i Īran* 26: 2–15. [CrossRef]

Petsche, Johanna. 2016. The Sacred Dance of the Enneagram: The History and Meanings Behind G. I. Gurdjieff's Enneagram Movements. *Fieldwork in Religion* 11: 53–75. [CrossRef]

Riso, Don Richard, and Russ Hudson. 1999. *The Wisdom of the Enneagram: The Complete Guide to Psychological and Spiritual Growth for the Nine Personality Types*. New York: Bantam Books.

Rohr, Richard, and Andreas Ebert. 2001. *The Enneagram: A Christian Perspective*. Chestnut Ridge: Crossroad Publishing.

Schwarz, Erik, and Shayesteh Zarrabi. 2017. DAS2-Theory of Personality: A Cognitive Approach to the Enneagram. *Psychology* 8: 1802–15. [CrossRef]

Tlemsani, Issam. 2010. Co-Evolution and Reconcilability of Islam and The West: The Context of Global Banking. *Journal of Education, Business and Society: Contemporary Middle Eastern Issues* 3: 262–76. Available online: http://www.emeraldinsight.com/10.1108/17537981011089569 (accessed on 18 December 2022). [CrossRef]

Tlemsani, Issam. 2020. Stock Returns Indicator: Case of Tadawul. *International Journal of Monetary Economics and Finance* 13: 1–15. [CrossRef]

Tlemsani, Issam, and Robin Matthews. 2010. The Financial Tower of Babel: Roots of Crisis. *International Journal of Islamic and Middle Eastern Finance & Management* 3: 334–50.

Tlemsani, Issam, Mohamed Ashmel Mohamed Hashim, and Robin Matthews. 2022. Games theory and strategic alliances: Applications to British Russian partnership. *Higher Education, Skills and Work-Based Learning* 12: 689–704. [CrossRef]

administrative sciences

Article

"Entrepreneurizing" College Programs to Increase Entrepreneurial Intentions: A Mediation Framework

Mahmoud Ibrahim Fallatah [1,]* and Tahar Lazhar Ayed [2,3]

1 Department of Business Administration, College of Business, Umm Al-Qura University, Makkah 24382, Saudi Arabia
2 School of Business and Law, Dar Al-Hekma University, Jeddah 22246, Saudi Arabia
3 School of Commerce, University of Sfax, Sfax 3025, Tunisia
* Correspondence: mifallatah@uqu.edu.sa

Abstract: The impact of entrepreneurship on the development of emerging economies is widely recognized. Research has focused on studying factors that increase entrepreneurship in societies, including the role of education in increasing entrepreneurial intentions among students. In this paper, we contribute to the entrepreneurship and education literature by examining the impact of entrepreneurial college programs on entrepreneurial intentions. Further, we study the mediating roles of perceived benefits and individual creativity. Using a sample of 438 students from a public university in Saudi Arabia, our findings reveal that students enrolled in entrepreneurial programs have higher levels of entrepreneurial intentions that those enrolled in non-entrepreneurial programs, and that perceived benefits and individual creativity partially mediate the aforementioned relationship. The paper opens the door for future research in the entrepreneurship and education literature and provides several managerial implications.

Keywords: entrepreneurship; entrepreneurial intentions; education; Saudi Arabia; individual creativity; perceived benefits

Citation: Fallatah, Mahmoud Ibrahim, and Tahar Lazhar Ayed. 2023. "Entrepreneurizing" College Programs to Increase Entrepreneurial Intentions: A Mediation Framework. *Administrative Sciences* 13: 50. https://doi.org/10.3390/admsci13020050

Received: 14 November 2022
Revised: 9 January 2023
Accepted: 15 January 2023
Published: 8 February 2023

1. Introduction

The impact of entrepreneurship on the development of emerging economies has been much discussed in the literature (Bruton et al. 2008, 2021). Specifically, research suggests that entrepreneurship has a positive impact on economic growth, employment, and productivity (Acs 2006; Audretsch et al. 2011; Lu et al. 2021). Thus, it is paramount for countries to create environments in which entrepreneurship is encouraged. One driver of entrepreneurship that has received much attention in the literature is education (e.g., Aronsson 2004; Hägg and Jones 2021; Honig 2004; Ndou 2021; Liñán et al. 2011; Potter 2008; Rauch and Hulsink 2015; Warhuus et al. 2021), which explains why countries around the world have invested heavily in entrepreneurship education, especially at the university level (Brush et al. 2003; Katz 2003; Lu et al. 2021; Ndou et al. 2018; Zhou and Xu 2012). Education has been related positively to entrepreneurial intentions (EI), defined as the intention to engage into an entrepreneurial activity in order to create a new business (Barba-Sánchez and Atienza-Sahuquillo 2018; Krueger et al. 2000; Lee et al. 2011; Liñán and Chen 2009; Liñán et al. 2011). Thus, for entrepreneurship researchers, it is essential to study entrepreneurial intentions since intention is largely recognized as the best predictor of behavior (Ajzen 1991).

Ample research has examined the relationship between education and the likelihood of an individual to become an entrepreneur (e.g., Amofah and Saladrigues 2022; Dickson et al. 2008; Elnadi and Gheith 2021; Lu et al. 2021; Van der Sluis et al. 2005), with varying results. On the one hand, for example, Acs and Armington (2005) detected a positive relationship between college education and the formation of new ventures. Similarly, Rauch and Hulsink (2015) determined education, particularly that related to entrepreneurship, to be influential in the intentions of students to be entrepreneurs. Recent studies also determined that entrepreneurship education increased entrepreneurial Intentions among business graduates (Anjum et al. 2022) and engineering students (Asimakopoulos et al. 2019). On the other hand, studies such as that of Van der Sluis et al. (2008) conclude that the relationships between education and EI is insignificant. Further, a meta-analytic study determined entrepreneurship education to have a negative impact on EI (Oosterbeek et al. 2010). These conflicting results suggest that context matters (Liñán et al. 2016; Walter and Block 2016). Specifically, there must be some contexts and factors where education encourages the formation of new ventures, and other contextual factors that have no impact on producing potential entrepreneurs. Among those factors are cognitive factors that research has determined to be significant in explaining behavioral decisions in the field of entrepreneurship (Baron 2004; Liñán and Chen 2009).

In this paper, our objective is to examine the impact of education on entrepreneurial intentions. Specifically, consistent with previous research (e.g., Foote and Hysa 2022; Boldureanu et al. 2020), we combine the Theory of Human Capital (Becker 1975) and the Entrepreneurial Self-Efficacy theory (Chen et al. 1998) with the Theory of Planned Behavior (TPB) (Ajzen 1991) and the theory of Entrepreneurship Event Model (EEM) (Shapero and Sokol 1982) to study the nature of educational programs (i.e., the degree to which they are entrepreneurial) and their effect on EI among college students in Saudi Arabia. Further, we build on the aforementioned theories to explore the mediating roles of individual creativity and perceived benefits.

Thus, our contribution in this paper is threefold. First, we contribute to the current debate on whether there is a relationship between education and entrepreneurial intentions. Some scholars have determined that education, in general, contributes to individuals' human capital (e.g., Ahn and Winters 2022; Parker and Praag 2006). Thus, educated individuals possess knowledge and skills that enable them to choose careers in a distinct way compared to the less educated ones, including the choice of being entrepreneurs (Lofstrom et al. 2014). Importantly, entrepreneurship education where individuals learn technical skills in areas such as strategic planning and developing business plans has been determined to be significantly related to entrepreneurial intentions (Martin et al. 2013; Rauch and Hulsink 2015; Ayed 2020). On the contrary, other studies determined education to be ineffective in predicting the likelihood of an individual becoming an entrepreneur (e.g., Dickson et al. 2008; Oosterbeek et al. 2010). In this paper, we extend the debate on the relationship between education and entrepreneurship by focusing not only on the degree to which students have been exposed to entrepreneurship education per se (e.g., Souitaris et al. 2007; Walter and Dohse 2012), but on the impact of the content and delivery of a collegiate program on the entrepreneurial intention of students.

Second, we contribute to the literature by building a model that links entrepreneurs' individual characteristics to their intention on engaging in entrepreneurial activities. The Theory of Planned Behavior (TPB) (Ajzen 1991) suggests that several personality traits act as motivational antecedents to entrepreneurial intentions, such as personal attitude, subjective norms, and perceived behavioral control. Further, the theory of Entrepreneurship Event Model (EEM) (Shapero and Sokol 1982) suggests that perceived desirability and perceived feasibility are also critical in prompting entrepreneurial behaviors (Dickel and Eckardt 2021; Krueger et al. 2000). Research has also defined other determinantal individual characteristics that positively impact entrepreneurial behaviors, such as alertness (Kirzner 1997), ego resilience (Block and Block 1980; Block and Kremen 1996; Chadwick and Raver 2020; Pérez-Nordtvedt and Fallatah 2022), sustainability traits (Joensuu-Salo et al. 2022), and spirituality (Pérez-Nordtvedt and Fallatah 2022). In this paper, our model seeks to examine the mediating roles of two relevant and important factors, individual creativity and individual's perceived benefits, in the relationship between Entrepreneurial Programs (EP), defined as graduate and undergraduate programs where technical knowledge and personal entrepreneurial skills are embedded in the program's courses and activities, and EI.

Third, we study the aforementioned relationship in the context of Saudi Arabia, an emerging yet wealthy economy. While several studies examined EI in Saudi Arabia (e.g., Aloulou 2015), few studies have examined the topic recently (e.g., Ayed 2020; Elnadi and Gheith 2021; Hoda and Fallatah 2022). Studying entrepreneurship in the context of Saudi Arabia at this time is vital since the country has set its Vision 2030, which put innovation and entrepreneurship at the forefront of its objectives (Fallatah 2021). Additionally, research has emphasized the role of culture in entrepreneurship (Liñán and Chen 2009), highlighting the fact that societies differ in their level of support and encouragement towards entrepreneurship (Busenitz and Lau 1996). Thus, our paper should shed some light on the role that education could play in helping the Kingdom achieve its entrepreneurial aspirations.

The rest of the paper is structured as follows. In the next section, we develop arguments on the relationship between education and EI. Then, we discuss the mediating roles of individual creativity and perceived benefits. The methodology section follows, where we describe our data collection process and analytical technique and present the results of our study. The paper concludes with the discussion and the conclusion sections.

2. Theoretical Background and Hypotheses Development

2.1. *Entrepreneurial Programs and Entrepreneurial Intention*

Several theories have been utilized to study EI; chief among them is the Theory of Planned Behavior (TPB) (Ajzen 1991), which emphasizes the role of intention as the main predictor of behavior. In the entrepreneurship literature, the theory explains that personal attitude (PA), subjective norms (SN), and perceived behavior control (PCB) are determinantal in influencing EI. Another major theory that has been employed to study EI is the Entrepreneurial Event Model (EEM) (Shapero and Sokol 1982), which indicates that desirability, feasibility, and propensity to act are key in regard to individuals' intentions to create a venture. Therefore, we expect that having the necessary knowledge and skills will increase an individual's capability and confidence to start their own business.

In regard to education and its impact on EI, scholars have relied on the Human Capital Theory (Becker 1975), which suggests that societies derive economic benefits by investing in people, particularly through education (Sweetland 1996). Chiefly, research asserts that education has a profound impact on the economic capability of individuals (Schultz 1971). Thus, research concludes that the knowledge and skills that individuals acquire through education and the various types of training are positively related to their intention to be entrepreneurs (Liñán 2004; Liñán and Chen 2009; Ndou 2021). Previous research also determined that education increased student awareness of entrepreneurship (Bae et al. 2014; Garavan and O'Cinneide 1994).

Additionally, scholars have utilized the Entrepreneurial Self-Efficacy theory (Chen et al. 1998) to explain EI. Entrepreneurial self-efficacy refers to an individual's belief in their ability to perform the entrepreneurship-related tasks effectively (Chen et al. 1998; McGee et al. 2009). Research argues that education should focus not only on providing knowledge, but also on entrepreneurial skills such as innovation, facing challenges and risk-taking, and more importantly on the belief system of potential entrepreneurs (Chen et al. 1998; Colombelli et al. 2022).

Thus, consistent with the self-efficacy theory, we argue that education increases the individual capability to perform entrepreneurial tasks. Specifically, education has an important role in developing technical, personal, and relational skills that are necessary for entrepreneurs to succeed (Baron 2006). While teaching students the technical aspects of entrepreneurship such as strategic planning and building business models is important (Rasmussen and Sørheim 2006), developing other necessary entrepreneurial skills that focus on the entrepreneur as an individual such as risk-taking and alertness is equally important. Those skills are necessary to increase student awareness of entrepreneurship (Chen et al. 1998), even among non-business students who are not necessarily exposed to technical knowledge about entrepreneurship (Asimakopoulos et al. 2019; Gilmartin et al. 2019; Vodă and Florea 2019) As put by education and curriculum scholars, entrepreneurial skills should be implemented in the social process of schooling (Gibb 2008; Giroux and Penna 1979). Education scholars emphasize that pedagogical models should be built upon a theoretical framework which situates schools within a societal context (Apple 1975). Thus, we believe that entrepreneurship, as a vital element of society, should be promoted heavily in the classroom, not only in business-related majors, but in all specialties.

Therefore, we argue that EPs are likely to produce potential entrepreneurs. As explained by the TBP (Ajzen 1991), we believe that the personal attitude and the perceived control behavior of students, as major determinants of EI, should also be evident in students who possess entrepreneurial skills (Liñán 2004; Liñán and Chen 2009). Additionally, as emphasized by EEM (Shapero and Sokol 1982), we argue that possessing such skills increases the individual desirability, feasibility, and propensity to act. In general, EP should develop a mindset among students and equip them with knowledge and skills that encourage students to think of entrepreneurship as a career option after their graduation (Colombelli et al. 2022). Put differently, we predict that students enrolled in such programs are more likely to have entrepreneurial intentions than their counterparts.

H1. *There is a positive relationship between enrolling in entrepreneurial college programs and entrepreneurial intentions among students.*

2.2. Perceived Benefits

As discussed above, enrolling in entrepreneurial programs should increase student awareness of entrepreneurship (Bae et al. 2014; Garavan and O'Cinneide 1994). Specifically, such programs should introduce students to the nuances of entrepreneurship and the required steps to becoming entrepreneurs. Students would be exposed to the benefits that entrepreneurship entails. Thus, we assert that being in an entrepreneurial program in college will provide students with more opportunities to appreciate the "perceived benefits" (PB) of entrepreneurship.

In turn, we argue that acknowledging the perceived benefits of entrepreneurship will lead students to form entrepreneurial intentions (Wu and Li 2011). More specifically, while enrolling in entrepreneurial programs might help increase entrepreneurial intentions among students, we believe that this will not materialize unless students believe in the benefits that might accrue to them from such an endeavor, whether they are economic (Parker 2008) or psychological benefits in the form of personal achievements (Delmar 2000). That is, students must perceive that being an entrepreneur is a favorable option compared to other options such as being employed by a public or a private organization. Thus, we hypothesize that perceived benefits will mediate the relationship between enrolling in entrepreneurial programs and having entrepreneurial intentions among students.

H2. *Perceived benefits mediate the relationship between entrepreneurial college programs and Entrepreneurial Intentions among students.*

2.3. Individual Creativity

Defined as the production of novel and useful ideas (Amabile 1996), creativity has been much discussed as a major component of entrepreneurship (Schumpeter 1934). Indeed, creative individuals are the ones who discover entrepreneurial opportunities and generate new ideas to exploit them (Baron 2006; Shane and Venkataraman 2000). While creativity is often thought of as a trait that individuals are born with, research asserts that creativity could be learned. To illustrate, Gundry et al. (2014) confirm that pedagogical approaches in education are very important in strengthening the students' ability to generate ideas. Thus, in entrepreneurial programs where brainstorming, problem-solving and role-playing along with other teaching methods that stimulate creative thinking are implemented, students are expected to develop several skills that encourage creativity (Osborn 1957; Ward 2004).

On the other hand, research suggests that creative individuals are more likely to discover opportunities and to exploit them (Gundry et al. 2014). Thus, since entrepreneurship is based on discovering new opportunities (Shane and Venkataraman 2000; Schumpeter 1934), and consistent with other scholars (e.g., Bellò et al. 2018), we argue that individual creativity will lead to higher levels of entrepreneurial intentions. Hence, we hypothesize that individual creativity (IC) will act as a mechanism through which entrepreneurial programs impact EI.

H3. *Individual creativity mediates the relationship between entrepreneurial college programs and Entrepreneurial Intentions among students.*

Figure 1 depicts our proposed model describing the impact of entrepreneurial programs on EI and the mediating roles of perceived benefits and individual creativity.

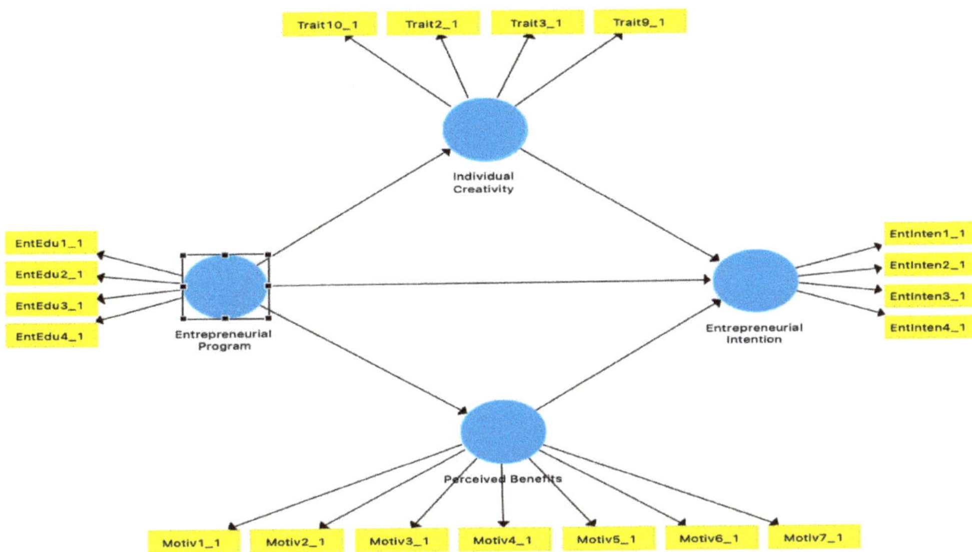

Figure 1. Research model and hypotheses.

3. Method

3.1. Sample

We followed a snowball sampling approach to collect data from graduate and undergraduate students in the college of business, where there is a mandatory course in entrepreneurship in one department, and the college of engineering, a college that has been determined to support entrepreneurship (Gilmartin et al. 2019), in a large public university in Saudi Arabia that has been emphasizing entrepreneurship in their recent strategic plan, with various degrees of response from its colleges. We collected data through surveys. The survey was translated from English to Arabic by a Saudi working professional fluent in both languages. Then, one of the authors fluent in both languages translated the survey back to ensure there were no changes in the meanings of the questions (Brislin 1986). Due to WhatsApp being the number one communication mode and because it is ubiquitously used for conducting business in Saudi Arabia (Saudi Arabia Social Media Statistics 2020), we sent a link of the online survey to students via email or a WhatsApp text. This is a common practice in research in the context of Saudi Arabia (e.g., Pérez-Nordtvedt and Fallatah 2022).

The survey was sent to 700 students from both colleges. The total number of surveys completed was 438, indicating a 62.57% response rate. Of our sampled students, approximately 52% were males, approximately 81% were undergraduate students, 82.4% majored in business, and 86.5% were under the age of 30. Additionally, 62.1% of our sample participants had working experience, 38.6% indicated that they have started or co-started a business in the past, and the majority stated that neither their parents (67.4%) nor any of their friends (58.4%) have started a business. Table 1 summarizes our sample.

Table 1. Description of the sample.

Profile	Measures	Frequency	Percentage
Distributed	-	700	-
Received	-	438	-
Valid	-	438	63.57
Gender	Male	229	52.3
	Female	209	47.7
Age	≤30	379	86.5
	>30	59	13.5
Education level	≤2 years	134	30.6
	>2 and ≤4 years	221	50.5
	>4	83	18.9
Field of study	Bus.	361	82.4
	N-Bus.	77	17.6
Parent who previously started a business	No	295	67.4
	Yes	143	32.6
Many friends have created their own business	No	256	58.4
	Yes	182	41.6
Worked for a small or new company	No	166	37.9
	Yes	272	62.1
Started a business	No	269	61.4
	Yes	169	38.6

3.2. *Measures*

3.2.1. Dependent Variable

Consistent with previous studies, we used the scale developed and validated by (Liñán and Chen 2009) to measure *EI*. Using a 5-level Likert scale, students were asked to evaluate the level of their agreement with several statements, such as "*my principal professional goal is to be an entrepreneur*" and "*I have very seriously thought about finding a firm*".

3.2.2. Independent Variables

To measure *Entrepreneurial programs*, we used a pre-validated scale adopted from Walter and Block (2016) to measure the degree to which the college programs were entrepreneurial. Using a 5-level Likert scale, students were asked to evaluate the level of their agreement with several statements such as "*my program helped me to understand the role of entrepreneurship in society*" and "*my program provided me with skills and competences that enable me to run a business*".

The measure used to assess *perceived benefits* was based on the scale developed by Amabile's Work Preference Inventory (WPI) (Amabile et al. 1994), which was later validated and adopted in several studies (e.g., Barba-Sánchez and Atienza-Sahuquillo 2012, 2018). The scale included items such as "*entrepreneurship will permit me to develop professionally and personally*" and "*entrepreneurship will permit me to be satisfied with my work*".

The measure of *individual creativity* was an adopted version of the scale developed by Bandera et al. (2018). Students were asked to evaluate the level of their agreement with the following two statements: "*I am creative when asked to work with limited resources*" and "*I often make novel connections and perceive new relationships between various pieces of information*".

3.2.3. Control Variables

As typical in EI studies with student samples, we controlled for *gender* (0 = female, 1 = male), *age*, and the *field of study* (0 = business, 1 = engineering), as well as *level* of students (0 = 2 years, 1 = between 2 and 4 years; 2 = 5 years and more). In addition, because prior experiences affect the desirability and the feasibility of starting a new venture

(Krueger 1993), we accounted for students' prior experience in entrepreneurship, asking if *students*, their *parents*, or any of their *friends* have ever started a business (0 = yes, 1 = no).

4. Analysis and Results

To test the hypotheses of our model, we used Structural Equation Modeling (SEM), a technique that is commonly used to estimate complex models with many constructs, indicator variables and structural paths without distributional assumptions imposed on the data (Hair et al. 2019). SEM is also a technique that has been widely used in studies related to EI (e.g., Farooq et al. 2018) and studies related to education and academic programs (e.g., Holtbrügge and Engelhard 2016). We used path analysis to test the direction and significance of the direct effect hypotheses.

4.1. PLS-SEM Algorithm: The Measurement Model Evaluation

To establish a valid and reliable measurement model, we followed the recommendation of Kline (2015) and used CR and AVE to test internal reliability and convergent validity, respectively (Bagozzi and Yi 1988; Hair et al. 2019; Nunkoo and Ramkissoon 2012; Nunkoo et al. 2013). The values of CR were all above 0.7, and all AVEs were above the cutoff point of 0.5 (Fornell and Larcker 1981) (Table 2).

Table 2. Outer Loadings, Composite Reliability, and Average Variance Extracted.

Variables and Items	OL	CR	AVE	Adj R^2
Entrepreneurial Program (EP)				
(1) My school education program helped me develop my sense of initiative—a sort of entrepreneurial attitude.	0.823	0.905	0.705	
(2) My school education program helped me to better understand the role of entrepreneurs in society.	0.810			
(3) My school education program made me interested to become an entrepreneur.	0.866			
(4) My school education program gave me skills and competences that enable me to run a business.	0.858			
Individual Creativity (IC)				
(1) I am creative when asked to work with limited resources.	0.591	0.841	0.573	3.9%
(2) I often make novel connections and perceive new relationships between various pieces of information.	0.732			
(3) I can produce a large number of ideas (fluidity).	0.844			
(4) I can produce new and unusual ideas (originality).	0.834			
Perceived benefits (PB): Entrepreneurship will permit me to:				
(1) Be the best at everything I do.	0.701	0.893	0.546	5.3%
(2) Develop professionally and personally.	0.786			
(3) Feel satisfied with my work.	0.706			
(4) Cover my personal needs.	0.743			
(5) Have good work relations.	0.801			
(6) Contribute to social well-being.	0.772			
(7) Gain social prestige.	0.650			
Entrepreneurial Intention (EI)				
(1) My principle professional goal is to be an entrepreneur.	0.700	0.875	0.638	31.4%
(2) I will make every effort to start and run my own enterprise.	0.873			
(3) I am determined to create a firm in the future.	0.755			
(4) I have very seriously thought of starting a firm.	0.853			

Additionally, to test the discriminant validity of our model, we followed the Fornell–Larker criterion (Fornell and Larcker 1981), which requires that the square root of the AVE of each construct must be greater than its highest correlation with the other constructs (Hair et al. 2017), a criterion that our findings fulfill. We also checked the VIF values and determined that all of them were lower than 5, which confirms the nonexistence of multicollinearity (Shirokova et al. 2016).

4.2. PLS-SEM Bootstrapping: The Structural Model Analysis

Performing the structural model analysis, results of the R^2 value show that our proposed model explains 31% of total variance in EI. Path coefficient values and t-values suggest that all relations in our model are significant and positive (p-value < 0.01). Particularly, Hypothesis 1, which predicted a positive relationship between EP and EI, is supported (β = 0.105, t = 2.663, p-value = 0.008). Figure 2 and Table 3 present the results of the path analysis. While all relationships were positively significant, our findings show that all independent variables contribute weakly to the explaining of their relative dependent variables as f^2 values were less than 0.15 (Cohen 1988).

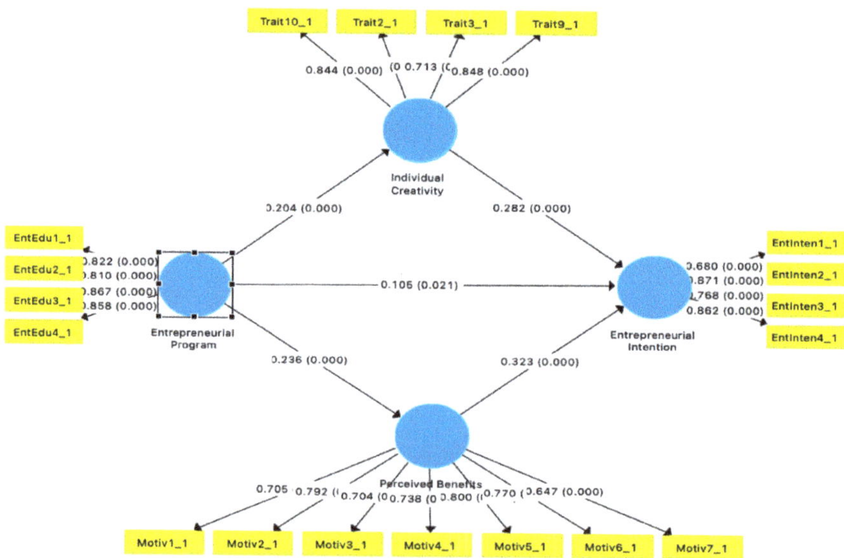

Figure 2. Measurement and structural model.

Table 3. Path coefficients.

Paths	Path Coefficients (β)	T Statistics (|O/STDEV|)	p-Values	f^2	VIF	Decision
EP -> EI	0.105	2.663	0.022	0.017	1.071	Supported
EP -> IC	0.204	4.764	0.000	0.045	1.000	Supported
EP -> PB	0.236	4.901	0.000	0.060	1.000	Supported
IC -> EI	0.282	5.923	0.000	0.081	1.393	Supported
PB -> EI	0.323	5.981	0.000	0.109	1.412	Supported

To test Hypothesis 2 and Hypothesis 3, which suggested mediating roles for IC and PB in the relationship between EP and EI, we followed the approach of Zhao et al. (2010). First, we could confirm that the mediation exists because all the indirect effects in the model were significant (Table 4). The findings show that EP has a significant impact on EI through PB (β = 0.077, t = 3.590; p-value = 0.000), which supports Hypothesis 2. In addition, the results show that EP has a significant impact on EI through IC (β = 0.058, t = 3.679;

p-value = 0.000), which supports Hypothesis 3. Then, we concluded that the relationship between EP and EI is partially mediated by IC and PB, as the direct effect between EP and EI was also significant (Preacher et al. 2007; Preacher and Hayes 2000, Ayed 2020)

Table 4. Mediation analysis.

Indirect Paths	Path Coefficients (β)	T Statistics	*p*-Values
EP -> PB -> EI	0.077	3.590	0.000
EP -> IC -> EI	0.058	3.679	0.000

5. Discussion

We argued that the more entrepreneurial the college program is, the more likely it is that the students have EI. Informed by the Human Capital Theory (Becker 1975) and the Entrepreneurial Self-Efficacy theory (Chen et al. 1998), we suggested that the knowledge and skills that students are exposed to in their formal education and the accompanying activities will positively impact their EI (Liñán 2004; Liñán and Chen 2009). Further, the study suggested mediating roles for PB and IC in the relationship between EP and EI.

Our findings supported the idea that collegiate programs that contain necessary entrepreneurial knowledge and skills in their contents and activities were more likely to produce students with EI. This is in line with previous research that detected a positive relationship between education and EI (Ahn and Winters 2022; Martin et al. 2013; Rauch and Hulsink 2015). Specifically, our findings confirmed the findings of recent research that detected a positive relationship between entrepreneurship education and EI in Saudi Arabia (Ayed 2020). On the other hand, while several research have determined insignificant (e.g., Van der Sluis et al. 2008) or negative (e.g., Oosterbeek et al. 2010) relationship between entrepreneurship education and EI, we believe that as universities are more aware of their role in the entrepreneurship ecosystem, college programs nowadays are more entrepreneurial in their curriculum and various activities. Thus, education had maybe not been impactful in the past, but our study, along with other recent ones, provides evidence that in cases where universities emphasize entrepreneurship, we are more likely to witness an increase in EI among students.

Further, our findings provided evidence that PB is a mediating mechanism through which EP impacts EI. More specifically, while enrolling in EP should increase EI among students directly, our findings illustrate that students tend to develop EI once they perceive the potential benefits of being entrepreneurs compared to other options (Delmar 2000; Parker 2008; Wu and Li 2011).

Likewise, our findings supported our hypothesis that EI is higher among students partially due to their individual creativity that they developed while in the program. Research has shown that creative individuals are more likely to discover entrepreneurial opportunities (Baron 2006; Shane and Venkataraman 2000; Schumpeter 1934). Our findings are also consistent with previous research that highlighted the role of education in increasing learners' creativity (Gundry et al. 2014).

5.1. Contributions, Limitations, and Future Research

Our study offers several contributions to the education and EI literature. First, it extends the conflicting debate over the role of education in promoting EI among students. Our study realizes that contextual factors matter in the Education–EI relationship. Thus, it suggests and determines that education per se might not be enough to increase EI, but that it is rather the entrepreneurial nature of the program that actually impacts EI among students. Second, our study develops and tests a model that links the entrepreneurs' individual characteristics to their intention in engaging in entrepreneurial activities. Our model examines the mediating roles of perceived benefits and individual creativity in the relationship between education and EI. As the debate continues over the role of education in EI, it is critical to explore the roles of other factors that might act as mechanisms through which the relationship functions. Third, acknowledging the importance of entrepreneurship in emerging countries (Schumpeter 1934), we contribute to the literature of entrepreneurship in these countries (Bruton et al. 2008; Tracey and Phillips 2011) by examining our model in Saudi Arabia, as the country takes on a major transformation plan that focuses on promoting entrepreneurship.

While our study provides several contributions to research, there are some limitations that should be noted. First, our data were collected from a single university, which might have an impact on the outcomes of our study, given that different universities have different cultures and priorities. Thus, future research could provide a more comprehensive study that includes different public and private universities in Saudi Arabia. Second, as typical with cross-sectional studies, our findings provide evidence of correlation between EP and EI, but we cannot confirm causation. Therefore, we recommend future studies to collect longitudinal data to offer a more accurate explanation of the relationship between EP and EI. In addition, as conflicting results continue to emerge in the relationship between EP and EI, interested scholars could find it appealing to examine the role of other contextual factors that affect the relationship.

5.2. Practical Implications

Along with our theoretical contributions, our study also provides several practical implications for policy-makers and university administrators. First, policy-makers in emerging economies can utilize education to promote entrepreneurship. In a country such as Saudi Arabia, where entrepreneurship is at the forefront of a national vision, policy-makers can contribute significantly to the vision by fostering entrepreneurship education in universities. Second, for university administrators, graduate and undergraduate programs should be designed to be more entrepreneurial by embedding nuances and entrepreneurial skill training in their course contents and extracurricular activities. While it is expected to include some elements of entrepreneurship education in business-related programs, our findings suggest that all non-business programs should also have activities that instill entrepreneurial skills in their students.

6. Conclusions

Entrepreneurship has been shown to have a huge role in emerging economies, and research has shown inconsistent results about the contribution of education to EI among students. Our study focuses on the help of the nature of educational programs in universities in increasing EI among students. We detect evidence that students enrolled in entrepreneurial programs are more likely to have EI than those in non-entrepreneurial programs. We also determine perceived benefits and individual creativity to have mediating roles between EP and EI. More studies are needed to further understand the ways in which universities can increase EI among its students.

Author Contributions: Conceptualization, M.I.F. and T.L.A.; methodology, T.L.A.; software, T.L.A.; validation, M.I.F. and T.L.A.; formal analysis, T.L.A.; investigation, T.L.A.; resources, M.I.F.; data curation, T.L.A.; writing—original draft, M.I.F. and T.L.A.; writing—review & editing, M.I.F.; visualization, T.L.A.; supervision, M.I.F. and T.L.A.; project administration, M.I.F.; funding acquisition, M.I.F. All authors have read and agreed to the published version of the manuscript.

Funding: This research was funded by the MBL Chair for Creativity and Entrepreneurship grant number [DSR-UQU-BLIE-010].

Institutional Review Board Statement: Not applicable.

Informed Consent Statement: Informed consent was obtained from all subjects involved in the study.

Data Availability Statement: The data presented in this study are available on request from the corresponding author. The data are not publicly available due to institutional restrictions.

Conflicts of Interest: The authors declare no conflict of interest.

References

Acs, Zoltan. 2006. How is entrepreneurship good for economic growth? *Innovations: Technology, Governance, Globalization* 1: 97–107. [CrossRef]

Acs, Zoltan J., and Catherine Armington. 2005. *Using Census BITS to Explore Entrepreneurship, Geography, and Economic Growth*. Washington, DC: SBA Office of Advocacy.

Ahn, Kunwon, and John V. Winters. 2022. Does education enhance entrepreneurship? *Small Business Economics*, 1–27. [CrossRef]

Ajzen, Icek. 1991. The theory of planned behavior. *Organizational Behavior and Human Decision Processes* 50: 179–211. [CrossRef]

Aloulou, Wassim. J. 2015. Entrepreneurial intention among freshmen students—Application of the theory of planned behaviour in Saudi context. *Journal of Enterprising Culture* 23: 473–500. [CrossRef]

Amabile, Teresa M. 1996. *Creativity in Context: Update to the Social Psychology of Creativity*. Boulder: Westview.

Amabile, Teresa M., Karl G. Hill, Beth A. Hennessey, and Elizabeth M. Tighe. 1994. The Work Preference Inventory: Assessing intrinsic and extrinsic motivational orientations. *Journal of Personality and Social Psychology* 66: 950. [CrossRef] [PubMed]

Amofah, Kwaku, and Ramon Saladrigues. 2022. Impact of attitude towards entrepreneurship education and role models on entrepreneurial intention. *Journal of Innovation and Entrepreneurship* 11: 1–30. [CrossRef]

Anjum, Temoor, Azadeh Amoozegar, Muhammad Farrukh, and Petra Heidler. 2022. Entrepreneurial intentions among business students: The mediating role of attitude and the moderating role of university support. *Education+ Training*. *ahead-of-print*. [CrossRef]

Apple, Michael W. 1975. The hidden curriculum and the nature of conflict. In *Curriculum Theorizing: The Reconceptualists*. Edited by William F. Pinar. Berkeley: McCutchan, pp. 95–119.

Aronsson, Magnus. 2004. Education matters—But does entrepreneurship education? An interview with David Birch. *Academy of Management Learning & Education* 3: 289–92.

Asimakopoulos, Grigorios, Virginia Hernández, and Javier Peña Miguel. 2019. Entrepreneurial intention of engineering students: The role of social norms and entrepreneurial self-efficacy. *Sustainability* 11: 4314. [CrossRef]

Audretsch, David B., Isabel Grilo, and A. Roy Thurik. 2011. Globalization, entrepreneurship, and the region. In *Handbook of Research on Entrepreneurship and Regional Development*. Cheltenham: Edward Elgar, pp. 11–32.

Ayed, Tahar Lazhar. 2020. Extending the debate over entrepreneurial education effectiveness: The case of a Saudi university. *Education+ Training* 62: 805–23. [CrossRef]

Bae, Tae Jun, Shanshan Qian, Chao Miao, and James O. Fiet. 2014. The relationship between entrepreneurship education and entrepreneurial intentions: A meta-analytic review. *Entrepreneurship Theory and Practice* 38: 217–54. [CrossRef]

Bagozzi, Richard P., and Youjae Yi. 1988. On the evaluation of structural equation models. *Journal of the Academy of Marketing Science* 16: 74–94. [CrossRef]

Bandera, Cesar, Regina Collins, and Katia Passerini. 2018. Risky business: Experiential learning, information and communications technology, and risk-taking attitudes in entrepreneurship education. *The International Journal of Management Education* 16: 224–38. [CrossRef]

Barba-Sánchez, Virginia, and Carlos Atienza-Sahuquillo. 2012. Entrepreneurial behavior: Impact of motivation factors on decision to create a new venture. *Investigaciones Europeas de Dirección y Economía de la Empresa* 18: 132–38. [CrossRef]

Barba-Sánchez, Virginia, and Carlos Atienza-Sahuquillo. 2018. Entrepreneurial intention among engineering students: The role of entrepreneurship education. *European Research on Management and Business Economics* 24: 53–61. [CrossRef]

Baron, Robert A. 2004. The cognitive perspective: A valuable tool for answering entrepreneurship's basic "why" questions. *Journal of Business Venturing* 19: 221–39. [CrossRef]

Baron, Robert A. 2006. Opportunity recognition as pattern recognition: How entrepreneurs "connect the dots" to identify new business opportunities. *Academy of Management Perspectives* 20: 104–19. [CrossRef]

Becker, Gary S. 1975. *Human Capital: A Theoretical and Empirical Analysis with Special Reference to Education*, 2nd ed. New York: Columbia University Press.

Bellò, Benedetta, Veronica Mattana, and Michela Loi. 2018. The power of peers: A new look at the impact of creativity, social context and self-efficacy on entrepreneurial intentions. *International Journal of Entrepreneurial Behavior & Research* 24: 214–33.

Block, Jack, and Adam M. Kremen. 1996. IQ and ego-resiliency: Conceptual and empirical connections and separateness. *Journal of Personality and Social Psychology* 70: 349–61. [CrossRef]

Block, Jeanne H., and Jack Block. 1980. The role of ego-control and ego-resiliency in the organization of behavior. In *Development of Cognition, Affect, and Social Relations. The Minnesota Symposia on Child Psychology*. Hillsdale: Lawrence Erlbaum, vol. 13, pp. 39–101.

Boldureanu, Gabriela, Alina Măriuca Ionescu, Ana-Maria Bercu, Maria Viorica Bedrule-Grigoruță, and Daniel Boldureanu. 2020. Entrepreneurship education through successful entrepreneurial models in higher education institutions. *Sustainability* 12: 1267. [CrossRef]

Brislin, Richard W. 1986. The wording and translation of research instruments. In *Field Methods in Cross-Cultural Research*. Berry. Beverly: Sage Publications, pp. 137–64.

Brush, Candida G., Irene M. Duhaime, William B. Gartner, Alex Stewart, Jerome A. Katz, Michael A. Hitt, Sharon A. Alvarez, G. Dale Meyer, and Sankaran Venkataraman. 2003. Doctoral education in the field of entrepreneurship. *Journal of Management* 29: 309–31. [CrossRef]

Bruton, Garry, Christopher Sutter, and Anna-Katharina Lenz. 2021. Economic inequality—Is entrepreneurship the cause or the solution? A review and research agenda for emerging economies. *Journal of Business Venturing* 36: 106095. [CrossRef]

Bruton, Garry D., David Ahlstrom, and Krzysztof Obloj. 2000. Entrepreneurship in emerging economies: Where are we today and where should the research go in the future. *Entrepreneurship Theory and Practice* 32: 1–14. [CrossRef]

Busenitz, Lowell W., and Chung-Ming Lau. 1996. A cross-cultural cognitive model of new venture creation. *Entrepreneurship Theory and Practice* 20: 25–40. [CrossRef]

Chadwick, Ingrid C., and Jana L. Raver. 2020. Psychological resilience and its downstream effects for business survival in nascent entrepreneurship. *Entrepreneurship Theory and Practice* 44: 233–55. [CrossRef]

Chen, Chao C., Patricia Gene Greene, and Ann Crick. 1998. Does entrepreneurial self-efficacy distinguish entrepreneurs from managers? *Journal of Business Venturing* 13: 295–316. [CrossRef]

Cohen, Jacob. 1988. *Statistical Power Analysis for the Behavioral Sciences*, 2nd ed. Hillsdale: Lawrence Erlbaum Associates, Publishers.

Colombelli, Alessandra, Shiva Loccisano, Andrea Panelli, Orazio Antonino Maria Pennisi, and Francesco Serraino. 2022. Entrepreneurship Education: The Effects of Challenge-Based Learning on the Entrepreneurial Mindset of University Students. *Administrative Sciences* 12: 10. [CrossRef]

Delmar, Frédéric. 2000. The psychology of the entrepreneur. In *Enterprise and Small Business*. Upper Saddle River: FT/Prentice-Hall.

Dickel, Petra, and Gordon Eckardt. 2021. Who wants to be a social entrepreneur? The role of gender and sustainability orientation. *Journal of Small Business Management* 59: 196–218. [CrossRef]

Dickson, Pat H., George T. Solomon, and K. Mark Weaver. 2008. Entrepreneurial selection and success: Does education matter? *Journal of Small Business and Enterprise Development* 15: 239–58. [CrossRef]

Elnadi, Moustafa, and Mohamed Hani Gheith. 2021. Entrepreneurial ecosystem, entrepreneurial self-efficacy, and entrepreneurial intention in higher education: Evidence from Saudi Arabia. *The International Journal of Management Education* 19: 100458. [CrossRef]

Fallatah, Mahmoud. 2021. Innovating in the Desert: A Network Perspective on Knowledge Creation in Developing Countries. *Journal of the Knowledge Economy* 12: 1533–51. [CrossRef]

Farooq, Muhammad Shoaib, Maimoona Salam, Saif ur Rehman, Alain Fayolle, Norizan Jaafar, and Kartinah Ayupp. 2018. Impact of support from social network on entrepreneurial intention of fresh business graduates: A structural equation modelling approach. *Education+ Training* 60: 335–53. [CrossRef]

Foote, Russell, and Eglantina Hysa. 2022. Improving Operational–Developmental Connections: Creating Positive "But-terfly Effects" in Universities. In *Governance and Management in Higher Education*. Bingley: Emerald Publishing Limited.

Fornell, Claes, and David F. Larcker. 1981. Structural equation models with unobservable variables and measurement error: Algebra and Statistics. *Journal of Marketing Research* 18: 382–88. [CrossRef]

Garavan, Thomas N., and Barra O'Cinneide. 1994. Literature review of problems associated with entrepreneurship education and training programmes. *Journal of European Industrial Training* 18: 3–12. [CrossRef]

Gibb, Allan. 2008. Entrepreneurship and enterprise education in schools and colleges: Insights from UK practice. *International Journal of Entrepreneurship Education* 6: 48.

Gilmartin, Shannon K., Marissa E. Thompson, Emily Morton, Qu Jin, Helen L. Chen, Anne Colby, and Sheri D. Sheppard. 2019. Entrepreneurial intent of engineering and business undergraduate students. *Journal of Engineering Education* 108: 316–36. [CrossRef]

Giroux, Henry A., and Anthony N. Penna. 1979. Social education in the classroom: The dynamics of the hidden curriculum. *Theory and Research in Social Education* 7: 21–42. [CrossRef]

Gundry, Lisa K., Laurel F. Ofstein, and Jill R. Kickul. 2014. Seeing around corners: How creativity skills in entrepreneurship education influence innovation in business. *The International Journal of Management Education* 12: 529–38. [CrossRef]

Hägg, Gustav, and Colin Jones. 2021. Educating towards the prudent entrepreneurial self—An educational journey including agency and social awareness to handle the unknown. *International Journal of Entrepreneurial Behavior & Research* 27: 82–103.

Hair, Joseph F., Jr., Marko Sarstedt, Christian M. Ringle, and Siegfried P. Gudergan. 2017. *Advanced Issues in Partial Least Squares Structural Equation Modeling*. Tousand Oaks: Sage.

Hair, Joseph F., Jeffrey J. Risher, Marko Sarstedt, and Christian M. Ringle. 2019. When to use and how to report the results of PLS-SEM. *European Business Review* 31: 2–24. [CrossRef]

Hoda, Najmul, and Mahmoud Fallatah. 2022. Social Networking Site Usage, Social Capital and Entrepreneurial Intention: An Empirical Study from Saudi Arabia. *The Journal of Asian Finance, Economics and Business* 9: 421–29.

Holtbrügge, Dirk, and Franziska Engelhard. 2016. Study abroad programs: Individual motivations, cultural intelligence, and the mediating role of cultural boundary spanning. *Academy of Management Learning & Education* 15: 435–55.

Honig, Benson. 2004. Entrepreneurship education: Toward a model of contingency-based business planning. *Academy of Management Learning & Education* 3: 258–73.

Joensuu-Salo, Sanna, Anmari Viljamaa, and Elina Varamäki. 2022. Sustainable Entrepreneurs of the Future: The Interplay between Educational Context, Sustainable Entrepreneurship Competence, and Entrepreneurial Intentions. *Administrative Sciences* 12: 23. [CrossRef]

Katz, Jerome. 2003. The chronology and intellectual trajectory of American entrepreneurship education: 1876–999. *Journal of Business Venturing* 18: 283–300. [CrossRef]

Kirzner, Israel. 1997. Entrepreneurial discovery and the competitive market process: An Austrian approach. *Journal of Economic Literature* 35: 60–85.

Kline, Rex. 2015. *Principles and Practice of Structural Equation Modeling*. New York: Guilford Publications.

Krueger, Norris. 1993. The impact of prior entrepreneurial exposure on perceptions of new venture feasibility and desirability. *Entrepreneurship Theory and Practice* 18: 5–21. [CrossRef]

Krueger, Norris F., Jr., Michael D. Reilly, and Alan L. Carsrud. 2000. Competing models of entrepreneurial intentions. *Journal of Business Venturing* 15: 411–32. [CrossRef]

Lee, Lena, Poh Kam Wong, Maw Der Foo, and Aegean Leung. 2011. Entrepreneurial intentions: The influence of organizational and individual factors. *Journal of Business Venturing* 26: 124–36. [CrossRef]

Liñán, Francisco. 2004. Intention-based models of entrepreneurship education. *Piccolla Impresa/Small Business* 3: 11–35.

Liñán, Francisco, and Yi-Wen Chen. 2009. Development and cross-cultural application of a specific instrument to measure entrepreneurial intentions. *Entrepreneurship Theory and Practice* 33: 593–617.

Liñán, Francisco, Juan A. Moriano, and Inmaculada Jaén. 2016. Individualism and entrepreneurship: Does the pattern depend on the social context? *International Small Business Journal* 34: 760–76. [CrossRef]

Liñán, Francisco, Juan Carlos Rodríguez-Cohard, and José M. Rueda-Cantuche. 2011. Factors affecting entrepreneurialintention levels: A role for education. *International Entrepreneurship and Management Journal* 7: 195–218. [CrossRef]

Lofstrom, Magnus, Timothy Bates, and Simon C. Parker. 2014. Why are some people more likely to become small-businesses owners than others: Entrepreneurship entry and industry-specific barriers. *Journal of Business Venturing* 29: 232–51. [CrossRef]

Lu, Genshu, Yaping Song, and Bingchao Pan. 2021. How university entrepreneurship support affects college students' entrepreneurial intentions: An empirical analysis from China. *Sustainability* 13: 3224. [CrossRef]

Martin, Bruce, Jeffrey McNally, and Michael Kay. 2013. Examining the Formation of Human Capital in Entrepreneurship: A Meta-Analysis of Entrepreneurship Education Outcomes. *Journal of Business Venturing* 28: 211–24. [CrossRef]

McGee, Jeffrey, Mark Peterson, Stephen Mueller, and Jennifer Sequeira. 2009. Entrepreneurial self-efficacy: Refining the measure. *Entrepreneurship Theory and Practice* 33: 965–88. [CrossRef]

Ndou, Valentina. 2021. Social entrepreneurship education: A combination of knowledge exploitation and exploration processes. *Administrative Sciences* 11: 112. [CrossRef]

Ndou, Valentina, Giustina Secundo, Giovanni Schiuma, and Giuseppina Passiante. 2018. Insights for Shaping Entrepreneurship Education: Evidence from the European Entrepreneurship Centers. *Sustainability* 10: 4323.

Nunkoo, Robin, and Haywantee Ramkissoon. 2012. Structural equation modelling and regression analysis in tourism research. *Current Issues in Tourism* 15: 777–802. [CrossRef]

Nunkoo, Robin, Haywantee Ramkissoon, and Dogan Gursoy. 2013. Use of structural equation modeling in tourism research: Past, present, and future. *Journal of Travel Research* 52: 759–71. [CrossRef]

Oosterbeek, Hessel, Mirjam Van Praag, and Auke Ijsselstein. 2010. The impact of entrepreneurship education on entrepreneurship skills and motivation. *European Economic Review* 54: 442–54. [CrossRef]

Osborn, Alex. 1957. *Applied Imagination*, revised ed. New York: Scribner.

Parker, Simon. 2008. Entrepreneurship, self-employment and the labour market. In *The Oxford Handbook of Entrepreneurship*. Oxford: Oxford University Press.

Parker, Simon, and Mirjam Van Praag. 2006. Schooling, capital constraints, and entrepreneurial performance: The endogenous triangle. *Journal of Business & Economic Statistics* 24: 416–31.

Pérez-Nordtvedt, Liliana, and Mahmoud Fallatah. 2022. Social innovation in Saudi Arabia: The role of entrepreneurs' spirituality, ego resilience and alertness. *Journal of Small Business Management* 60: 1080–121. [CrossRef]

Potter, Jonathan. 2008. Entrepreneurship and higher education: Future policy directions. *Local Economic & Employment* 27: 422–48.

Preacher, Kristopher, and Andrew Hayes. 2008. Asymptotic and resampling strategies for assessing and comparing indirect effects in multiple mediator models. *Behavior Research Methods* 40: 879–91. [CrossRef] [PubMed]

Preacher, Kristopher, Derek Rucker, and Andrew Hayes. 2007. Addressing moderated mediation hypotheses: Theory, methods, and prescriptions. *Multivariate Behavioral Research* 42: 185–227. [CrossRef] [PubMed]

Rasmussen, Einar, and Roger Sørheim. 2006. Action-based entrepreneurship education. *Technovation* 26: 185–94. [CrossRef]

Rauch, Andreas, and Willem Hulsink. 2015. Putting entrepreneurship education where the intention to act lies: An investigation into the impact of entrepreneurship education on entrepreneurial behavior. *Academy of Management Learning & Education* 14: 187–204.

Saudi Arabia Social Media Statistics. 2020. Global Media Insight. Available online: https://www.globalmediainsight.com/blog/saudi-arabia-social-media-statistics (accessed on 23 December 2021).

Schultz, Theodore. 1971. *Investment in Human Capital. The Role of Education and of Research*. New York: The Free Press.

Schumpeter, Joseph. 1934. *The Theory of Economic Development*. Cambridge: Harvard University Press.

Shane, Scott, and Sankaran Venkataraman. 2000. The promise of entrepreneurship as a field of research. *Academy of Management Review* 25: 217–26. [CrossRef]

Shapero, Albert, and Lisa Sokol. 1982. Social dimensions of entrepreneurship. In *The Encyclopedia of Entrepreneurship*. Edited by C. A. Kent, D. L. Sexton and K. H. Vesper. Englewood Cliffs: Prentice Hall, pp. 72–90.

Shirokova, Galina, Oleksiy Osiyevskyy, and Karina Bogatyreva. 2016. Exploring the intention-behavior link in student entrepreneurship: Moderating effects of individual and environmental characteristics. *European Management Journal* 34: 386–99. [CrossRef]

Souitaris, Vangelis, Stefania Zerbinati, and Andreas Al-Laham. 2007. Do entrepreneurship programmes raise entrepreneurial intention of science and engineering students? The effect of learning, inspiration and resources. *Journal of Business Venturing* 22: 566–91. [CrossRef]

Sweetland, Scott. 1996. Human capital theory: Foundations of a field of inquiry. *Review of Educational Research* 66: 341–59. [CrossRef]

Tracey, Paul, and Nelson Phillips. 2011. Entrepreneurship in emerging markets. *Management International Review* 51: 23–39. [CrossRef]

Van der Sluis, Justin, Mirjam Van Praag, and Wim Vijverberg. 2005. Entrepreneurship selection and performance: A meta-analysis of the impact of education in developing economies. *The World Bank Economic Review* 19: 225–61. [CrossRef]

Van der Sluis, Justin, Mirjam Van Praag, and Wim Vijverberg. 2008. Education and entrepreneurship selection and performance: A review of the empirical literature. *Journal of Economic Surveys* 22: 795–841. [CrossRef]

Vodă, Ana Iolanda, and Nelu Florea. 2019. Impact of personality traits and entrepreneurship education on entrepreneurial intentions of business and engineering students. *Sustainability* 11: 1192. [CrossRef]

Walter, Sascha, and Dirk Dohse. 2012. Why mode and regional context matter for entrepreneurship education. *Entrepreneurship & Regional Development* 24: 807–35.

Walter, Sascha, and Jörn Block. 2016. Outcomes of entrepreneurship education: An institutional perspective. *Journal of Business Venturing* 31: 216–33. [CrossRef]

Ward, Thomas. 2004. Cognition, creativity, and entrepreneurship. *Journal of Business Venturing* 19: 173–88. [CrossRef]

Warhuus, Jan, Franziska Günzel-Jensen, Sarah Robinson, and Helle Neergaard. 2021. Teaming up in entrepreneurship education: Does the team formation mode matter? *International Journal of Entrepreneurial Behavior & Research* 27: 1913–35.

Wu, Lingfei, and Jun Li. 2011. Perceived value of entrepreneurship: A study of the cognitive process of entrepreneurial career decision. *Journal of Chinese Entrepreneurship* 3: 134–46. [CrossRef]

Zhao, Xinshu, John Lynch, Jr., and Qimei Chen. 2010. Reconsidering Baron and Kenny: Myths and truths about mediation analysis. *Journal of Consumer Research* 37: 197–206. [CrossRef]

Zhou, Mansheng, and Haixia Xu. 2012. A review of entrepreneurship education for college students in China. *Administrative Sciences* 2: 82–98. [CrossRef]

administrative sciences

MDPI

Article

Entrepreneurship Education, a Challenging Learning Process towards Entrepreneurial Competence in Education

Heliona Miço [1,*] and Jonida Cungu [2]

1 Department of Law, Epoka University, 1032 Tirana, Albania
2 Department of Linguistics, University of Elbasan "Aleksandër Xhuvani", 3001 Elbasan, Albania
* Correspondence: hmico@epoka.edu.al

Abstract: Entrepreneurship education is a recent field in education. From a field mainly related to small business, it is extended towards enhancement of students' entrepreneurial attitudes and skills. It can support students in developing an independent and versatile way by growing the spirit of entrepreneurship. Developing entrepreneurship competence among students requires the mastery of concepts by teachers. Training teachers in entrepreneurship education helps them apply specific competences, methods, and tools to encourage confidence in learners' own capabilities and to stimulate flexibility, leadership, and initiative. To understand the teachers' entrepreneurial competence, an online survey was developed to assess the level of mastery of such competence in Albanian teachers. The survey was designed to analyze the development of entrepreneurial competence of teachers, and their entrepreneurship education and training. The research goal of the survey is to evaluate the influence mechanisms of gaining entrepreneurship education competence and the way of implementing this competence in pre-university education schools. Results from the questionnaire highlighted the teachers' need for the acquisition of entrepreneurial competence at every professional level, starting from initial teacher education. The findings from the survey are analyzed by taking into consideration the European policies regarding entrepreneurship education, such as the European Entrepreneurship Competence Framework (EntreComp), and their implementation in the Albanian educational sector.

Keywords: entrepreneurship education; university; teachers; competences; EntreComp framework; learning; Albania

Citation: Miço, Heliona, and Jonida Cungu. 2023. Entrepreneurship Education, a Challenging Learning Process towards Entrepreneurial Competence in Education. *Administrative Sciences* 13: 22. https://doi.org/10.3390/admsci13010022

Received: 30 November 2022
Revised: 5 January 2023
Accepted: 6 January 2023
Published: 11 January 2023

1. Introduction

Entrepreneurship education is a new field (Bueckmann Diegoli et al. 2018), which is spreading and developing beyond the business sphere. It was initially treated in the academy at the subject level in the late 1970s and early 1980s. (Wilson 2008; Hägg and Kurczewska 2021). This period coincided with the period of increased interest in entrepreneurship and small business. Unemployment encouraged different countries to find incentive mechanisms to influence the self-employment of citizens making entrepreneurship education to be considered as a contributing factor for economic growth. Entrepreneurship education was appraised as an academic tool, as well as an important model for the development of the society (Fayolle 2013). Based on these circumstances, the concept of entrepreneurship education is appearing in the educational curricula of many countries (Deveci and Seikkula-Leino 2018).

What does entrepreneurship mean? It is defined as the individual ability to find a business idea and transform it into practice (European Commission 2011). According to Shane and Venkataraman (2000) entrepreneurship is defined as the study of opportunity sources that consist of processes in which opportunities are discovered, evaluated, and used. In order to master these processes, proper education is needed, which has thus given rise to entrepreneurship education (Deveci and Seikkula-Leino 2018).

There are different meanings and various definitions applied in different countries regarding entrepreneurship education. Entrepreneurship education is concerned with learning for entrepreneurship, learning through entrepreneurship and learning about entrepreneurship (Gibb 2005). It refers to an individual's ability to turn ideas into action and includes creativity, innovation, risk taking, and the ability to plan and manage projects in order to achieve objectives (Commission of the European Communities 2006; Gautam and Singh 2015; Raposo and do Paço 2011; Secundo et al. 2021). Moreover, entrepreneurship education has been defined as a research focused process enabling us to investigate the most favorable education process to produce graduates in order to transform them into individuals who have life skills (Abiogu 2011). The Eurydice (2016) used the following definition for the entrepreneurship education: "Entrepreneurship education is about learners developing the skills and mind-set to be able to turn creative ideas into entrepreneurial action. This is a key competence for all learners, supporting personal development, active citizenship, social inclusion and employability. It is relevant across the lifelong learning process, in all disciplines of learning and to all forms of education and training (formal, non-formal and informal) which contribute to an entrepreneurial spirit or behavior, with or without a commercial objective".

Moreover, it is estimated that connecting entrepreneurship education only with business world is not enough, as it limits the opportunities of learners and teaching community. Entrepreneurship education has taken on dimensions related to social, psychological and pedagogical aspects, making it a process through which learners acquire a broad set of competencies by bringing individual, social, and economic benefits (European Commission 2011; Hägg and Gabrielsson 2019; Ndou et al. 2019).

It is accepted that there are two main approaches regarding entrepreneurship education (Eurydice 2016). A questionnaire organized by the European Commission with the participation of European countries showed that entrepreneurship education is recognized and embedded entrepreneurship education in a policy document in 21 countries of the European Union (European Commission 2014). The data showed that one of the approaches treats entrepreneurship education in a broad sense related to European key competencies, where learning outcomes related to employability, active citizenship and entrepreneurial skills for life and work are emphasized. The second approach has a narrower goal, focusing entrepreneurship education on learning outcomes linked with entrepreneurial and business activity.

Different studies have supported these approaches (Wach 2014). Firstly, it is accepted by Béchard and Grégoire (2005) that, in terms of educational theories, research on entrepreneurial education is rather incomplete and mainly focuses on the economic and business content of the teaching. Bae et al. (2014, as cited in Welsh et al. 2016) looked at entrepreneurship education as "education for entrepreneurial attitudes and skills" versus entrepreneurial intentions defined as "desires to own or start a business". According to Deveci and Seikkula-Leino (2018), the first approach is about creating a company or job. The other approach focuses on the individual and aims to improve the entrepreneurial attitudes and behaviors of students (European Commission 2004; Rizza and Varum 2011). Ndou (2021) highlights that even though entrepreneurship education has been mainly a priority for management and business students, in today's environment, characterized by the rapid development of new technologies and the complexity of society, it has emerged as a relevant competence to be created at all levels of education and for different disciplines. While for Akhuemonkhan et al. (2013) entrepreneurship training involves identifying "the sources of opportunities, the process of discovery, evaluation, and exploitation of opportunities; and the set of individuals who discover, evaluate and exploit them".

By being a process of providing students with ability to generate ideas and shaping the skills necessary to implement these ideas, entrepreneurship education has specific objectives (European Commission 2014):

1. promoting the development of personal qualities such as creativity, spirit of initiative, risk-taking and responsibility that are relevant to entrepreneurship;

2. raising students' awareness of self-employment as a career option (the message being that you can become not only an employee, but also an entrepreneur);
3. providing the business skills that are needed in order to start a new venture.

These objectives connect entrepreneurship education with pedagogy focusing on students' activity in learning (Gibb 2005). The learning situations are flexible, interactive, and based on multidimensional knowledge development (Ikävalko et al. 2008). Being a non-traditional teaching method, many studies take a pedagogical approach in defining entrepreneurship education (Ratten and Jones 2021; Fayolle et al. 2016; Igwe et al. 2022), giving teachers a primary role in the implementation of entrepreneurship education in schools (Ruskovaara et al. 2016; Oksanen et al. 2021).

The European approach to entrepreneurship education has also influenced the educational policies and legal framework in Albania. Starting from 2012, there was a legislative framework regarding the fundamental competences on education (On Pre-University Education System in the Republic of Albania Law of 2012, Pub. L. No. 69/2012 2012, Pub. L. No. 69/2012 2012). The educational system aims to create conditions and opportunities for students in order to build and develop knowledge, skills, attitudes and values according to the society's requirements; to develop in an independent and comprehensive manner; as well as to contribute to the construction and well-being of their own by helping them to face life's challenges constructively. The development of the spirit of entrepreneurship is presented as one of the qualities that the student receives at the level of pre-university education. This is attributed to the competence of entrepreneurship, which is included for the first time, in the Albanian legal and policy framework of education in 2012.

Despite steps taken towards entrepreneurial competences in the Albanian education system, there is a lack of adequate teacher training and practices to ensure effective entrepreneurship education. The inclusion of life, entrepreneurship and the environment competence in the Albanian education system, not just in the curriculum, needs a multifaceted analysis. Such an analysis will evaluate the knowledge obtained by teachers in pre-service education and training through the entrepreneurship pedagogy, continuous professional development, cooperation of the school with business and start-up companies and other forms of implementation related to curricular aspects and its implementation. Such an analysis should be accompanied by the legal framework and the necessary standards compatible with European education policies to enable its implementation by both teachers and students.

This study contributes to a portion of this larger analysis by assessing the entrepreneurial competencies of teachers, the curriculum in pre-service education and continuous educational training, and the approach of schools towards the distribution of entrepreneurship education in Albania. An online survey was completed by 233 teachers of different pedagogical profiles employed in different schools located in urban areas as well as in rural areas. The aim of this study was to assess the need of educators to be equipped with entrepreneurial characteristics and to identify approaches toward meeting those needs.

The article is divided into six sections. Section 1 is the introductory part and provides a concise overview on the article. Section 2 attempts a review of the literature in order to provide insight into the role of entrepreneurship education to the acquisition of learners' knowledge as an approach of European policies, focusing on the European Entrepreneurship Competence Framework. This section also highlights the gaps in the research regarding the assessment of the level of mastery of entrepreneurship education in teaching staff. Section 3 deals with methodology used for the research with justification for its adoption. Section 4 presents the findings of data analyzes gathered by the survey. Section 5 interprets the findings in the framework of a literature review. Section 6 concludes with policy recommendations for effective mastery of teachers' entrepreneurship competence in order to promote students' personal and social development.

2. Theoretical Background of Literature

2.1. The Approach of the European Commission on the Need for Entrepreneurship Education

The efforts of the European Commission for including entrepreneurship competence into education have been concretized by including entrepreneurship as one of the eight European Key Competences in 2006 in the educational systems of the countries of the European Union (European Parliament and Council of European Union 2006). The Eurydice study identified different attitudes of the European states for the inclusion of entrepreneurship competence in the mainstream education system (Eurydice 2016), by accepting a fragmented approach for this policy (Seikkula-Leino et al. 2021; Apostu et al. 2022). The launch of Europe 2020 strategy through education and training turned attention to entrepreneurship competences emphasizing the need for the development of entrepreneurship mindset for all learners (European Commission 2010). On the other hand, Sustainable Development Goals adopted by United Nations have highlighted in the framework of quality education the entrepreneurial learning as a means to equip young people with relevant skills, for employment, decent jobs and entrepreneurship (United Nations 2015). The main approach through which the learners can acquire entrepreneurship skills and knowledge is accepted to be through the education system (Jónsdóttir 2007; Commission of the European Communities 2006; Eurydice 2016).

Recent research has accepted that entrepreneurship education contributed to risk taking attitudes, to the formation of new business and to the proclivity to be self-employed (Alvarez Marques and Albuquerque 2012; Ndou et al. 2018). The two-folded approach sees the purpose of entrepreneurship education as related to education of students to take more responsibility for themselves and their learning, to try to achieve their goals, to be creative, to discover existing opportunities, and to cope in a complicated society. While, another aim is for them to take an active role in the labour market and consider entrepreneurship as a natural career choice (Seikkula-Leino et al. 2019; Gibb 2005).

To better understand and integrate entrepreneurship across the education system, the European Entrepreneurship Competence Framework (EntreComp) is one of the responses of the European Commission (Bacigalupo et al. 2016) for the promotion of entrepreneurial learning towards social, cultural, or financial value creation (Seikkula-Leino et al. 2021). Considering entrepreneurship as a competence with transversal nature, EntreComp was launched as a common reference framework for entrepreneurship to help citizens to develop their ability to actively participate in society, to manage their own lives and careers and to start value-creating initiatives. The EntreComp conceptual model is made up of three main competence areas: 'Ideas and opportunities', 'Resources' and 'Into Action' and 15 competences that, together, make up the building blocks of entrepreneurship as a competence for all citizens (see Table 1) (Bacigalupo et al. 2016; Eurydice 2016). In the EntreComp framework, the entrepreneurship is both an individual and collective competence, defined as the capacity to act upon opportunities and ideas and transform them into value for others. The value that is created can be financial, cultural or social (European Commission 2018). The framework describes the development of 15 competences along a progression model in learning outcomes, over eight levels.

Table 1. EntreComp Framework.

EntreComp Framework		
No.	**Area**	**Competence**
1.	Ideas and opportunities	Spotting opportunities Creativity Vision Valuing ideas Ethical and sustainable thinking

Table 1. *Cont.*

EntreComp Framework		
No.	**Area**	**Competence**
2.	Resources	Self-awareness & self-efficacy Motivation & perseverance Mobilising resources Financial & economic literacy Mobilising others
3.	Into action	Taking the initiative Planning & management Coping with uncertainty, ambiguity & risk Working with others Learning through experience

2.2. The Benefits of Entrepreneurship Education

It is accepted that entrepreneurship education is not just about teaching someone to run a business, it is about encouraging creative thinking and promoting a strong sense of self-worth and empowerment (Garavan and O'Cinneide 1994; Gautam and Singh 2015; European Commission 2021a). According to European Commission (2021a) entrepreneurship education gives responsibility to learners by encouraging them to do things themselves, guides them towards identifying and seizing opportunities, supports inventive learning, encourages confidence in the learner's own capabilities, pushes students to take risks and encourages them to cooperate with others. On the other hand, entrepreneurship education cannot achieve its objectives without the involvement of businesspeople, who play an important role in the acquisition of entrepreneurial competence (Apostu et al. 2022). Their involvement changes the role of the teacher. However, teachers are the ones who will play the main role in realizing the goals of entrepreneurship education and will equip students with entrepreneurial competencies (Ruskovaara and Pihkala 2013; Birdthistle et al. 2007). Yet, recent research has confirmed that the implementation of entrepreneurship education has been a challenge for teachers (Fayolle 2013; Oksanen et al. 2022; European Commission 2021a).

2.3. The Pedagogy Applied to Entrepreneurship Education

Teachers face difficulties in finding contents and methods to implement entrepreneurship education (Ruskovaara and Pihkala 2013; Seikkula-Leino 2008). Entrepreneurship education research is mainly based on the theoretical and conceptual understanding of entrepreneurship and learning, while entrepreneurship education is a question of learning for entrepreneurship, about entrepreneurship and through entrepreneurship (Gibb 2005; European Commission 2021a). According to Seikkula-Leino et al. (2010), entrepreneurship education is not an established part of teachers' pre-service education and training, and continuous professional development. For this reason, it is necessary to change the curricula and develop new teaching and learning methods towards equipping teachers with the entrepreneurship competence.

There are different theories about the competences and previous experience of teachers, whether having entrepreneurial experience in order to pass it on to students (Bueckmann Diegoli et al. 2018), or whether having the possibility to acquire the entrepreneurial competence during initial teacher education and continuous professional development (Ibáñez-Cubillas and Gijón Puerta 2021; Peltonen 2015). Research demonstrates that collaborative learning can help teachers to adopt a more entrepreneurial teaching approach, rather than traditional education methods which do not provide for proper entrepreneurial skills and competences (Alvarez Marques and Albuquerque 2012).

The pedagogy applied to entrepreneurship education should be built on the active role of learners in the learning process which leads towards the non-traditional route of education (Gibb 2005). According to Johannisson et al. (1997) as cited in Alvarez Marques and Albuquerque (2012), the entrepreneurial pedagogy is adopted to the educational legal framework and to different levels of education within the mainstream educational system. This implies a progressive strategy by considering the age of the students and their previously gained knowledge, starting with "field game" type activities and continuing with "field projects". This requires the teacher to play an active role in the acquisition of entrepreneurial skills and in mentorship of students in the way to acquire entrepreneurial skills and attitudes. It is necessary for teachers to receive training in either or both the experiential pedagogy, and the business content, to enable students to think independently and have the opportunity to learn through "errors" (European Commission 2021a).

In the Albanian education system, the concept of entrepreneurship competence has recently been included in the curricular framework of education. The eight key competences for lifelong learning of the European Union (Council of the European Union 2018) are transferred in the seven basic competences of the Albanian Pre-University Education Curricular Framework. Entrepreneurship competence was mentioned for the first time in 2012 in the legal framework, being later adopted by the curriculum framework, with the terminology "Competence for life, entrepreneurship and the environment", to enable individuals to manage different situations that they encounter in life (Albanian Ministry of Education and Sport 2014). Through this competence, the school tends to prepare the individual to demonstrate skills in entrepreneurship, in organizing initiatives and group works (The Organisation for Economic Cooperation and Development (OECD 2015)). Through the data gathered by the survey, it is emphasized that theoretically teachers possess the entrepreneurship competence, while practically they have the possibility to share with students only the teaching process of entrepreneurship.

2.4. The Gaps in the Research

It has been widely argued that the most relevant challenges of entrepreneurship education consist in the different dimensions of competencies to be created, the pedagogical approaches, learning strategies, and knowledge creation processes (Fayolle 2013; Ndou 2021; Secundo et al. 2016). The research on the entrepreneurship education affects educational programs, syllabi, and adapted methodologies to focus on students' capacity i.e., the ability to benefit from the opportunities offered which they will be faced with (Banha et al. 2022). According to Volkmann et al. (2009), multiple approaches and interactive teaching methods are necessary to promote creativity, innovation, critical thinking, opportunity recognition and social awareness. Since there are required collaborative learning and a connection with experiential pedagogy and business content to enable students to think independently, it is necessary for the teacher to have acquired entrepreneurship education competence. Akpan (2021) considered entrepreneurship education a lifelong process, starting at elementary school and progressing through all levels of education, including adult education. Nevertheless, Akpan (2021) features the dearth of teaching staff equipped with knowledge on entrepreneurship education and therefore the lack of entrepreneurial competencies of teachers. Huang et al. (2020) identified the factors that influence the formation of teachers' competence in entrepreneurship education, linking them to professional training, new modes of teaching, entrepreneurial practice, entrepreneurial culture, and policy guarantee. Regardless of the volume in research on the importance and the role that entrepreneurship education plays in the development of the student, making him capable of "turning ideas into actions, ideas that generate values for someone other than for yourself", there is a gap in research on the formation of the teacher with competencies and skills as a bridge between entrepreneurship education and the student (Colombelli et al. 2022). The assessment of teacher competence on entrepreneurship education, the factors that influence its mastery and the mechanisms of implementing teachers' skills on entrepreneurship education have not been treated sufficiently.

The article intends to evaluate the entrepreneurship competencies of teachers by focusing on the factors that lead to obtaining this competence, such as university studies, in-service training in the framework of the professional development of teachers and support by policies and legislation. Albania was chosen not only for reasons of practicality in data collection, but to demonstrate that the inclusion of entrepreneurship competencies in policies and legislation, rather than in teachers' knowledge and skills, is not sufficient. Based on the above discussion, the research question is posed: What is the level of mastery of entrepreneurial competences of teachers, the factors that influence this competence, and its implementation in the teaching process? For this research question, the following hypotheses are presented:

Hypothesis 1. *Teachers have received insufficient knowledge in entrepreneurship education from university studies or training.*

Hypothesis 2. *Teachers do not sufficiently exercise entrepreneurship competence due to lack of knowledge and skills.*

3. Methodology

The researchers employed the survey method to gain insight into the level of entrepreneurial competence of Albanian teachers and their need to develop competence in initial teacher education, continuous professional development and teachers' professional networks. As noted in the literature, the survey research is used to gather information about population groups to "learn about their characteristics, opinions, attitudes, or previous experiences" (Leedy and Ormrod 2005; Brewer 2009). Providing information, the survey research has a specific purpose: to improve the effectiveness of educational research (Ebel 1980; Walston et al. 2017). The study collected data through a questionnaire designed to measure the level of recognition of entrepreneurial competence by the teaching staff of the Albanian education system. The questionnaire was administered online using Google Forms and the teachers' answers are recorded. The questionnaire was organized with 21 multiple choice questions, four of which were open-ended questions. The questionnaire was sent in the form of a link to the heads of 30 pre-university education schools, who were invited to cooperate in completing the survey together with their teachers. The schools are randomly selected and belong to the following types: elementary school, secondary school and vocational school. Educational institutions belong to the district of Elbasan and are located in rural and urban areas. In total, the number of teachers of all schools where the survey way sent reached approximately 600 teachers, but the questionnaire was completed only by 233 of them. There was no sampling selection of teachers. They belong to all subject profiles developed in pre-university education, grouped into: social sciences, natural sciences, computer science and vocational education. The online survey guaranteed the anonymity of the teachers enabling them to provide true opinions, and to share ideas in a safe and comfortable environment. The teachers were free to complete the questionnaire without interference, and this is proven by the number of teaching staff who completed the questionnaire, which is different from the total number of teachers, to whom the questionnaire was sent for completion. The answers to these questions provided information about the teaching staff's level of knowledge regarding entrepreneurial competence. The questionnaire completed by 233 teachers and school leaders who belonged to the nine-year education cycle, secondary education and vocational education would provide data on the mastery of entrepreneurship competence in pre-university education.

Data analysis starts with the presentation of descriptive statistics of the data gathered from google form survey. Detailed information is obtained using crosstabs and correlations analyses of different variables. After gathering the data from survey, it is used cross-tabulation for investigating the relationship between teachers' approach of getting the knowledge on entrepreneurship education and the ways of implementing this knowledge through various teaching methods. The information provided by cross-tabulations serves to investigate the relationship between variables such as: "*Did you gain knowledge of entrepreneurship education during your initial teacher education and training?*" and "*If you received training on entrepreneurial competence, it was provided by:*", as well as between "*Do you develop projects at school that affect the development of entrepreneurial competence"* and *"Do you try to develop the entrepreneurial culture in students through your ideas during teaching?"*. To support the hypothesis, correlation analysis is used to measure the strength of the direct correlation of different variables such as knowledge obtained at university or training programs and entrepreneurial pedagogy.

Data collection took place during November 2022. The collected data from the questionnaire served to reflect on the policy and legal framework of Albanian education regarding entrepreneurial competence. They were also used to provide recommendations for changes needed in order for the Albanian education system to embrace entrepreneurship education.

3.1. Participants and Their Demographics

The questionnaire was completed by 233 Albanian teachers working in pre-university education. Participants were randomly selected from basic, secondary and vocational education schools. The schools were located both in urban and rural areas. According to the demographic data collected through the questionnaire, participants included young and experienced teachers of different genders. The participants held various positions within the schools.

The questionnaire was filled out individually by teachers online, without any possibility of intervention. The questionnaire ensured the preservation of confidentiality and informed participants' that the data of the questionnaire would be valid only for study purposes. In the sections below, questions from the questionnaire are indicated by the abbreviations Q1 for Question 1, Q2 for Question 2, etc.

Demographically, the participants varied. With regard to years of experience (Q1), 12.1% of respondents had five years of work experience or fewer, 12.1% had 5–10 years, 43.7% had 10–20 years, 11.7% had 20–25 years, and 20.3% had over 25 years of work experience. In terms of gender (Q2), 183 (79.2%) participants were female and 48 (20.8%) were male. Regarding the location of the participants' educational institutions (Q3), 130 (56.3%) participants reported working in urban areas, and 101 (43.7%) in administrative units and rural areas. Regarding the educational cycle where they teach (Q4), 38 (16.5%) participants taught in primary education, 147 (63.6%) participants taught in lower secondary education; 44 (19%) participants taught in secondary education and only 2 (0.9%) participants in vocational education. Finally, the participants were involved in different subject areas (Q5), with 148 (64.1%) participants in social sciences, 59 (25.5%) in the natural sciences, 22 (9.5%) vocational education, and two (0.9%) computer sciences.

3.2. Instrumentation

A questionnaire consisting of 21 questions, developed by the research team for this study, was administered to identify teachers' level of knowledge of entrepreneurial competence. In addition, the questionnaire assessed participants' understanding of the role entrepreneurial competence plays in the teaching and learning process. There were multiple-choice questions, where participants had the possibility of selecting more than one option and provide comments on the answer in case they were required to interpret their position. The questionnaire was sent to teachers via email invitation. Completing the questionnaire created the possibility for each participant to be identified by email address. Consequently, the data obtained from the questionnaire were individual, identifiable, and non-repetitive.

4. Results

4.1. Survey's Components and Results

The purpose of the questionnaire was to collect reliable and concrete data from 233 educational workers who participated in the survey. The quantitative data were used for the empirical interpretation of the study. In the questionnaire, the collected data helped to define the theoretical and explanatory framework of the study, on the level of entrepreneurship competence in pre-university education in Albania.

The second set of questions is related to the collection of qualitative data, focusing on data regarding teachers' knowledge about entrepreneurship education and the level of knowledge applicability in the institutions where they teach. Data interpretation aims to highlight whether the curriculum used nowadays in education meets the requirements of entrepreneurship education. Hence, the researchers addressed this issue in the questionnaire by asking (Q6), "*Did you gain knowledge of entrepreneurship education during your initial teacher education and training?*" From the received responses, 93 (39.9%) participants reported that they did not receive entrepreneurship education during university training, 72 (30.9%) participants have received partial knowledge while 68 (29.2%) participants reported receiving this knowledge. This is explained by the fact that some of the participants classify the knowledge of pedagogy, psychology, and professional practice related to entrepreneurship education; some others specify the economics subjects as complementary to this knowledge and a significant part of the participants state that the university did not offer special modules on entrepreneurship education during their study period. Such an indicator is sufficient to create the possibility that the teachers apart from not receiving knowledge on entrepreneurship education, were not clear if the knowledge received belonged to this field.

Another question (Q7) sought to obtain information on the needs of teachers for additional knowledge in the area of entrepreneurship education. Specifically, the question asked: "*Have you participated in entrepreneurship training after finishing higher education?*" Response to this question varied and 103 (44.2%) of participants claimed to have participated in training programs on entrepreneurial competence, 42 (18%) of participants planned to participate in training programs in the framework of continuous professional development and 88 (37.8%) of individuals did not participate in such training after initial teacher education. Such indicators expressed the variation in teachers' knowledge of entrepreneurial competence.

In 2014, the Albanian curricular framework of education changed, including the competence for life, entrepreneurship and environment and shifted the attention towards learning based on key competences. The answer to (Q9) provided data on the extent of development of entrepreneurship competence during training sessions on the development of the curriculum based on core competences: "*Have you acquired knowledge of entrepreneurship competence during training for the curriculum based on core competences?*" Only 104 (44.6%) of the participants received trainings for entrepreneurship competence which enabled them to integrate it with the subject they taught. Another 79 (33.9%) participants responded that they received partial training. While only 50 (21.5%) responded negatively. From the data, it is noticed that less than half of the participants have knowledge about entrepreneurship competence. This means that only a part of teachers may use entrepreneurship knowledge for implementation during the teaching process.

In order to evaluate the interest of teachers for participating in training programs for improving entrepreneurial competence, (Q13) asked: "*Would you be interested in being involved in training that focuses on improving entrepreneurial competence?*" Most of the participants 153 (65.7%) are very interested in participating in trainings towards improving entrepreneurial competence, 65 (27.9%) participants are unsure about participating in these trainings and 15 (6.4%) do not show interest in improving such knowledge. These data are of particular importance as a positive indicator of the legitimacy of teachers towards entrepreneurial competence (Foliard et al. 2019).

The answer to question (Q14) provided different results: "Have your school established links and collaborative structures with businesses and community organizations to support the entrepreneurship curriculum?" From the data, 106 (45.5%) of the participants stated that they try to create connections and cooperation structures with businesses and community organizations to support the entrepreneurship curriculum in their schools. Another 97 (41.6%) of participants make partial efforts for such collaborations for the benefit of the entrepreneurship curriculum, while 30 (12.9%) of them do not make such an effort. The teachers were asked to provide their experiences related to possible tools used for the implementation of entrepreneurship education and they specified: fairs, collaborations within the subject "Education for Career", "field projects" for the creation of a business company, various activities within the initiative "School as Community center", crafts, activities supported by Junior Achievement, improvisation of entrepreneurial situations. When combined with data from (Q6), 164 (71%) participants reported that they did not receive entrepreneurship education during university training or have partially received it, which shows the need to improve and gain additional knowledge in the field of entrepreneurship education.

In order to guarantee a sustainable quality in education, teachers are organized in professional networks where they exchange experiences, knowledge, and materials for entrepreneurial learning. The answers to (Q15) provided data on the inclusion of teachers in professional networks related to entrepreneurial education: *Are you part of a professional network related to entrepreneurial learning and teaching?* Only 94 (40.3%) participants stated they were part of educational professional networks, 91 (39.1%) participants reported that they were not organized in any professional network, and 48 (20.8%) stated that such a professional network does not exist.

The answer to (Q16), "Do you have discussions/roundtables on specific topics for entrepreneurial education at school, in order to develop your professionalism?" provided similar results. The responses show that 58 (24.9%) of the participants answered yes and another 95 (40.8%) answered sometimes. On the other hand, 28 (12%) participants stated these roundtables are held rarely, where 52 (22.3%) of the participants stated the round tables regarding entrepreneurship education were never organized.

Another question (Q17) asked participants about the implementation of school projects which orient students towards implementation of entrepreneurship competence. This question asked, "*Do you plan or implement school projects that guide students towards the implementation of entrepreneurial competence?*". According to the provided answers, 96 (41.2%) participants confirmed their engagement in such projects, 88 (37.8%) indicated their partial participation in these projects; 19 (8.2%) of participant stated their engagement as "Very rarely", while 30 (12.9%) participants responded negatively.

When asked about *the involvement of the school in different organizations for training/projects related to entrepreneurship education* (Q18), 73 (31.3%) participants reported involvement of school in different projects related to entrepreneurship education and another 90 (38.6%) participants answered sometimes. On the other hand, 18 (7.7%) participants answered "very rarely" and 52 (22.3%) participants confirmed that the school in never involved in different organizations for entrepreneurship education.

Teachers were also asked *about the concepts that school curriculum addresses regarding entrepreneurship education* (Q19). For this question, 88 (37.8%) participants confirmed the inclusion of entrepreneurship education concepts in the school curriculum, 84 (36.1%) pointed partial inclusion and 61 (26.2%) participants responded negatively. They identified as well different subjects related to entrepreneurial competence such as "Finance in your hands", the elective module "Entrepreneur for a day", economy, and cultural heritage. From the data, it is noticed different levels of development of entrepreneurship competence, depending on the different subjects.

When asked about the participants' opinion on the level of entrepreneurial competences in the role of supporter and facilitator of students' knowledge (Q20), only 69 (29.6%) participants consider their competences as sufficient. Another 128 (54.9%) participants indicated partial possession of this competence and 36 (15.5%) stated that they did not possess the entrepreneurship competencies. The participants were asked to express themselves about their experience and the answers expressed different considerations regarding the activities related to goals of entrepreneurship education, such as: "I have no competence; it is an interdisciplinary connection; development through projects, fairs, group work; work with student government; computer entrepreneurship, management of recyclables, personal budget management, cooperation with organizations, visits to business locations. The data highlighted the struggle of educators to identify the content and methods needed for implementation of entrepreneurship education.

In the final question (Q21), participants were asked *to list the forms of knowledge learning for entrepreneurial education*. The results reveal that, 86 (39.6%) participants acknowledged that they have received this knowledge through qualifications or training programs. Sixty (27.6%) participants reported initial teacher education as a form through which they learnt regarding entrepreneurship education, 51 (23.5%) participants reported both continuous professional development and initial teacher educations as possible routes for this knowledge, and 20 (9.3%) participants stated other forms of knowledge learning (self-taught, personal skills, school activities).

4.2. Statistical Analysis of Main Indicators

The following questions shown in Table 2, highlight the lack of formal learning on entrepreneurship education received from universities and training programs, which is reflected in the perception of entrepreneurship competence by the teaching staff and the way of exercise the entrepreneurship competence due to insufficient knowledge and skills (information presented in Table 3). From the received qualitative data, it is debatable whether teachers are prepared and able to cultivate the culture of entrepreneurship in students.

Table 2. Crosstabulation of (Q6) and (Q8).

Did You Gain Knowledge of Entrepreneurship Education during Your Initial Teacher Education and Training? * If You Received Training on Entrepreneurial Competence, It Was Provided by: Crosstabulation

		If you received training on entrepreneurial competence, it was provided by:			Total
		Training programs	Educational institutions	By the individual	
Did you gain knowledge of entrepreneurship education during your initial teacher education and training?	No	35.5%	20.4%	44.1%	100.0%
	Partly	38.9%	23.6%	37.5%	100.0%
	Yes	47.1%	38.2%	14.7%	100.0%
Total		39.9%	26.6%	33.5%	100.0%

Table 3. Crosstabulation of (Q8) and (Q10).

If You Received Training on Entrepreneurial Competence, It Was Provided by: * If Entrepreneurship Education Was Part of the Training Program, Were Approaches Such as: Project-Based Learning, Active Learning or Independent Learning, Part of the Training Pedagogy? Crosstabulation

		If entrepreneurship education was part of the training program, were approaches such as: project-based learning, active learning or independent learning, part of the training pedagogy?			Total
		No	Partly	Yes	
If you received training on entrepreneurial competence, it was provided by:	Training programs	7.5%	31.2%	61.3%	100.0%
	Educational institutions	11.3%	30.6%	58.1%	100.0%
	By the individual	29.5%	39.7%	30.8%	100.0%
Total		15.9%	33.9%	50.2%	100.0%

Since the p-value is less than our chosen significance level $\alpha = 0.01$ as shown in Table 4, we can reject the null hypothesis, and conclude that there is an association between the two above-mentioned variables.

Although almost half of the respondents expressed that they could manage to cultivate the culture of entrepreneurship as shown in Table 5, they still feel the need to attend training in this field in order to improve their competences.

Table 4. Chi-Square Tests of (Q8) and (Q10).

Chi-Square Tests			
	Value	**df**	**Asymptotic Significance (2-Sided)**
Pearson Chi-Square	24.092 [a]	4	0.000
Likelihood Ratio	24.065	4	0.000
Linear-by-Linear Association	20.734	1	0.000
N of Valid Cases	233		

[a] 0 cells (0%) have expected count less than 5. The minimum expected count is 9.85.

Table 5. Crosstabulation of (Q11) and (Q12).

Do You Develop Projects at School That Affect the Development of Entrepreneurial Competence? * Do You Try to Develop the Entrepreneurial Culture in Students through Your Ideas during Teaching? Crosstabulation					
		Do you try to develop the entrepreneurial culture in students through your ideas during teaching?			Total
		It's not important	Sometimes	Often	
Do you develop projects at school that affect the development of entrepreneurial competence?	No	21.1%	56.1%	22.8%	100.0%
	Yes	0.6%	44.3%	55.1%	100.0%
Total		5.6%	47.2%	47.2%	100.0%

Since the *p*-value is less than our chosen significance level $\alpha = 0.01$ as shown in Table 6, we can reject the null hypothesis, and conclude that there is an association between "Do you develop projects at school that affect the development of entrepreneurial competence" and "Do you try to develop the entrepreneurial culture in students through your ideas during teaching?".

Table 6. Chi-Square Tests of (Q11) and (Q12).

Chi-Square Tests			
	Value	**df**	**Asymptotic Significance (2-Sided)**
Pearson Chi-Square	43.175 [a]	2	0.000
Likelihood Ratio	39.640	2	0.000
N of Valid Cases	233		

[a] 1 cells (16.7%) have expected count less than 5. The minimum expected count is 3.18.

This is illustrated by taking the following questions as an example. "If entrepreneurship education was part of the training program, were approaches such as: project-based learning, active learning or independent learning, part of the training pedagogy?" and "Did you gain knowledge of entrepreneurship education during your initial teacher education and training", showing a positive correlation of 0.322. According to Table 7, this result emphasizes the connection between the acquisition of knowledge on entrepreneurship education and their exercise by teachers in activities related to this competence.

Table 7. Correlation matrix of variables Q8, Q10, Q11, Q12.

Correlations					
		If You Received Training on Entrepreneurial Competence, It Was Provided by:	**Do You Try to Develop the Entrepreneurial Culture in Students through Your Ideas during Teaching?**	**Did You Gain Knowledge of Entrepreneurship Education during Your Initial Teacher Education and Training?**	**If Entrepreneurship Education Was Part of the Training Program, Were Approaches Such as: Project-Based Learning, Active Learning or Independent Learning, Part of the Training Pedagogy?**
If you received training on entrepreneurial competence, it was provided by:	Pearson Correlation	1	−0.116	−0.193 **	−0.299 **
	Sig. (2-tailed)		0.077	0.003	0.000
	N	233	233	233	233
Do you try to develop the entrepreneurial culture in students through your ideas during teaching?	Pearson Correlation	−0.116	1	0.196 **	0.388 **
	Sig. (2-tailed)	0.077		0.003	0.000
	N	233	233	233	233
Did you gain knowledge of entrepreneurship education during your initial teacher education and training?	Pearson Correlation	−0.193 **	0.196 **	1	0.322 **
	Sig. (2-tailed)	0.003	0.003		0.000
	N	233	233	233	233
If entrepreneurship education was part of the training program, were approaches such as: project-based learning, active learning or independent learning, part of the training pedagogy?	Pearson Correlation	−0.299 **	0.388 **	0.322 **	1
	Sig. (2-tailed)	0.000	0.000	0.000	
	N	233	233	233	233

** Correlation is significant at the 0.01 level (2-tailed). All combinations showing two asterisks (**) have a significant correlation among each other. Negative signs show that they have a negative correlation.

Overall, the results of the questionnaire show that teachers may not be prepared for entrepreneurship competence. Enterprise education is usually conceived more broadly, seeking to foster self-esteem and confidence by drawing on the individual's talents and creativity, while building the relevant skills and values that will assist students in expanding their perspectives on schooling and opportunities beyond. Methodologies are based on the use of personal, behavioral, motivational, attitudinal, and career planning activities (UNESCO/ILO 2006). The data obtained from the questionnaire provide the level of knowledge on the entrepreneurship competence possessed by teachers, as well as the methods of implementation during the teaching process. From the data, it is noticeable that a significant number of participants admit that the knowledge obtained in university studies for entrepreneurship education was at minimal levels or absent.

During preparation of teachers in initial teacher education, students receive knowledge on "Psycho-pedagogical training" and "Curricula training". The knowledge related to entrepreneurship education is integrated in the educational programs of "Psycho-Pedagogical Training". However, there is no real courses in higher education studies to equip future teachers with the competence of entrepreneurship education. The lack of obtaining this knowledge, highlighted as well in the questionnaire, creates a lack of connection between the competence of entrepreneurship and the individual who is preparing for this competence.

Entrepreneurship education programs can significantly change the entrepreneurial intentions of participants (Almahry et al. 2018), which can only happen if future teachers are prepared for this competence. Teachers mostly perceive entrepreneurship education as knowledge that is supplemented by pedagogy and psychology modules and not as an attitude and ability that is converted into individual initiatives for entrepreneurial activities with individual and social benefit.

Entrepreneurship education allows the student to develop different aspects of entrepreneurial self-efficacy, since the knowledge gained equips them with perceptions about their entrepreneurial skills (Porfírio et al. 2022). The data show that the development of modules, such as "Career Education", "Finance in your hands", and the projects developed by Junior Achievement (Q14), are indicators that the introduction of entrepreneurship education has a positive impact on the entrepreneurial attitudes of students (Junior Achievement of Albania 2014), despite the lack of teachers' knowledge on entrepreneurship education during their university studies.

Entrepreneurial culture is achieved and enlightened through education and training, processes which were found to be lacking in the group of respondents. If we have entrepreneurial competencies applicable, then we will believe that entrepreneurs are not born but they are made through their life experience (Gautam and Singh 2015).

5. Discussion

Entrepreneurship education is one of the fastest growing subject areas in the world due to increased interest in its ability to connect current business practices with academic theory (Ratten and Usmanij 2021). The main goal of entrepreneurship education is to develop certain levels of entrepreneurial competence. This is also reflected in the curricular framework of education in Albania, where entrepreneurship competence is defined as an attitude that orients students towards the future, to develop the spirit of entrepreneurship and to be motivated to meet the objectives (Albanian Ministry of Education and Sport 2014).

Yet, challenges with adapting education to entrepreneurship and better-preparing teacher education students is a long-standing issue (Arruti and Panos-Castro 2020). To help address this issue, the European Entrepreneurship Competence Framework (EntreComp) aims to identify the competencies that make someone entrepreneurial, focusing on three competence areas with five competences each of them (Bacigalupo et al. 2016). These areas are summarized in Table 8.

Table 8. Study Findings in the light of EntreComp Framework.

EntreComp Framework			
No.	**Area**	**Competence**	**Study Findings**
1.	Ideas and opportunities	Spotting opportunities Creativity Vision Valuing ideas Ethical and sustainable thinking	49.4% of participants acknowledge the lack of training on project-based learning, active learning or independent learning, as part of the entrepreneurship pedagogy. (Q10) 60% of the participants acknowledge that they need further knowledge on entrepreneurship education (Q21).
2.	Resources	Self-awareness & self-efficacy Motivation & perseverance Mobilising resources Financial & economic literacy Mobilising others	(Q19): Only 37.7% of the participants confirmed the inclusion of entrepreneurship education concepts in the school curriculum, particularly in specific subjects such as "Finance in your hands", the elective module "Entrepreneur for a day", economy, and cultural heritage. 60% of the participants acknowledge that they need further knowledge on entrepreneurship education (Q21).
3.	Into action	Taking the initiative Planning & management Coping with uncertainty, ambiguity & risk Working with others Learning through experience	105 (45.5%) of the participants stated that they try to create connections and cooperation structures with businesses and community organizations to support the entrepreneurship curriculum in their schools, while 54.5% of them make partial or no efforts at all (Q14). 60% of the participants acknowledge that they need further knowledge on entrepreneurship education (Q21).

Entrepreneurial competence is both individual and collective capacity (McCallum et al. 2018). Entrepreneurship is a competence for life, which must be developed through training programs or special modules in educational institutions. The results collected from the data analysis support Hypothesis 1: Teachers have received insufficient knowledge in entrepreneurship education from university studies or training. The data collected from variables Q6 and Q8 show that only 38.2% of teachers received training on entrepreneurial competence from educational institutions and 47.1% of teachers have provided such knowledge from training programs. While the cross tabulation of variable (Q8) *"If you received training on entrepreneurial competence, it was provided by: training programs; educational institutions; by individuals"* with the variable (Q10) *"If entrepreneurship education was part of the training program, were approaches such as: project-based learning, active learning or independent learning, part of the training pedagogy?"* supports the Hypothesis 2: Teachers will not sufficiently exercise entrepreneurship competence due to lack of knowledge and skills. Moreover, the correlation emphasizes that the lack of knowledge in entrepreneurship competence on the part of the teachers affects the lack of expansion of the competence in specific activities distributed in the curriculum and in the pedagogical aspect.

The unsatisfactory level of training and knowledge obtained on entrepreneurship education has resulted in the poor level of individual and collective capacities. The teacher needs to understand the entrepreneurship process and then to implement his/her ideas, in collaboration with others. This seems difficult to achieve, for as long as the teacher in the role of the student has a lack of knowledge on entrepreneurial education, the possibility to engage with the approach to entrepreneurial education becomes more difficult (Q6; Q7; Q8; Q21). Entrepreneurship education requires sound education for the educators (European Commission 2021a). The data obtained from this study identified shortcomings in the professional engagement of teachers in order to meet the competences of each area of EntreComp Framework.

During initial teacher education and training, the student performs mainly the professional practice and course assignments in the form of mini-projects. These courses try to fulfill the minimum needs on entrepreneurship education. This learning method is mainly related to knowledge about the concept of entrepreneurship rather than the process of explaining the concept and implementing it in a work process. Even after initial teacher education, it seems that only some educators receive training on entrepreneurial education, where knowledge is mainly related to economics and business subjects. While entrepreneurship education should be included in all study programs, not only in business study programs, since one of the key competencies of the curricular framework in education is related to the competence of entrepreneurship (Q10; Q19).

Since 2012, the non-profit organization Junior Achievement Albania has been operating in Albania for the development of entrepreneurship education, which is dedicated to preparing teachers and youngsters for the world of entrepreneurship. Teachers get to experience a non-traditional way of teaching, getting closer to "the entrepreneurial teachers" who challenge young students to think outside the box and learn by doing (Junior Achievement of Albania 2014). Consequently, entrepreneurship education cannot be treated as knowledge only in economics, but in any other educational program, starting from primary education.

For instance, in the "Albanian language and literature" study program, a special module which enables students to develop computer platforms on the use of standard language can be implemented during initial teacher education (Q6). Such a platform can also serve as a start-up business. In this way, real-life situations are used to help students formulate their own ideas about engaging in or creating a new venture (Brawer 1997). Regardless of the fact that teacher training in entrepreneurial education within the framework of the Junior Achievement program is offered to a limited number of teachers, it is considered necessary to extend the activities of this program to every local educational institution (Albanian Ministry of Education and Sport and Albanian Ministry of Finance and Economy 2022).

It is necessary to develop a common framework to evaluate and improve the teacher training curriculum with involvement of clear methodologies to build entrepreneurship competencies through pedagogical processes and learning context (Fayolle et al. 2006). Entrepreneurship education is the study of the source of opportunities and the process of discovery (Gautam and Singh 2015), therefore knowledge of entrepreneurship education in second cycle study programs that prepare future teachers (Miço 2019), will enable students to return ideas into action and acquire entrepreneurial skills for implementing them during the teaching profession (Q19; Q20; Q21). In this way, it is possible to cultivate the culture of entrepreneurship, which starts with knowledge, feeds with ideas and is implemented with actions in practice.

Moreover, the regulatory framework should be expanded to include entrepreneurship education and pedagogy (Q6; Q7; Q8; Q9). Since various studies have recognized the provision of entrepreneurship education through non-traditional teaching methods (Gibb 2005), it is necessary that teaching methods be included in the education act and regulatory legal framework, to be a reference part of policy implementation (European Commission 2021b). In this way, educators will have the sense of identity and will play an active engagement in the learning methods of entrepreneurship education (Kelchtermans 2005).

Considering that entrepreneurship education requires a collaborative role of school with business community and other social organizations, an increased autonomy of school is required for a better implementation of the activities related to entrepreneurship competence (Q10; Q14). This will increase the knowledge of the teacher in entrepreneurship education, by serving as a boost for entrepreneurship competence simultaneously for teachers and students. On the other hand, since the concept of autonomy has remained unimplemented from the legal aspect, this requires the implementation of the Albanian law on the pre-university education system, regarding the financial and administrative autonomy of schools (Miço and Cungu 2022; On Pre-University Education System in the Republic of Albania Law of 2012, Pub. L. No. 69/2012 2012, Pub. L. No. 69/2012 2012).

6. Conclusions

The need to address entrepreneurship education in terms of teacher education has become evident not only from international policies with the support of international organizations, but also from the implementation of the competence based curriculum in Albanian education. For a variety of reasons, Albanian educators encountered lack of knowledge about entrepreneurship education. This study indicates the need to strengthen entrepreneurship competence in education in terms of teaching competence, pre-service teaching programs, in-service training, curricula, and cooperation with the school with stakeholders. The goal is a holistic approach towards an educational system that guides learners to identity and seize opportunities, supports incentive learning, and encourages confidence in the learners' own capabilities. The study highlighted the fact that the entrepreneurial competence of teachers is interconnected with a vivid role of school in community. In Albania, research in the framework of the connection between entrepreneurship education and teachers is insufficient. Entrepreneurship education competence is perceived to be offered to students mainly through one of the elective subjects at pre-university education level, despite being a key competence of the curriculum framework. For this reason, this paper is innovative and carries values that serve to address the suggestions made to change this perception. The limitation of this study is that more factors should be considered to further explore teachers' competencies in entrepreneurship education and the relationship between teachers' knowledge and skills and entrepreneurship competencies. An analysis according to the teacher's professional profile should be added to the questionnaire to highlight different levels of entrepreneurship education obtained in different profiles of the teaching profession. Furthermore, further study on how to improve teachers' competency through university studies and training would serve the best implementation of entrepreneurship competence.

Entrepreneurship education embraces a non-traditional teaching method that requires specific competencies, methods, and tools. It increases teachers and school principals' competence by providing them with new knowledge through the learning-by-doing method. Since knowledge about pedagogy, didactics and psychology are part of Master programs in the field of teaching, it is necessary that entrepreneurship education to be part of pre-service training. Teachers should be provided with the theoretical content of entrepreneurship education, the importance of this methodology for students, and methods on how to implement this knowledge via in-class practice.

Given that the Albanian education system continues to be centralized both politically and financially, the need for investment in entrepreneurship education must be met by the government in parallel with curricular changes in pre-service and in-service training. To facilitate this, mechanisms should be built not only to promote and improve teachers' knowledge in entrepreneurship competence, but to introduce ways of incorporating entrepreneurial knowledge in the legal framework and in Albanian educational policies. The results of this study further reveal the need to adopt the European Entrepreneurship Competence Framework (EntreComp), to include it into educational policy and legislation, and to break it down into concrete areas of the teaching profession. The results of this study can provide a focal point for policy reflection on the part of the state, with the goal of increasing investment in education to facilitate the entrepreneurship education.

In addition, this study may increase awareness among faculties and universities that offer pre-service teaching programs to enable different courses with a focus on improving entrepreneurship competence of graduating teachers about to enter future classrooms. Finally, the study seeks to provide a voice for governmental institutions developing educational policies in Albania to provide ongoing training for teachers in the field of entrepreneurship education.

Author Contributions: The authors have collaborated together in every phase of the article. Conceptualization, H.M. and J.C.; methodology, H.M. and J.C.; software, H.M. and J.C.; validation, H.M. and J.C.; formal analysis: H.M. and J.C.; resources, H.M. and J.C.; data curation, H.M. and J.C.; writing—original draft preparation, H.M. and J.C.; writing—review and editing, H.M. and J.C.; visualization, H.M. and J.C.; All authors have read and agreed to the published version of the manuscript.

Funding: This research received no external funding.

Institutional Review Board Statement: Ethical review and approval were waived for this study due to the anonymity of the survey participants.

Informed Consent Statement: Informed consent was obtained from all subjects involved in the study.

Data Availability Statement: Data supporting reported results are gathered via google form, created from the email address jonidacungu21@gmail.com. The data are available in every moment they are required.

Conflicts of Interest: The authors declare no conflict of interest.

References

Abiogu, Godwin. C. 2011. Entrepreneurship education: A veritable means of reconstructing tertiary institutions in Nigeria. In *The 29 Thannual Conference of Philosophy of Education Association of Nigeria (PEAN)*. Lagos: Lagos State University.

Akhuemonkhan, I. A., L. Raimi, and A. O. Sofoluwe. 2013. Entrepreneurship Education and Employment Stimulation in Nigeria. *Journal of Studies in Social Sciences* 3. Available online: https://infinitypress.info/index.php/jsss/article/view/66 (accessed on 18 December 2022).

Akpan, Okon Edem. 2021. Entrepreneurship Studies Education in Tertiary Institutions in Nigeria: Theories and Practices. *International Journal of Vocational and Technical Education Research* 7. Available online: https://www.eajournals.org/wp-content/uploads/Entrepreneurship-Studies-Education-in-Tertiary-Institutions-in-Nigeria.pdf (accessed on 18 December 2022).

Albanian Ministry of Education and Sport. 2014. Korniza Kurrikulare e Arsimit Parauniversitar të Republikës së Shqipërisë. (Curricular Framework of Pre-University Education of the Republic of Albania). Available online: https://ascap.edu.al/wp-content/uploads/2017/03/Korniza-Kurrikulare.pdf (accessed on 18 December 2022).

Albanian Ministry of Education and Sport, and Albanian Ministry of Finance and Economy. 2022. *On the 2022–2023 School Year in the Pre-University Education System*; Ministry of Education and Sport and Ministry of Finance and Economy, Instruction no. 22, Dated 27.07.2022. Official Gazette no. 114. 2022. Available online: https://www.qbz.gov.al/eli/fz/2022/114/3fe28bad-febb-4c60-943f-82bd9609344f (accessed on 18 December 2022).

Almahry, Fatima Fouad, Adel M. Sarea, and Allam M. Hamdan. 2018. A review paper on entrepreneurship education and entrepreneurs' skills. *Journal of Entrepreneurship Education* 21. Available online: https://www.abacademies.org/articles/a-review-paper-on-entrepreneurship-education-and-entrepreneurs-skills-7734.html (accessed on 18 December 2022).

Alvarez Marques, L., and C. Albuquerque. 2012. Entrepreneurship education and the development of young people life competencies and skills. *ACRN Journal of Entrepreneurship Prespectives* 1. Available online: http://www.acrn.eu/resources/Journals/Joe022012/201202c.pdf (accessed on 18 December 2022).

Apostu, Simona Andreea, Lindita Mukli, Mirela Panait, Iza Gigauri, and Eglantina Hysa. 2022. Economic Growth through the Lenses of Education, Entrepreneurship, and Innovation. *Administrative Sciences* 12: 74. [CrossRef]

Arruti, Arantza, and Jessica Panos-Castro. 2020. International entrepreneurship education for pre-service teachers: A longitudinal study. *Education + Training* 62: 825–41. [CrossRef]

Bacigalupo, Margherita, Panagiotis Kampylis, Yves Punie, and Godelieve Van den Brande. 2016. *EntreComp: The Entrepreneurship Competence Framework*. EUR 27939 EN. Luxembourg: Publication Office of the European Union. [CrossRef]

Banha, Francisco, Luís Serra Coelho, and Adão Flores. 2022. Entrepreneurship Education: A Systematic Literature Review and Identification of an Existing Gap in the Field. *Education Sciences* 12: 336. [CrossRef]

Béchard, Jean-Pierre, and Denis Grégoire. 2005. Entrepreneurship education research revisited: The case of higher education. *Academy of Management Learning and Education* 4: 22–43. [CrossRef]

Birdthistle, Naomi, Briga Hynes, and Patricia Fleming. 2007. Entreprise education program in secondary schools in Irland: A multi-stakeholder perspective. *Education + Training* 49: 265–76. Available online: https://eric.ed.gov/?id=EJ767292 (accessed on 18 December 2022). [CrossRef]

Brawer, Florence B. 1997. *Simulation as a Vehicle in Entrepreneurship Education*; Digest Number 97-1. Ewing Marion Kauffman Foundation. Kansas City: Center for Entrepreneurial Leadership. Available online: https://files.eric.ed.gov/fulltext/ED433468.pdf (accessed on 18 December 2022).

Brewer, Ernest W. 2009. Conduction Survey Research in Education. In *Handbook of Research on E-Learning Applications for Career and Technical Education: Technologies for Vocational Training*. Edited by Viktor Wang. Hershey: IGI Global, pp. 519–533. [CrossRef]

Bueckmann Diegoli, Rafaela, Héctor San Martín Gutiérrez, and María del Mar García de los Salmones. 2018. Teachers as Entrepreneurial Role Models: The impact of a Teachers Entrepreneurial Experience and Student Learning Styles in Entrepreneurial Intentions. *Journal of Entrepreneurship Education* 21. Print ISSN 1098-8394; Online ISSN 1528-2651. Available online: https://www.abacademies.org/articles/teachers-as-entrepreneurial-role-models-the-impact-of-a-teachers-entrepreneurial-experience-and-student-learning-styles-in-entrepr-6991.html (accessed on 18 December 2022).

Colombelli, Alessandra, Shiva Loccisano, Andrea Panelli, Orazio Antonino Maria Pennisi, and Francesco Serraino. 2022. Entrepreneurship Education: The Effects of Challenge-Based Learning on the Entrepreneurial Mindset of University Students. *Administrative Sciences* 12: 10. [CrossRef]

Commission of the European Communities. 2006. Communication from the Commission to the Council, the European Parliament, the European Economic and Social Committee and the Committee of the Regions. *Implementing the Community Lisbon Programme: Fostering Entrepreneurial Mindsets through Education and Learning*. Brussels, 13.02.2006. COM (2006) 33 final. Available online: https://eur-lex.europa.eu/LexUriServ/LexUriServ.do?uri=COM:2006:0033:FIN:en:PDF (accessed on 18 December 2022).

Council of the European Union. 2018. *Recommendation of 22 May 2018. On Key Competences for Lifelong Learning (2018/C 189/01)*. Official Journal of the European Union. C 189/1. 04.06.2018. Available online: https://eur-lex.europa.eu/legal-content/EN/TXT/PDF/?uri=CELEX:32018H0604(01)andrid=7 (accessed on 18 December 2022).

Deveci, I., and Jaana Seikkula-Leino. 2018. A review of Entrepreneurship Education in Teacher Education. *Malaysian Journal of Learning and Instruction* 15: 105–48. Available online: https://files.eric.ed.gov/fulltext/EJ1185795.pdf (accessed on 18 December 2022). [CrossRef]

Ebel, Robert L. 1980. Survey research in education: The need and the value. *Peabody Journal of Education* 57: 1980. [CrossRef]

European Commission. 2004. Helping to Create an Entrepreneurial Culture a Guide on Good Practices in Promoting Entrepreneurial Attitudes and Skills through Education. Unit B.1: Entrepreneurship (SC27 3/4) Directorate-General for Enterprise, B-1049 Brussels. Available online: https://op.europa.eu/en/publication-detail/-/publication/98912089-1cff-4f51-ab71-a3d422804a5c (accessed on 18 December 2022).

European Commission. 2010. *Communication from the Commission. Europe 2020 A Strategy for Smart, Sustainable and Inclusive Growth*. Brussels 3.3.2010. Available online: https://ec.europa.eu/eu2020/pdf/COMPLET%20EN%20BARROSO%20%20%20007%20-%20Europe%202020%20-%20EN%20version.pdf (accessed on 18 December 2022).

European Commission. 2011. *Entrepreneurship Education: Enabling Teachers as a Critical Success Factor "A Report on Teacher Education and Training to Prepare Teachers for the Challenge of Entrepreneurship Education"*. Final Report. Bruxelles: Entrepreneurship Unit. Available online: https://www.tesguide.eu/policy-strategy/enabling-teachers-as-a-critical-success-factor.htm (accessed on 18 December 2022).

European Commission. 2014. Thematic Working Group on Entrepreneurship Education: Final Report. Available online: https://www.tesguide.eu/policy-strategy/itemid/40911/ (accessed on 18 December 2022).

European Commission. 2018. *EntreComp: The European Entrepreneurship Competence Framework*. Luxembourg: Publications Office of the European Union. Available online: https://iefp.eapn.pt/docs/AE1_Quadro_Europeu_competencias_digitais.pdf (accessed on 19 December 2022). [CrossRef]

European Commission. 2021a. *A Guide to Fostering Entrepreneurship Education. Five Key Actions towards a Digital, Green and Resilient Europe*. European Innovation Council and SMEs Executive Agency (EISMEA). B-1049, Brussels. Available online: http://www.ee-hub.eu/component/attachments/?task=downloadandid=1560:EA0921266ENN (accessed on 18 December 2022).

European Commission. 2021b. *Albania 2021 Report*. Accompanying the Document Communication from the Commission to the European Parliament, the Council, the European Economic and Social Committee and the Committee of the Regions 2021 Communication on EU Enlargement Policy. Strasbourg 19.10.2021. Available online: https://ec.europa.eu/neighbourhood-enlargement/albania-report-2021_en (accessed on 18 December 2022).

European Commission/EACEA/Eurydice. 2016. *Entrepreneurship Education at School in Europe*. Eurydice Report. Luxemburg: Publications Office of the European Union. Available online: https://publications.europa.eu/resource/cellar/74a7d356-dc53-11e5-8fea-01aa75ed71a1.0001.02/DOC_1 (accessed on 21 December 2022).

European Parliament, and Council of European Union. 2006. *On Key Competences for Lifelong Learning*. Recommendation 2006/962/EC. Official Journal of the European Union L 394/10 30.12.2006. Available online: https://eur-lex.europa.eu/LexUriServ/LexUriServ.do?uri=OJ:L:2006:394:0010:0018:en:PDF (accessed on 18 December 2022).

Fayolle, Alain. 2013. Personal views on the future of entrepreneurship education. *Entrepreneurship and Regional Development* 25: 692–701. [CrossRef]

Fayolle, Alain, Benoît Gailly, and Narjisse Lassas-Clerc. 2006. Assessing the impact of entrepreneurship education programmes: A new methodology. *Journal of European Industrial Training* 30: 701–20. Available online: https://www.emerald.com/insight/content/doi/10.1108/03090590610715022/full/html (accessed on 18 December 2022). [CrossRef]

Fayolle, Alain, Caroline Verzat, and Robert Wapshott. 2016. In Quest of Legitimacy, the Theoretical and Methodological Foundations of Entrepreneurship Education Research. *International Small Business Journal* 34. Available online: https://journals.sagepub.com/doi/10.1177/0266242616649250 (accessed on 18 December 2022). [CrossRef]

Foliard, Stéphane, Sandrine Le Pontois, Alain Fayolle, and Isabell Diermann. 2019. The Legitimacy of Teachers in Entrepreneurship Education: What We can Learn from a Literature Review. In *Creating Entrepreneurial Space: Talking through Multi-Voices, Reflections on Emerging Debates Contemporary Issues in Entrepreneurship Research*. Bingley: Emerald Publishing Limited, vol. 9A, pp. 7–23. ISSN 2040-7246. [CrossRef]

Garavan, Thomas, and Barra O'Cinneide. 1994. Entrepreunership Education and Training Programs: A review and evaluation—Part 2. *Journal of European Industrial Training* 18: 13–21. Available online: https://www.emerald.com/insight/content/doi/10.1108/03090599410073505/full/html (accessed on 18 December 2022). [CrossRef]

Gautam, Manish, and Sunil Singh. 2015. Entrepreneurship Education: Concept, Characteristics and Implications for Teacher Education. *An International Journal of Education. SPIJE* 5: 21–35, ISSN 2231-2323 (Print), 2231-2404 (Online). Available online: https://ir.ucc.edu.gh/xmlui/bitstream/handle/123456789/4308/ENTREPRENEURSHIP%20EDUCATION.pdf?sequence=1andisAllowed=y (accessed on 18 December 2022).

Gibb, Alain. 2005. The future of entrepreneurship education. Determining the basis for coherent policy and practice. In *The Dynamics of Learning Entrepreneurship in a Cross-Cultural University Context*. Edited by P. Kyrö and C. Carrier. Entrepreneurship Education Series 2/2005; Hämeenlinna: University of Tampere, Research Centre for Vocational and Professional Education.

Hägg, Gustav, and Jonas Gabrielsson. 2019. A systematic literature review of the evolution of pedagogy in entrepreneurial education research. *International Journal of Entrepreneurial Behavior and Research* 26: 829–61. [CrossRef]

Hägg, Gustav, and Agnieszka Kurczewska. 2021. The Brief History of Entrepreneurial Education. In *Entrepreneurship Education*, 1st ed. London: Routledge. ISBN 9781003194972.

Huang, Yangjie, Lanyijie An, Lanying Liu, Zelin Zhuo, and Peng Wang. 2020. Exploring Factors Link to Teachers' Competencies in Entrepreneurship Education. *Frontiers in Psychology* 11: 563381. [CrossRef]

Ibáñez-Cubillas, Pilar, and José Gijón Puerta. 2021. Competences for the initial training of teachers in the teaching of entrepreneurship. *Education Policy Analysis Archives* 29: 138. [CrossRef]

Igwe, Paul Agu, Nnamdi O. Madichie, Okechukwu Chukwuemeka, Mahfuzur Rahman, Nonso Ochinanwata, and Ikenna Uzuegbunam. 2022. Pedagogical Approaches to Responsible Entrepreneurship Education. *Sustainability* 14: 9440. [CrossRef]

Ikävalko, Markku, Elena Ruskovaara, and Jaana Seikkula-Leino. 2008. Rediscovering Teacher's Role in Entrepreneurship Education. Paper presented at the EFMD 2009 Entrepreneurship Conference, Barcelona, Spain, February 26–27; Barcelona: Institución Superior de Formación Universitaria S.L. Available online: https://www.researchgate.net/publication/228684793_Rediscovering_teachers_role_in_entrepreneurship_education (accessed on 18 December 2022).

Johannisson, Bengt, Dan Halvarsson, and E. Lövstål. 1997. Stimulating and Fostering Entrepreneurship through University Training—Learning within an Organizing Context. Paper presented at the Conference Internationalizing Entrepreneurship Education and Training, IntEnt97, Monterey Bay, CA, USA, June 25–27.

Jónsdóttir, Svanborg Rannveig. 2007. *Analyses of Entrepreneurship Education in Vocational Education and Training in Iceland*. Reykjavik: Leonardo National Agency. Available online: https://www.researchgate.net/publication/260421576_Analysis_of_entrepreneurship_education_in_vocational_education_and_training_in_Iceland/link/00b7d53622bb598b55000000/download (accessed on 18 December 2022).

Junior Achievement of Albania. 2014. *Annual Report. Academic Year 2013–2014*. Available online: https://junior-albania.org/wp-content/uploads/2021/05/Annual-Report-JA-2013-2014.pdf (accessed on 18 December 2022).

Kelchtermans, Geert. 2005. Teachers' emotions in educational reforms: Self-understanding, vulnerable commitment and micropolitical literacy. *Teaching and Teacher Education* 21: 995–1006. [CrossRef]

Leedy, Paul D., and Jeanne Ellis Ormrod. 2005. *Practical Research: Planning and Design*. Upper Saddle River: Prentice Hall. Available online: http://www.worldcat.org/title/practical-research-planning-and-design/oclc/53831701 (accessed on 18 December 2022).

McCallum, Elin, Rebecca Weicht, Lisa McMullan, and Alison Price. 2018. *EntreComp into Action—Get Inspired, Make It Happen: A User Guide to the European Entrepreneurship Competence Framework*. Edited by M. Bacigalupo and W. O'keeffe. Luxembourg: Publications Office of the European Union. [CrossRef]

Miço, Heliona. 2019. The Teaching Profession in Albania and the Continuous Need for Improvement through Teacher Training Reforms. In *Rethinking Teacher Education for the 21st Century Trends, Challenges and New Directions*. Opladen, Berlin and Toronto: Verlag Barbara Budrich, p. 150. ISBN 978-3-8474-2241-9. Available online: https://shop.budrich-academic.de/wp-content/uploads/2019/09/9783847412571.pdf?v=e892e780304d (accessed on 18 December 2022).

Miço, Heliona, and Jonida Cungu. 2022. The Need for Digital Education in the Teaching Profession: A Path Toward Using the European Digital Competence Framework in Albania. *IAFOR Journal of Education: Technology in Education* 10. [CrossRef]

Ndou, Valentina. 2021. Social Entrepreneurship Education: A Combination of Knowledge Exploitation and Exploration Processes. *Administrative Sciences* 11: 112. [CrossRef]

Ndou, Valentina, Giustina Secundo, Giovanni Schiuma, and Giuseppina Passiante. 2018. Insights for shaping entrepreneurship education: Evidence from the European entrepreneurship centers. *Sustainability* 10: 4323. [CrossRef]

Ndou, Valentina, Gioconda Mele, and Pasquale Del Vecchio. 2019. Entrepreneurship education in tourism: An investigation among European Universities. *Journal of Hospitality, Leisure, Sport & Tourism Education* 25: 100175.

OECD. 2015. *Entrepreneurship at a Glance 2015*. Paris: OECD Publishing. [CrossRef]

Oksanen, Lea, Felicity Healey-Benson, and Elin McCallum. 2021. Take a Chance on CPD! How One School Put Its Faith in the EntreCompEdu CPD Programme and Developed Whole-School Collective Entrepreneurial Education. *Kwartalnik Pedagogiczny Rok* 2021: 262. Available online: https://repository.uwtsd.ac.uk/id/eprint/2033/ (accessed on 18 December 2022). [CrossRef]

Oksanen, Lea, Elena Oikkonen, and Timo Pihkala. 2022. Adopting Entrepreneurship Education—Teachers Professional Development. *Entrepereneurship Education and Pedagogy*, 1–23. [CrossRef]

On Pre-University Education System in the Republic of Albania Law of 2012, Pub. L. No. 69/2012. 2012 June 21. Available online: https://arsimi.gov.al/wp-content/uploads/2019/02/Permbledhje-parauniversitari-8-Tetor-2020.pdf (accessed on 18 December 2022).

Peltonen, Katariina. 2015. How can teachers' entrepreneurial competences be developed? A collaborative learning perspective. *Education and Training* 7. [CrossRef]

Porfírio, José, Tiago Carrilho, Jacinto Jardim, and Volker Wittberg. 2022. Fostering Entrepreneurship Intentions: The Role of Entrepreneurship Education. *Journal of Small Business Strategy* 32: 1–10. [CrossRef]

Raposo, Mário Lino Barata, and Arminda Maria Finisterra do Paço. 2011. Entrepreneurship education: Relationship between education and entrepreneurial activity. *Psicothema* 23: 453–57. Available online: https://www.researchgate.net/publication/51507462_Entrepreneurship_education_Relationship_between_education_and_entrepreneurial_activity (accessed on 18 December 2022).

Ratten, Vanessa, and Paul Jones. 2021. Covid-19 and entrepreneurship education: Implications for advancing research and practice. *International Journal of Management Education* 19: 100432. Available online: https://www.sciencedirect.com/science/article/pii/S1472811720303992 (accessed on 18 December 2022). [CrossRef]

Ratten, Vanessa, and Petrus Usmanij. 2021. Entrepreneurship education: Time for a change in research direction? *The International Journal of Menagement Education* 19: 100367. [CrossRef]

Rizza, Caroline, and Celeste Amorim Varum. 2011. Directions in entrepreneurship education in Europe. *Investigaciones de Economía de la Educación* 6: 517–38. Available online: https://econpapers.repec.org/bookchap/aecieed06/06-32.htm (accessed on 18 December 2022).

Ruskovaara, Elena, and Timo Pihkala. 2013. Teachers implementing entrepreneurship education: Classroom practices. *Education + Training* 55: 204–16. [CrossRef]

Ruskovaara, Elena, Minna Hämäläinen, and Timo Pihkala. 2016. Head teachers managing entrepreneurship education: Empirical evidence from general education. *Teaching and Teacher Education* 55: 155–64. Available online: https://www.sciencedirect.com/science/article/abs/pii/S0742051X1630004X (accessed on 18 December 2022). [CrossRef]

Secundo, Giustina, Valentina Ndou, and Pasquale Del Vecchio. 2016. Challenges for instilling entrepreneurial mindset in scientists and engineers: What works in European universities? *International Journal of Innovation and Technology Management* 13: 1640012. [CrossRef]

Secundo, Giustina, M. E. L. E. Gioconda, Pasquale Del Vecchio, Elia Gianluca, Alessandro Margherita, and Ndou Valentina. 2021. Threat or opportunity? A case study of digital-enabled redesign of entrepreneurship education in the COVID-19 emergency. *Technological Forecasting and Social Change* 166: 120565. [CrossRef]

Seikkula-Leino, Jaana. 2008. Advancing entrepreneurship education in the Finnish basic education—The prospect of developing local curricula. In *The Dynamics between Entrepreneurship, Environment and Education*. Edited by A. Fayolle and P. Kyro. Cheltenham: Edward Elgar, pp. 168–90.

Seikkula-Leino, Jaana, Elena Ruskovaara, Markku Ikavalko, Johanna Mattila, and Tiina Rytkölä. 2010. Promoting entrepreneurship education: The role of the teacher? *Education + Training* 52: 117–27. [CrossRef]

Seikkula-Leino, Jaana, Elena Ruskovaara, Timo Pihkala, Iván Diego Rodríguez, and Jane Delfino. 2019. Developing Entrepreneurship Education in Europe: Teachers' Commitment to Entrepreneurship Education in the UK, Finland and Spain. In *The Role and Impact of Entrepreneurship Education: Methods, Teachers and Innovative Programmes*. Cheltenham: Edward Elgar Publishing. [CrossRef]

Seikkula-Leino, Jaana, Maria Salomaa, Svanborg Rannveig Jónsdóttir, Elin McCallum, and Hazel Israel. 2021. EU Policies Driving Entrepreneurial Competences—Reflections from the Case of EntreComp. *Sustainability* 13: 8178. [CrossRef]

Shane, Scott, and Sankaran Venkataraman. 2000. The Promise of Entrepreneurship as a Field of Research. *The Academy of Management Review* 25. Available online: https://www.jstor.org/stable/259271 (accessed on 18 December 2022). [CrossRef]

UNESCO/ILO. 2006. *Towards an Entrepreneurial Culture for the Twenty-first Century. Stimulating Entrepreneurial Spirit through Entrepreneurship Education in Secondary Schools*. Geneva: International Labour Organization and United Nations Educational, Scientific and Cultural Organization. ISBN 978-92-3-104034-4.

United Nations. 2015. Transforming Our World: The 2030 Agenda for Sustainable Development. A/RES/70/1. Available online: https://sdgs.un.org/goals (accessed on 18 December 2022).

Volkmann, Christine, Karen E. Wilson, Steve Mariotti, Daniel Rabuzzi, Shai Vyakarnam, and A. Sepulveda. 2009. Educating the Next Wave of Entrepreneurs. Unlocking Entrepreneurial Capabilities to Meet the Global Challenges of the 21st Century. In *A Report of the Global Education Initiative*. Geneva: World Economic Forum. Available online: https://www.gvpartners.com/web/pdf/WEF_EE_Full_Report.pdf (accessed on 18 December 2022).

Wach, Krzysztof. 2014. Europeanisation of Entrepreneurship Education in Europe—Looking Back and Looking Forward. *Horizons of Education* 13: 11–31. Available online: http://cejsh.icm.edu.pl/cejsh/element/bwmeta1.element.ojs-issn-1643-9171-year-2014-volume-13-issue-26-article-oai_horyzonty_ignatianum_edu_pl_article_56 (accessed on 18 December 2022).

Walston, Jill, Jeremy Redford, and Monica P. Bhatt. 2017. *Workshop on Survey Methods in Education Research: Facilitator's Guide and Resources (REL 2017–214)*; Washington, DC: U.S. Department of Education, Institute of Education Sciences, National Center for Education Evaluation and Regional Assistance, Regional Educational Laboratory Midwest. Available online: http://ies.ed.gov/ncee/edlabs (accessed on 18 December 2022).

Welsh, Dianne H. B., William L. Tullar, and Hamid Nemati. 2016. Entrepreneurship Education: Process, method or both? *Journal of Innovation and Knowledge* 1: 125–32. [CrossRef]

Wilson, Karen. 2008. Chapter 5: Entrepreneurship Education in Europe. In *Entrepreneurship and Higher Education, Chapter 5*. New York: OECD. Available online: https://papers.ssrn.com/sol3/papers.cfm?abstract_id=1392369 (accessed on 18 December 2022).

administrative sciences

MDPI

Article

Examining the Impact of COVID-19 on Entrepreneurial Intention through a Stimulus–Organism–Response Perspective

Gentjan Çera [1,*], Margarita Ndoka [2], Ines Dika [3] and Edmond Çera [4]

1 Faculty of Economics and Agribusiness, Agricultural University of Tirana, 1001 Tirana, Albania
2 Faculty of Economy, European University of Tirana, 1001 Tirana, Albania
3 Faculty of Economics, University of Tirana, 1001 Tirana, Albania
4 Faculty of Management and Economics, Tomas Bata University in Zlin, 760 01 Zlín, Czech Republic
* Correspondence: gcera@ubt.edu.al

Abstract: Among scholars, there is an interest in understanding how entrepreneurial behavior is influenced by the consequences of crises. The COVID-19 pandemic may negatively or positively affect individuals' behavior, including entrepreneurial intention. Thus, this paper seeks to study whether or not the economic shock caused by the pandemic reinforces the intention to start a business. The research was administered at the individual level by distributing a structured survey. The hypotheses were developed based on a unique conceptual framework integrating the planned behavior theory and a stimulus–organism–response perspective. The relationships were tested using the structural equation modeling method with an original dataset of more than 800 respondents from three post-communist transition countries. The results indicate that the COVID-19 pandemic, seen as an opportunity, positively influences both the antecedents of entrepreneurial intention and individuals' intention to start a business. The message that these findings convey is that, even in crises, there are opportunities from which one can benefit, including the individual's propensity to engage in startup activities. By examining the impact of the COVID-19 crisis on entrepreneurial behavior, educational institutions and policymakers can design effective policies to foster entrepreneurship and reduce unemployment, particularly among the youth.

Keywords: COVID-19; entrepreneurial intention; PLS-SEM; theory of planned behavior; Albania; Kosovo; North Macedonia

Citation: Çera, Gentjan, Margarita Ndoka, Ines Dika, and Edmond Çera. 2022. Examining the Impact of COVID-19 on Entrepreneurial Intention through a Stimulus–Organism–Response Perspective. *Administrative Sciences* 12: 184. https://doi.org/10.3390/admsci12040184

Received: 30 October 2022
Accepted: 1 December 2022
Published: 5 December 2022

1. Introduction

It is generally accepted among scholars that disasters and crises lead to economic and societal changes in people's behavior and lifestyles (Menter 2022; Rayburn et al. 2022). Such changes can manifest as negative and positive influences on entrepreneurial activity (Krichen and Chaabouni 2021; Meahjohn and Persad 2020). Therefore, an exogenous shock not only poses additional challenges to individuals, organizations, and economies, but can also offer them new opportunities for business innovation (Brown and Rocha 2020). According to Aly (2022), entrepreneurship is seen as a vital factor in achieving a resilient economy in times of crisis. Entrepreneurial activity can be fed by encouraging and motivating individuals to create new businesses. Prior research has shown that in order to avoid failure and to ensure sustainability, individuals and organizations must be provided with support during crises (Noelia and Rosalia 2020; Ratinho et al. 2020; Çera et al. 2019; Dvorský et al. 2019; Alshebami and Seraj 2022b).

The COVID-19 pandemic is an unprecedented event that spread quickly worldwide. Being a highly infectious illness, it has impacted global public health because of its high level of transmission and increased death rate—mostly among the elderly, people with impaired immune systems, and those with underlying medical conditions (Mueller et al. 2020). Today, even though most of the governmental measures have been removed globally, the infection is still present (Our World in Data n.d.). This crisis has definitely changed the behavior in terms of how individuals work and live (Hale et al. 2021; Ratten 2021).

Generally, practitioners and academics believe that fostering entrepreneurship in times of crisis and economic recession is an adequate response (Capella-Peris et al. 2020; Meahjohn and Persad 2020). The COVID-19 pandemic has threatened public health by putting it under pressure and forcing governments to implement measures such as lockdowns. Nevertheless, this pandemic has created new opportunities for entrepreneurs (Ketchen and Craighead 2020; McGee and Terry 2022; Usman and Sun 2022), and this may represent the right moment for individuals who want to carry on their career in entrepreneurship (Godswill et al. 2021; Krichen and Chaabouni 2021; Ruiz-Rosa et al. 2020).

Considering the benefits provided by entrepreneurial activity—including social and economic aspects (decreasing the unemployment rate), especially for young adults—researchers, educational institutions, and public officials (i.e., governments) are particularly interested in having a better view of the impact of various factors on individuals' entrepreneurial behavior, including the intention to start a business. Such interest is more present in times of crisis, including the COVID-19 pandemic. A better understanding of these determinants (particularly during a crisis) would make it possible to design new policies or adjust existing ones to boost entrepreneurial activity.

According to Ratten (2021), the pandemic should be seen not only as a cause of considerable havoc, but also as a crisis that created an environment suitable for new entrepreneurial opportunities to flourish. Hence, the adversity of COVID-19 may lead to a new way of doing business (Usman and Sun 2022). Therefore, it would be interesting to see the actual effect of the COVID-19 crisis on individuals' intention to start a business.

Even though there are a considerable number of papers covering entrepreneurial intention (Abebe and Alvarado 2018; Barba-Sánchez and Atienza-Sahuquillo 2018; Belas et al. 2017; Neneh 2019; Palalić et al. 2017; Perez-Quintana et al. 2017; Zarnadze et al. 2022; Çera et al. 2021), minimal research has focused on the role of the COVID-19 pandemic on increasing individuals' intention to start up a business (Godswill et al. 2021; Hernández-Sánchez et al. 2020; Li et al. 2022; Ratten 2021; Trif et al. 2022). Therefore, this paper seeks to shed light on the relationship mentioned above by introducing an integration of two theories: the theory of planned behavior (Ajzen 1991), and the stimulus–organism–response perspective (Mehrabian and Russell 1974). Such research will provide useful insights for the entrepreneurship literature and policymakers.

The rest of this paper is organized as follows: The article's next section is dedicated to theoretical lenses and the development of hypotheses. Then, the results are interpreted after the description of the methodological procedures. The fifth section of the article consists of a discussion of the findings, followed by the section dedicated to the conclusion.

2. Literature Review

2.1. Theoretical Lenses

The present study uses two theoretical lenses: the theory of planned behavior (Ajzen 1991), and a stimulus–organism–response framework (Mehrabian and Russell 1974). The literature on these theoretical views in the context of entrepreneurial intention is discussed below.

Scholars consider individuals' intentions towards startups to be a difficult topic to study (Liñán and Fayolle 2015; Maheshwari et al. 2022). The complexity of this topic lies in the fact that individuals' intention is affected by several factors (Shane et al. 2003; Murnieks et al. 2020; Lüthje and Franke 2003), including the mental process that underlies the intentional actions (Entrialgo and Iglesias 2020) and the sophisticated process based on perception (Krueger and Carsrud 1993; Krueger et al. 2000). One of the predominant models used to study this topic is the theory of planned behavior (Maheshwari et al. 2022), introduced by Ajzen (1991), which proposes that attitudes, subjective norms, and perceived behavioral control are three key determining factors of one's intention towards a particular action and, in turn, leading to that person's actual action or behavior. The efficacy of this theory has been tested, showing that the model works (Krueger and Carsrud 1993; Kautonen et al. 2015; Munir et al. 2019;

van Gelderen et al. 2008; Zampetakis et al. 2017). The majority of the papers that used this theory applied the model without the relationship between intention and action/behavior. However, there is evidence of a strong correlation between an individual's intention and their actual behavior toward starting a business (Neneh 2019). In a meta-analysis, Armitage and Conner (2001) found that the intention–behavior correlation was statistically significant, reflecting a medium-sized effect ($r = 0.47$). Therefore, studying entrepreneurial intention may provide insights into the actual behavior towards starting a business. Moreover, this model has been used in the context of the COVID-19 pandemic (Ruiz-Rosa et al. 2020; Godswill et al. 2021; Krichen and Chaabouni 2021).

As mentioned earlier, in this paper, a different theory is applied that complies with the theory of planned behavior: the stimulus–organism–response perspective. This theory was introduced by Mehrabian and Russell (1974), consisting of three elements: stimulus, organism, and response. In this framework, stimuli refer to a set of factors, including the environment and information load. The organism is the second element of this framework, and it refers to the organism's conditions, which consist of emotional reactions to environmental stimuli. The third and final element of this framework is labeled as "response", which represents an approach or avoidance action or behavior.

These two theoretical perspectives can be merged to provide a better view of the context of the present study. Hence, the COVID-19 pandemic is seen as a stimulus coming from the external environment, affecting an individual's organism conditions. In this study, the organism is represented by determinants of entrepreneurial intention (i.e., attitude, subjective norms, and perceived behavioral control). Lastly, entrepreneurial intention covers the response component of the stimulus–organism–response perspective.

2.2. *Development of Hypotheses*

2.2.1. Attitudes towards Behavior and Entrepreneurial Intention

Once the theoretical lenses used in this study were set, the development of the hypothesis could proceed. The following paragraphs discuss the relationships based on the two mentioned theories. The first four hypotheses deal with the theory of planned behavior, while the last set represents the relationships between COVID-19 and other factors.

An individual's attitude towards entrepreneurship is defined as the extent to which a person holds a negative or positive attitude towards becoming an entrepreneur (Liñán and Chen 2009). From this definition, one can say that people with a positive perception of being an entrepreneur are more likely to have a firm interest in engaging in startup activity, whereas people with a negative perception are more likely to have no interest in such activity. Prior research demonstrates that there is a positive association between attitude and entrepreneurial intention (Joensuu-Salo et al. 2015; Feola et al. 2019; Maes et al. 2014; Haus et al. 2013; Liñán and Chen 2009), including limited research covering the time of the COVID-19 pandemic (Ruiz-Rosa et al. 2020). Nevertheless, some studies do not report a significant influence of attitudes on entrepreneurial intention, even during COVID-19 (Godswill et al. 2021; Nguyen et al. 2020). Thus, it is not clear whether attitude's effect on entrepreneurial intention is positive. Therefore, there is a need to study this relationship. Thus, our first hypothesis is as follows:

Hypothesis 1 (H1). *Personal attitude towards entrepreneurship positively influences entrepreneurial intention.*

2.2.2. Subjective Norms and Entrepreneurial Intention

According to the theory of planned behavior, the second determinant of a person's intention is the subjective norm, which is known as the social influence on an individual to perform (or not) a particular behavior (Ajzen 1991). This is related to the belief that an important person, relatives, friends, or others will endorse (or not) a specific behavior, e.g., a decision to start up a business. Prior studies show a positive effect of subjective norms on entrepreneurial intention (Moriano et al. 2012; Rantanen and Toikko 2017; Mirjana et al.

2018; Maresch et al. 2016; Misoska et al. 2016). Moreover, it is difficult to find a paper reporting an insignificant relationship—for example, the study of Godswill et al. (2021), which was conducted in the context of the COVID-19 pandemic. The present study may offer additional evidence about this relationship in the context of the pandemic. Thus, subjective norms (i.e., social influence) are expected to positively predict one's intention to start a business. Therefore, our second hypothesis is as follows:

Hypothesis 2 (H2). *An individual's entrepreneurial intention is positively influenced by subjective norms.*

2.2.3. Perceived Behavioral Control and Entrepreneurial Intention

Based on the theory of planned behavior, perceived behavioral control is the third main determinant of an individual's intention (Ajzen 1991). In the context of entrepreneurship, this is seen as the belief and confidence that a person has in carrying out business activities as an entrepreneur. Based on this logic, the more opportunities and resources a person believes they have and the fewer constraints they foresee, the greater their perceived control over a particular action is expected to be, including startup activity. Previous studies confirm the positive effect of perceived behavioral control on entrepreneurial intention (Al-Jubari 2019; Joensuu-Salo et al. 2015; Kautonen et al. 2015; Liñán and Chen 2009; Nguyen et al. 2020), including those conducted during the COVID-19 pandemic (Ruiz-Rosa et al. 2020; Godswill et al. 2021). Although there is such evidence, there is a need to study this relationship in the context of COVID-19 in post-communist countries. Thus, our third hypothesis is as follows:

Hypothesis 3 (H3). *Perceived behavioral control positively influences entrepreneurial intention.*

2.2.4. The Role of COVID-19

Previous studies have tried to shed light on the impact of COVID-19 on different aspects of entrepreneurship, including the intention to start a business (Lopes et al. 2021; Botezat et al. 2022). Arve et al. (2022) conducted an experiment and found that the majority of prospective entrepreneurs either canceled or postponed their projects during the first months of the pandemic. Nevertheless, some studies see this crisis as a chance to implement a business idea by establishing a firm. Research found that most of the students from Erasmus University Rotterdam did not change their entrepreneurial intention due to COVID-19 (Wismans et al. 2022). In addition, the latter study demonstrated that the share of students who increased their entrepreneurial intention (19%) was higher than those who decreased such intention (16%). Hence, evidence supports the claim that COVID-19 offers new chances for entrepreneurship. Moreover, seeing COVID-19 as an opportunity to engage in entrepreneurial activity is more common than perceiving it as a threat (Lungu et al. 2021). This finding is supported by a prior study conducted in a war setting, which suggests that even under conditions of war, people develop entrepreneurial intentions in case they can grow from adversity and believe in their abilities (Bullough et al. 2014). Thus, one can say that crisis may create a suitable environment for individuals to see entrepreneurial opportunities. According to Krichen and Chaabouni's (2021) research, there is a positive and statistically significant impact of COVID-19 seen as an opportunity on students' likelihood to start a business. This finding is consistent with other research that highlights the pandemic's potential beneficial effects on entrepreneurship (Botezat et al. 2022; Lungu et al. 2021). Consequently, a positive effect of COVID-19 on entrepreneurial intention was also expected to be present in this study.

Recently published papers have utilized the theory of planned behavior to explore the impact of COVID-19 on behavioral changes, including the effects of COVID-19 on the determinants of behavioral intention (i.e., attitude, subjective norms, and perceived behavioral control) (Srisathan and Naruetharadhol 2022; Prasetyo et al. 2020; Han et al. 2020; Lucarelli et al. 2020). It is generally known that external factors influence individuals' attitudes towards particular actions. In this context, according to Rayburn et al. (2022), in response to the COVID-19 pandemic, individuals moved from fear to frugality, either by following new behaviors forced by the crisis, or by going back to their behavior prior to the crisis. Hence, attitudes towards different aspects change in a crisis setting, such as attitudes towards entrepreneurship in general and starting up a business. In the context of the COVID-19 pandemic, Gomes et al. (2021) demonstrated that the positive and significant influence of attitudes toward behavior and entrepreneurial intention was present in both situations: before and during the pandemic. Moreover, the latter study shows a slightly more significant effect during the COVID-19 pandemic than before it.

Similar to attitudes, evidence shows that subjective norms and perceived behavioral control increased due to COVID-19 (Botezat et al. 2022). According to prior research, people's lifestyles have changed due to COVID-19 (Rayburn et al. 2022; Ratten 2021). At the community level, to avoid the transmission of illness, individuals were recommended to take additional hygienic measures. Individuals are pursuing digitization more aggressively than ever before in order to respect social distancing norms, embracing new activities and interactions—including teleworking—and adjusting everyday habits to fit a new reality (Srisathan and Naruetharadhol 2022). Therefore, a person's friends and relatives may push them to take action to start a business, meaning that subjective norms are influenced by COVID-19. Indeed, previous research supports such an association (Prasetyo et al. 2020; Srisathan and Naruetharadhol 2022; Han et al. 2020).

Very few papers have discussed the impact of COVID-19 on perceived behavioral control. By definition, perceived behavioral control is the comfort level of a person in performing any particular behavior (Ajzen 1991). Its determinants are assumed to be the set of accessible control beliefs, such as beliefs about the presence of factors that can enable or constrain a certain behavior. This reasoning leads to the concept of resilience, which refers to the ability that a person has to recover from or adjust easily to change or misfortune (Sinclair and Wallston 2004; Alshebami and Seraj 2022a). Studies have shown that resilience is an important factor in crisis settings, including in entrepreneurship (Arve et al. 2022; Bullough et al. 2014; Sharma and Rautela 2021; Schepers et al. 2021; Alshebami 2022). Prior research has found that perceived behavioral control is affected by crises, including COVID-19, supporting the existence of this association (Prasetyo et al. 2020; Srisathan and Naruetharadhol 2022).

Based on the above discussion, one can conclude that COVID-19 influences attitudes toward entrepreneurship, subjective norms, and perceived behavioral control. Thus, our fourth hypothesis is as follows:

Hypothesis 4a–c (H4a–c). *The COVID-19 pandemic has a positive effect on attitudes to start a business (H4a), subjective norms (H4b), and perceived behavioral control (H4c).*

Hypothesis 4d (H4d). *Entrepreneurial intention is positively affected by the COVID-19 pandemic.*

The integration of the theory of planned behavior and the stimulus–organism–response perspective is illustrated in Figure 1. Additionally, the figure also shows the proposed linkages (i.e., hypotheses).

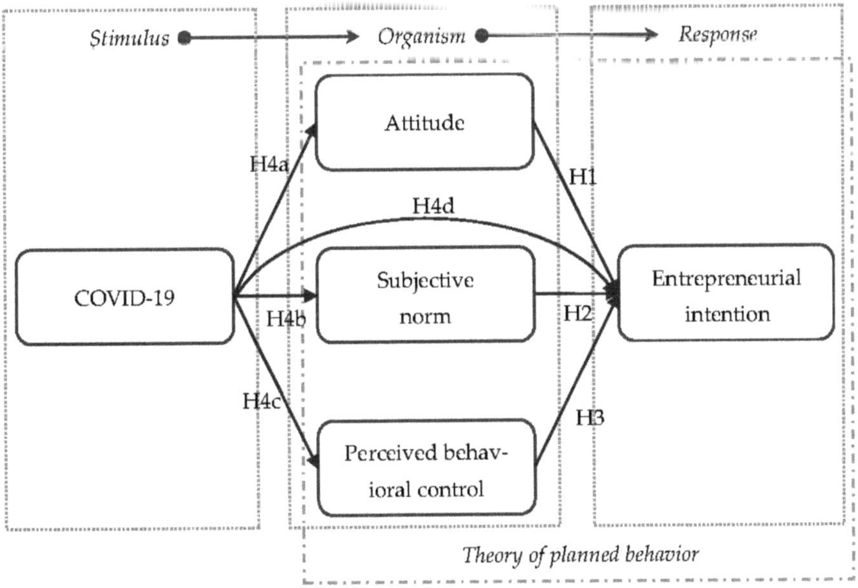

Figure 1. Conceptual framework and hypotheses.

3. Method and Procedures

3.1. Research Instrument and Sample

In order to meet the goals of this research, a survey was conducted to test the research model and indicate the significance of the relationships. The use of surveys is a quantitative method that can infer the population by studying a sample (Creswell and Creswell 2017). This type of method implies the need for primary data collection. Hence, a questionnaire was developed based on the literature review.

The research covered three countries: Albania, Kosovo, and North Macedonia. After the validation of the questionnaire, it was translated into the Albanian and Macedonian languages. The data were collected during the COVID-19 pandemic at the end of 2021.

The respondents were selected by following a two-stage sampling procedure: (i) selection of primary sampling unit, and (ii) selection of the respondents. The first stage was fulfilled by randomly selecting participants from among the voting centers. The second stage consisted of selecting the respondents following a methodology of starting from the voting center and then moving clockwise, always getting further from the starting point. More than 800 valid responses were collected, with more than 200 respondents from each country. Such a sample size is well above the recommendation of Hair et al. (2010).

Table 1 shows the sample profile (overall and per country). For the most part, the pattern of the subsample profiles reflects one of the overall samples. Three out of five respondents were 24 years old or less. The majority of the respondents were female. Almost 70% of the respondents were settled in urban areas (i.e., cities).

3.2. Measurement of Variables

The variables of this research were measured as proposed in the literature, with minor changes, including wording or adaptation to the context. The dependent variable in this paper is entrepreneurial intention. There are different ways in which this variable has been measured in the literature (Armitage and Conner 2001; Çera and Çera 2020; Franke and Lüthje 2004; Krueger and Carsrud 1993; Lim et al. 2016; Çera et al. 2020). However, as claimed by Thompson (2009), an individual's intention cannot be captured by considering only one item/statement; therefore, entrepreneurial intention in this work is measured by four items/statements, which can be found in the Appendix A. The source for this measurement was the work published by Liñán and Chen (2006).

Table 1. Sample profile.

Variable	Category	Country			
		Albania n = 412	Kosovo n = 207	North Macedonia n = 203	Total N = 822
Settlement	City	87.9%	48.8%	49.3%	68.5%
	Village	12.1%	51.2%	50.7%	31.5%
	Total	100%	100%	100%	100%
Gender	Male	26.7%	29.0%	25.1%	26.9%
	Female	73.3%	71.0%	74.9%	73.1%
	Total	100%	100%	100%	100%
Age	18–24 years old	66.5%	46.4%	58.1%	59.4%
	25–35 years old	33.5%	53.6%	41.9%	40.6%
	Total	100%	100%	100%	100%

Regarding the independent variables, excluding the COVID-19 variable, all of the others were measured similarly to the approach of García-Rodríguez et al. (2017). A single-item variable was used to measure the impact of COVID-19 on the antecedents of the individuals' intent to act and their intentions themselves. The statement reads "the COVID-19 pandemic situation has made me optimistic about starting a business". The respondents were asked to indicate their level of agreement with the statement (1 = strongly disagree, 5 = strongly agree). A similar type of measurement was used in a prior study (Krichen and Chaabouni 2021). Appendix A (Table A1) summarizes the list of items/indicators used to measure each variable included in this research.

3.3. Method

The partial least squares structural equation modelling (PLS-SEM) method was used to test the proposed conceptual framework. PLS-SEM was performed using SmartPLS 3.0 (Ringle et al. 2015) computer software. The PLS approach is a variance-based structural equation modeling (SEM) method (Hair et al. 2017). This approach enables assessment of the measurement model, including the reliability and validity of the constructs and the structural model. Therefore, it can test the formulated hypotheses by examining the standardized path coefficients. As recommended by the literature, the standardized coefficients were estimated using the bootstrap procedure, with 5000 iterations of resampling (Hair et al. 2019).

Since the three countries share similar cultures and levels of economic development, our analysis considered one dataset rather than three sub-datasets (one per country). According to Hofstede (2011), these countries share very similar cultural values (see Figure 2). Unfortunately, there are no reports for Kosovo. However, Kosovo is inhabited by Albanians and has many things in common not only with Albania, but also with North Macedonia. As the graph depicts, there are few differences between Albania and North Macedonia. Therefore, the three countries share similar cultural values. This leads to the suggestion of analyzing the data as a whole, rather than separately.

3.4. Checking Assumptions

A PLS-SEM method is an approach based on assumptions. Their violation (individually or collectively) leads to problems in the interpretation of the results that this method generates. Therefore, the violation of any of this approach's assumptions is an indication that its output is misleading. To avoid such issues there is a need to check some assumptions, which are mostly related to the measurement model, including the reliability and validity of the items and scales.

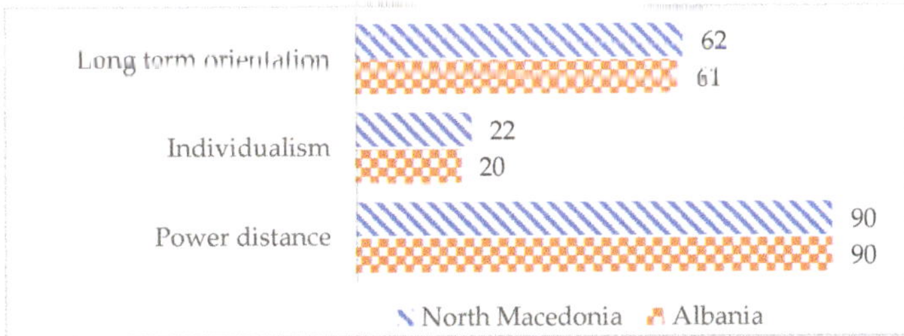

Figure 2. Hofstede's cultural dimensions for Albania and North Macedonia. *Source*: Hofstede Insights: https://www.hofstede-insights.com/ (accessed on 22 October 2022).

In order to assess the fitness of the model, a list of metrics can be examined. In this context, Cronbach's alpha, composite reliability (CR), and rho alpha provide information about scale reliability, while average variance extracted (AVE) reports the extent to which the scale reliability and convergent validity are satisfactory. These metrics are assessed and reported in Table 2. Since the values of Cronbach's alpha (above 0.70), composite reliability (above 0.60), and rho alpha are above the thresholds for all scales (Hair et al. 2019), it can be said that the data show satisfactory reliability and convergent validity of the constructs. In addition, item reliability can be assessed by examining the factor loadings, which should be above 0.708 (Hair et al. 2019). Indeed, as reported in Table 2, all loadings are above this threshold, leading to the conclusion that all constructs explain more than half of the indicator's variance, providing evidence to accept indicator reliability.

Table 2. Descriptive statistics and measurement model quality attributes.

Variable	Mean	Standard Deviation	Loadings	VIF	CA	rho_A	CR	AVE
COVID-19	2.20	1.21	1	1	1	1	1	1
EI	-	-	-	-	0.9079	0.9104	0.9354	0.7837
ei1	3.19	1.26	0.8639	2.4197				
ei2	3.25	1.22	0.8979	2.8824				
ei3	3.48	1.30	0.9018	3.1293				
ei4	3.50	1.28	0.8768	2.7662				
ATT	-	-	-	-	0.9349	0.9357	0.9505	0.8367
att1	3.29	1.27	0.8976	3.0892				
att2	3.40	1.32	0.9270	4.0227				
att3	3.56	1.35	0.9140	3.5009				
att5	3.28	1.30	0.9199	3.7092				
SN	-	-	-	-	0.8739	0.8919	0.9215	0.7966
sn1	3.74	1.22	0.8783	1.8999				
sn2	3.59	1.22	0.9223	3.4762				
sn3	3.33	1.21	0.8762	2.9185				
PBC	-	-	-	-	0.9047	0.9067	0.9265	0.6777
pbc1	3.62	1.18	0.8193	2.3296				
pbc2	3.53	1.11	0.8517	2.5354				
pbc3	3.68	1.14	0.8589	2.7262				
pbc5	3.33	1.17	0.7913	2.0059				
pbc6	3.37	1.12	0.7983	2.1293				
pbc7	3.28	1.12	0.8175	2.2221				

Note: VIF, variance influence factor; CA, Cronbach's alpha; CR, composite reliability; AVE, average variance extracted; ATT, attitude; EI, entrepreneurial intention; PBC, perceived behavioral control; SN, subjective norms; COVID-19, the COVID-19 pandemic.

Moreover, Table 2 shows the variance influence factor (VIF) for each indicator. In general, VIF indicates the presence of multicollinearity in a relationship. However, since the data show that the VIF values are below 5 (Hair et al. 2019), one can say that there is no multicollinearity issue within the measurement model.

Another crucial issue to consider in PLS-SEM deals with the discriminant validity, which indicates how distinct one construct is from others. Table 3 provides information on this issue, since it reports the correlations' heterotrait–monotrait ratio (HTMT). It is recommended to examine HTMT coefficients when using PLS-SEM as a measure of discriminant validity (Henseler et al. 2015). The rule of thumb is that the HTMT values should be below 0.85. In Table 3, all of the coefficients satisfy this rule. This test result indicates that the discriminant validity is set in this paper. Additionally, Table 3 reports the correlation coefficients among the measured constructs.

Table 3. Correlation matrix and discriminant validity—HTMT.

	ATT	COVID-19	EI	PBC	SN
ATT		0.2598	0.5888	0.6382	0.4566
COVID-19	0.2687		0.2196	0.2225	0.1657
EI	0.6370	0.2297		0.4711	0.3801
PBC	0.6918	0.2353	0.5186		0.5778
SN	0.4936	0.1717	0.4172	0.6374	

Note: Correlation coefficients are above the diagonal, while HTMT coefficients are below it. ATT, attitude; EI, entrepreneurial intention; PBC, perceived behavioral control; SN, subjective norms; COVID-19, the COVID-19 pandemic.

Figure 3 graphically illustrates the main results of the measurement model, as generated by SmartPLS 3.0.

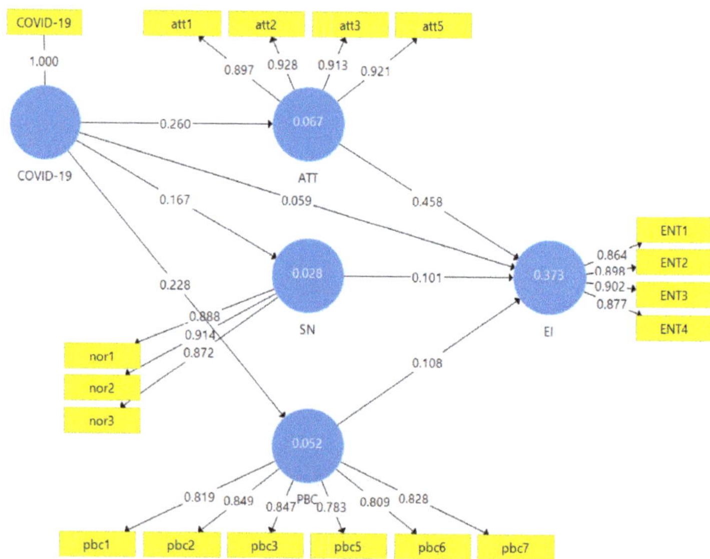

Figure 3. Measurement model. *Note*: ATT, attitude; EI, entrepreneurial intention; PBC, perceived behavioral control; SN, subjective norms; COVID-19, the COVID-19 pandemic.

4. Results

Upon checking the assumptions of the PLS-SEM method, the output of the analysis can be interpreted. This means that the satisfaction of the PLS-SEM's assumptions leads to the examination of the formulated hypotheses. The tested model explains 37.2% of the variation in entrepreneurship intention, 6.7% in attitude, 5.2% in perceived behavioral control, and almost 3% in subjective norms. These statistics are summarized in Table 4.

Table 4. R-squares.

Construct	R Squared	Adjusted R Squared
Attitude	0.067	0.066
Entrepreneurial intention	0.372	0.369
Perceived behavioral control	0.052	0.051
Subjective norms	0.028	0.027

According to the proposed conceptual framework, the entrepreneurial intention is determined by attitude, subjective norms, perceived behavioral control, and COVID-19. The results of the path analysis are summarized in Table 5. As indicated in the Method and Procedures section, the path coefficient's statistical significance was examined to conclude whether the hypotheses were supported or not.

Table 5. Results of hypotheses testing via bootstrapping (direct effect).

Hypothesis	Path	Coefficient	*t*-Value	VIF
H1	ATT → EI	0.459	11.98 ***	1.762
H2	SN → EI	0.101	2.902 **	1.532
H3	PBC → EI	0.108	2.672 **	2.052
H4a	COVID-19 → ATT	0.260	8.313 ***	1.028
H4b	COVID-19 → PBC	0.228	6.952 ***	1.028
H4c	COVID-19 → SN	0.165	5.108 ***	1.000
H4d	COVID-19 → EI	0.220 [a]	6.492 ***	1.762

Note: VIF, variance influence factor; ATT, attitude; EI, entrepreneurial intention; PBC, perceived behavioral control; SN, subjective norms; COVID-19, the COVID-19 pandemic; a, total effect; ** and *** imply that the test result is significant at the 99% and 99.9% levels, respectively.

Subjective norms positively influenced attitudes ($\beta = 0.426$, $t = 13.68$, $p < 0.001$) and perceived behavioral control ($\beta = 0.557$, $t = 21.49$, $p < 0.001$). These findings support H1a and H1b, meaning that subjective norms are a significant determinant of both an individual's attitude and their perceived behavioral control. The data show that an individual's intention toward startups is statistically significantly and positively affected by attitude ($\beta = 0.459$, $t = 11.98$ $p < 0.001$), subjective norms ($\beta = 0.101$, $t = 2.902$, $p < 0.01$), and perceived behavioral control ($\beta = 0.108$, $t = 2.672$, $p < 0.01$). Thus, there is evidence in support of H1, H2, and H3. These hypotheses deal with the standard model of the theory of planned behavior (Ajzen 1991). The remaining hypotheses link the impact of the COVID-19 pandemic with the theory of planned behavior variables.

In this paper, the role of stimulus in the stimulus–organism–response paradigm is played by the COVID-19 pandemic, which influences all factors mentioned in the theory of planned behavior (see Figure 1). The data show that COVID-19 statistically and positively influences attitude ($\beta = 0.260$, $t = 8.313$, $p < 0.001$), perceived behavioral control ($\beta = 0.228$, $t = 6.952$, $p < 0.001$), and subjective norms ($\beta = 0.165$, $t = 5.108$, $p < 0.001$). Based on these results, one can conclude that COVID-19 impacts the antecedents of individuals' intention to start a business, showing strong evidence in support of H4a–c. The last hypothesis deals with the impact of COVID-19 on an individual's entrepreneurial intention. Table 5 shows the total effect of the COVID-19 pandemic on entrepreneurial intention, which is statistically significant ($\beta = 0.220$, $t = 6.492$, $p < 0.001$). In addition, this influence is positive, meaning that an increase in the values of the variable that measures COVID-19 leads to an increase in individuals' entrepreneurial intentions.

Figure 4 graphically illustrates the path analysis generated by SmartPLS 3.0. Note that the total effect is not plotted in this figure. Instead, the figure provides information on the inner model by showing the path coefficients along with their statistical significance (*t*-statistics).

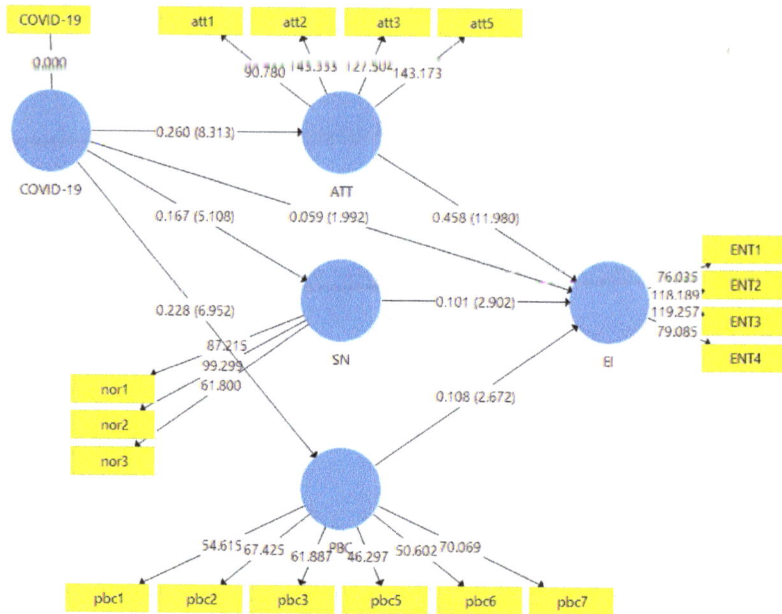

Figure 4. Path analysis—inner model: path coefficients (*t*-values). *Note*: ATT, attitude; EI, entrepreneurial intention; PBC, perceived behavioral control; SN, subjective norms; COVID-19, the COVID-19 pandemic.

5. Discussion

This paper aimed to examine the impact of the COVID-19 pandemic on individuals' intentions toward starting a business. The integration of two theories was proposed: the theory of planned behavior (Ajzen 1991), and the stimulus–organism–response perspective (Mehrabian and Russell 1974). The integration of these two theories offers a conceptual framework that can determine the impact of external stimuli (here represented by COVID-19) on entrepreneurial intention and its determinants.

The main finding of this work is that crisis, in addition to posing additional challenges to individuals and organizations, can also be seen as a generator of new opportunities. This finding is consistent with the limited research that has been conducted in this context (Ketchen and Craighead 2020; Krichen and Chaabouni 2021; Li et al. 2022; Ratten 2021; Usman and Sun 2022). Hence, due to the COVID-19 pandemic, individuals can find new business opportunities and a suitable situation to implement new ideas, which may lead to innovation (Brown and Rocha 2020). Such linkages can be seen with individuals' entrepreneurial behavior as well, including the intention to start a business. Thus, as this research demonstrates, entrepreneurial intention is positively affected by COVID-19 (seen as an opportunity). According to the findings of our work, individuals who perceive times of crisis as an opportunity may engage in startup activities to benefit from the situation, as their entrepreneurial intention is increased. This finding seems reasonable from the point of view of the entrepreneurial situation, which can form the perception of various risks that individuals face in a crisis context (Rayburn et al. 2022; Traczyk and Zaleskiewicz 2016). This is linked to the individuals' attitudes towards starting a business, which is an essential determinant of entrepreneurial intention and was found to be influenced by COVID-19. This result reinforces the positive impact that a crisis (seen as an opportunity and not as a threat) can have on entrepreneurial behavior, as shown in this study, which contradicts two prior studies (Godswill et al. 2021; Nguyen et al. 2020).

Nevertheless, Ruiz-Rosa et al. (2020) found similar results in a study on the social entrepreneurial intention of students from a university in Spain in the context of COVID-

19. Additionally, the data show that subjective norms and perceived behavioral control are positively influenced by COVID-19 which, in turn, affects entrepreneurial intention. These findings are consistent with the limited prior research carried out in the context of the COVID-19 pandemic (Botezat et al. 2022; Gomes et al. 2021; Nguyen et al. 2020; Ruiz-Rosa et al. 2020).

Such findings lead to the discussion on how to increase entrepreneurial activity. Various factors can influence entrepreneurial activity; however, one that all scholars agree on is that of education on entrepreneurship. Since entrepreneurship education has been found to be a significant determinant of individuals' intention towards engagement in startup activities (Çera et al. 2020; Dana et al. 2021; Durán-Sánchez et al. 2019; Hoppe 2016; Mwasalwiba 2010; Papagiannis 2018; Paray and Kumar 2020; Pedrini et al. 2017; Premand et al. 2016; Oo et al. 2018; Oosterbeek et al. 2010), it is unreasonable to doubt the role of education in this regard. Therefore, educational institutions are seen as critical actors in motivating students towards entrepreneurship since, through their curricula, they can be equipped with the knowledge and skills needed for starting and managing a business. Moreover, scholars claim that the entrepreneurial university environment is an essential factor that can increase entrepreneurial intention and actual behavior (García-Rodríguez et al. 2017; Ndou et al. 2018, 2019; Trif et al. 2022; Çera et al. 2021). Since COVID-19 has impacted the traditional means of providing entrepreneurship education (Hoti et al. 2022; Ndou 2021; Kripa et al. 2021), educational institutions should address the challenges and make use of innovative ways to deliver the best practices to equip students with adequate knowledge and skills (Cunningham 2022). Recently, there has been a discussion in the literature on the need to shift from the traditional means of offering entrepreneurship education to digital methods (Volkmann and Grünhagen 2022; Lehmann et al. 2022). This need to shift from the traditional approach to a new one is present due to COVID-19. Therefore, the COVID-19 pandemic has also created new challenges and opportunities for educational institutions.

6. Conclusions

6.1. Implications of the Study

Driven by the theory of planned behavior (Ajzen 1991) and the stimulus–organism–response perspective (Mehrabian and Russell 1974), this study provides a unique and improved research model for investigating the positive impact of COVID-19 on individuals' intentions to start a business in the context of three post-communist transition countries. Furthermore, the combination of these two theories provides the possibility of investigating the abovementioned relationship by seeing the COVID-19 pandemic as an external inducement (i.e., stimulus) that influences attitudes, subjective norms, perceived behavioral control, and entrepreneurial intention.

The findings of this research provide theoretical contributions and practical implications. Regarding this paper's contribution to the entrepreneurship literature, the authors believe that the integration of the two abovementioned theoretical lenses should be considered as a novelty of the paper. Putting the theory of planned behavior (Ajzen 1991) into a stimulus–organism–response paradigm (Mehrabian and Russell 1974) would be a useful approach that provides results. Therefore, this study adds to the existing literature by offering a new and unique conceptual framework, which may be useful for investigating the impacts of exogenous shocks on entrepreneurial intention and its determinants in a crisis context. In addition, in terms of theoretical contribution, the current paper demonstrates that a disaster or crisis that occurs, such as COVID-19, can not only pose additional challenges but also provide new opportunities which, in turn, lead to the increase in individuals' intentions towards starting a business. Therefore, our findings are valuable in strengthening the literature on entrepreneurial intention, which is ample in terms of research conducted in "normal times" but limited when a disaster or crisis occurs, such as the COVID-19 pandemic.

Regarding the practical implications of this research, from the policymakers' point of view, it is imperative to understand the effects of a crisis on individuals' intentions and behavior toward startup activity, because this can lead to a reduction in unemployment—especially among young adults. Therefore, according to this research, policymakers and educational institutions should adjust the existing policies, strategies, instruments, and curricula to face the challenges raised by COVID-19 and benefit from the new opportunities.

6.2. Limitations

Although our research's goal was met, this study is not free of limitations. Firstly, the study focuses on individuals' intentions rather than their actual behavior toward starting a business. Even though there is a significant correlation between entrepreneurial intention and behavior (Bae et al. 2014; Joensuu-Salo et al. 2020), it is still not certain that intention will turn into behavior in either the near or far future (Bogatyreva et al. 2019). Secondly, from a methodological perspective, a crisis's impact should be measured by applying a pre- and post-test research design. Finally, the generalization of the findings obtained by the presented research model is limited to the countries that this study covers. Therefore, scholars should be advised to use and test the proposed conceptual framework in different contexts, as further research could contribute to overcoming the abovementioned limitations.

Author Contributions: Conceptualization, G.Ç. and I.D.; methodology, E.Ç. and G.Ç.; software, G.Ç.; validation, M.N. and E.Ç.; formal analysis, G.Ç.; investigation, E.Ç.; resources, E.Ç. and I.D; data curation, E.Ç.; writing—original draft preparation, G.Ç. and M.N.; writing—review and editing, M.N. and I.D.; visualization, G.Ç.; supervision, G.Ç. and I.D.; project administration, M.N.; funding acquisition, E.Ç. and I.D. All authors have read and agreed to the published version of the manuscript.

Funding: This research received no external funding.

Institutional Review Board Statement: Not applicable.

Informed Consent Statement: Not applicable.

Data Availability Statement: Not applicable.

Conflicts of Interest: The authors declare no conflict of interest.

Appendix A

Table A1. Items and sources of the variables used in the research.

Code	Items and Sources
Indicate your level of agreement with the following statements for each (1 = strongly disagree, 5 = strongly agree)	
	COVID-19 (Krichen and Chaabouni 2021)
	The COVID-19 pandemic situation has made me optimistic in starting a business
	Entrepreneurial intention (Liñán and Chen 2006)
ei1	I am ready to do anything to be an entrepreneur
ei2	My professional goal is to become an entrepreneur
ei3	I will make every effort to start and run my own firm
ei4	I am determined to create a firm in the future
	Attitude (García-Rodríguez et al. 2017)
att1	Being an entrepreneur implies more advantages than disadvantages to me
att2	A career as entrepreneur is attractive for me
att3	If I had the opportunity and resources, I would become an entrepreneur
att4 *	Being an entrepreneur would entail great satisfaction for me
att5	Among various options, I would rather become an entrepreneur

Table A1. *Cont.*

Code	Items and Sources
	Perceived behaviour control (García-Rodríguez et al. 2017)
pbc1	I am usually able to protect my personal interests
pbc2	When I make plans, I am almost certain to make them work
pbc3	I can pretty much determine what will happen in my life
pbc4 *	For me, being an entrepreneur would be very easy
pbc5	If I wanted to, I could easily pursue a career as entrepreneur
pbc6	As entrepreneur, I would have complete control over the situation
pbc7	As an entrepreneur, the chances of success would be very high
	Subjective norms (García-Rodríguez et al. 2017)
Pursuing a career as an entrepreneur, how do people in your environment react? (1 = very negatively, 5 = very positively)	
sn1	Your close family
sn2	Your friends
sn3	Your fellow students/colleagues

* Removed from the analysis due to the violation of the PLS-SEM assumptions.

References

Abebe, Michael A., and David Alvarado. 2018. Blessing in Disguise? Social and Institutional Determinants of Entrepreneurial Intentions Following Involuntary Job Loss. *Journal of Small Business Management* 56: 555–72. [CrossRef]

Ajzen, Icek. 1991. The Theory of Planned Behavior. *Organizational Behavior and Human Decision Processes* 50: 179–211. [CrossRef]

Al-Jubari, Ibrahim. 2019. College Students' Entrepreneurial Intention: Testing an Integrated Model of SDT and TPB. *SAGE Open* 9: 2158244019853467. [CrossRef]

Alshebami, Ali Saleh. 2022. Psychological Features and Entrepreneurial Intention among Saudi Small Entrepreneurs during Adverse Times. *Sustainability* 14: 7604. [CrossRef]

Alshebami, Ali Saleh, and Abdullah Hamoud Ali Seraj. 2022a. Exploring the Influence of Potential Entrepreneurs' Personality Traits on Small Venture Creation: The Case of Saudi Arabia. *Frontiers in Psychology* 13: 1403–3. [CrossRef]

Alshebami, Ali Saleh, and Abdullah Hamoud Ali Seraj. 2022b. Investigating the Impact of Institutions on Small Business Creation Among Saudi Entrepreneurs. *Frontiers in Psychology* 13: 3346–46. [CrossRef]

Aly, Maha. 2022. COVID-19 Pandemic Lockdown: The Era of Connection and Creation. In *The COVID-19 Crisis and Entrepreneurship: Perspectives and Experiences of Researchers, Thought Leaders, and Policymakers*. Edited by David B. Audretsch and Iris A. M. Kunadt. International Studies in Entrepreneurship. Cham: Springer International Publishing, pp. 177–83. [CrossRef]

Armitage, Christopher J., and Mark Conner. 2001. Efficacy of the Theory of Planned Behaviour: A Meta-Analytic Review. *British Journal of Social Psychology* 40: 471–99. [CrossRef]

Arve, Malin, Claudine Desrieux, and Romain Espinosa. 2022. Entrepreneurial Intention and Resilience: An Experiment during the COVID-19 Lockdown. *Managerial and Decision Economics*, 1–18. [CrossRef]

Bae, Tae Jun, Shanshan Qian, Chao Miao, and James O. Fiet. 2014. The Relationship Between Entrepreneurship Education and Entrepreneurial Intentions: A Meta-Analytic Review. *Entrepreneurship Theory and Practice* 38: 217–54. [CrossRef]

Barba-Sánchez, Virginia, and Carlos Atienza-Sahuquillo. 2018. Entrepreneurial Intention among Engineering Students: The Role of Entrepreneurship Education. *European Research on Management and Business Economics* 24: 53–61. [CrossRef]

Belas, Jaroslav, Beáta Gavurová, J. Schonfeld, Katarína Zvaríková, and Tomas Kačerauskas. 2017. Social and Economic Factors Affecting the Entrepreneurial Intention of University Students. *Transformations in Business & Economics* 16: 220–39.

Bogatyreva, Karina, Linda F. Edelman, Tatiana S. Manolova, Oleksiy Osiyevskyy, and Galina Shirokova. 2019. When Do Entrepreneurial Intentions Lead to Actions? The Role of National Culture. *Journal of Business Research* 96: 309–21. [CrossRef]

Botezat, Elena-Aurelia, Alexandru Constăngioară, Anca-Otilia Dodescu, and Ioana-Crina Pop-Cohuţ. 2022. How Stable Are Students' Entrepreneurial Intentions in the COVID-19 Pandemic Context? *Sustainability* 14: 5690. [CrossRef]

Brown, Ross, and Augusto Rocha. 2020. Entrepreneurial Uncertainty during the COVID-19 Crisis: Mapping the Temporal Dynamics of Entrepreneurial Finance. *Journal of Business Venturing Insights* 14: e00174. [CrossRef]

Bullough, Amanda, Maija Renko, and Tamara Myatt. 2014. Danger Zone Entrepreneurs: The Importance of Resilience and Self–Efficacy for Entrepreneurial Intentions. *Entrepreneurship Theory and Practice* 38: 473–99. [CrossRef]

Capella-Peris, Carlos, Jesús Gil-Gómez, Manuel Martí-Puig, and Paola Ruíz-Bernardo. 2020. Development and Validation of a Scale to Assess Social Entrepreneurship Competency in Higher Education. *Journal of Social Entrepreneurship* 11: 23–39. [CrossRef]

Çera, Edmond, Gentjan Çera, and Engjell Skreli. 2021. The Relationship between Entrepreneurship Education and Entrepreneurial Intention: Evidence from a Transition Country. *International Journal of Entrepreneurship and Small Business* 43: 548–69. [CrossRef]

Çera, Gentjan, A Mlouk, Edomond Çera, and Arjan Shumeli. 2020. The Impact of Entrepreneurship Education on Entrepreneurial Intention. A Quasi-Experimental Research Design. *Journal of Competitiveness* 12: 39–56. [CrossRef]

Çera, Gentjan, and Edmond Çera. 2020. Intention to Start a Business and Entrepreneurship Education Programme: A Pre- and Post-Programme Research Design. *Journal of Enterprising Communities: People and Places in the Global Economy* 14: 603–19. [CrossRef]

Çera, Gentjan, Edmond Çera, Zoltan Rozsa, and Svitlana Bilan. 2021. Entrepreneurial Intention as a Function of University Atmosphere, Macroeconomic Environment and Business Support: A Multi-Group Analysis. *European Journal of Training and Development* 45: 706–24. [CrossRef]

Çera, Gentjan, Jaroslav Belas, and Eliska Zapletalikova. 2019. Explaining Business Failure through Determinist and Voluntarist Perspectives. *Serbian Journal of Management* 14: 257–75. [CrossRef]

Creswell, John W., and J. David Creswell. 2017. *Research Design: Qualitative, Quantitative, and Mixed Methods Approaches*. Thousand Oak: SAGE Pubications.

Cunningham, James A. 2022. COVID-19: Entrepreneurial Universities and Academic Entrepreneurship. In *The COVID-19 Crisis and Entrepreneurship: Perspectives and Experiences of Researchers, Thought Leaders, and Policymakers*. Edited by David B. Audretsch and Iris A. M. Kunadt. International Studies in Entrepreneurship. Cham: Springer International Publishing, pp. 209–27. [CrossRef]

Dana, Leo-Paul, Mehdi Tajpour, Aidin Salamzadeh, Elahe Hosseini, and Mahnaz Zolfaghari. 2021. The Impact of Entrepreneurial Education on Technology-Based Enterprises Development: The Mediating Role of Motivation. *Administrative Sciences* 11: 105. [CrossRef]

Durán-Sánchez, Amador, María de la Cruz Del Río-Rama, José Álvarez-García, and Diego Fernando García-Vélez. 2019. Mapping of Scientific Coverage on Education for Entrepreneurship in Higher Education. *Journal of Enterprising Communities* 13: 84–104. [CrossRef]

Dvorský, Ján, Zora Petráková, Gentjan Çera, and Andrea Folvarčna. 2019. Important Factors for the Entrepreneurship in Central Europe. *Innovative Marketing* 15: 71–83. [CrossRef]

Entrialgo, Montserrat, and Víctor Iglesias. 2020. Entrepreneurial Intentions among University Students: The Moderating Role of Creativity. *European Management Review* 17: 529–42. [CrossRef]

Feola, Rosangela, Massimiliano Vesci, Antonio Botti, and Roberto Parente. 2019. The Determinants of Entrepreneurial Intention of Young Researchers: Combining the Theory of Planned Behavior with the Triple Helix Model. *Journal of Small Business Management* 57: 1424–43. [CrossRef]

Franke, Nikolaus, and Christian Lüthje. 2004. Entrepreneurial Intentions of Business Students—A Benchmarking Study. *International Journal of Innovation and Technology Management* 1: 269–88. [CrossRef]

García-Rodríguez, Francisco J., Esperanza Gil-Soto, Inés Ruiz-Rosa, and Desiderio Gutiérrez-Taño. 2017. Entrepreneurial Potential in Less Innovative Regions: The Impact of Social and Cultural Environment. *European Journal of Management and Business Economics* 26: 163–79. [CrossRef]

Godswill, Agu Agu, Onwuka Onwuka Okwara, Ebere Rejoice Okocha, and Nnamdi O. Madichie. 2021. COVID-19 Pandemic and Entrepreneurial Intention among University Students: A Contextualisation of the Igbo Traditional Business School. *African Journal of Economic and Management Studies* 13: 89–104. [CrossRef]

Gomes, Sofia, Marlene Sousa, Tânia Santos, José Oliveira, Márcio Oliveira, and João M. Lopes. 2021. Opening the "Black Box" of University Entrepreneurial Intention in the Era of the COVID-19 Pandemic. *Social Sciences* 10: 181. [CrossRef]

Hair, Joseph F., G. Tomas M. Hult, Christian M. Ringle, and Marko Sarstedt. 2017. *A Primer on Partial Least Squares Structural Equation Modeling*, 2nd ed. Los Angelos: Sage.

Hair, Joseph F., Jeffrey J. Risher, Marko Sarstedt, and Christian M. Ringle. 2019. When to Use and How to Report the Results of PLS-SEM. *European Business Review* 31: 2–24. [CrossRef]

Hair, Joseph F., William C. Black, Barry J. Babin, Rolph E. Anderson, and Ronald L. Tatham. 2010. *Multivariate Data Analysis*. Hoboken: Prentice Hall.

Hale, Thomas, Noam Angrist, Rafael Goldszmidt, Beatriz Kira, Anna Petherick, Toby Phillips, Samuel Webster, Emily Cameron-Blake, Laura Hallas, Saptarshi Majumdar, and et al. 2021. A Global Panel Database of Pandemic Policies (Oxford COVID-19 Government Response Tracker). *Nature Human Behaviour* 5: 529–38. [CrossRef]

Han, Heesup, Amr Al-Ansi, Bee-Lia Chua, Beenish Tariq, Aleksandar Radic, and Su-hyun Park. 2020. The Post-Coronavirus World in the International Tourism Industry: Application of the Theory of Planned Behavior to Safer Destination Choices in the Case of US Outbound Tourism. *International Journal of Environmental Research and Public Health* 17: 6485. [CrossRef]

Haus, Inga, Holger Steinmetz, Rodrigo Isidor, and Rüdiger Kabst. 2013. Gender Effects on Entrepreneurial Intention: A Meta-analytical Structural Equation Model. *International Journal of Gender and Entrepreneurship* 5: 130–56. [CrossRef]

Henseler, Jörg, Christian M. Ringle, and Marko Sarstedt. 2015. A New Criterion for Assessing Discriminant Validity in Variance-Based Structural Equation Modeling. *Journal of the Academy of Marketing Science* 43: 115–35. [CrossRef]

Hernández-Sánchez, Brizeida Raquel, Giuseppina Maria Cardella, and José Carlos Sánchez-García. 2020. Psychological Factors That Lessen the Impact of COVID-19 on the Self-Employment Intention of Business Administration and Economics' Students from Latin America. *International Journal of Environmental Research and Public Health* 17: 5293. [CrossRef]

Hofstede, Geert. 2011. Dimensionalizing Cultures: The Hofstede Model in Context. *Online Readings in Psychology and Culture* 2: 1–26. [CrossRef]

Hoppe, Magnus. 2016. Policy and Entrepreneurship Education. *Small Business Economics* 46: 13–29. [CrossRef]

Hoti, Irida, Blerta Dragusha, and Valentina Ndou. 2022. Online Teaching during the COVID-19 Pandemic: A Case Study of Albania. *Administrative Sciences* 12: 116. [CrossRef]

Joensuu-Salo, Sanna, Anmari Viljamaa, and Elina Varamäki. 2020. Do Intentions Ever Die? The Temporal Stability of Entrepreneurial Intention and Link to Behavior. *Education + Training* 62: 325–38. [CrossRef]

Joensuu-Salo, Sanna, Elina Varamäki, and Anmari Viljamaa. 2015. Beyond Intentions—What Makes a Student Start a Firm? *Education + Training* 57: 853–73. [CrossRef]

Kautonen, Teemu, Marco van Gelderen, and Matthias Fink. 2015. Robustness of the Theory of Planned Behavior in Predicting Entrepreneurial Intentions and Actions. *Entrepreneurship Theory and Practice* 39: 655–74. [CrossRef]

Ketchen, David J., and Christopher W. Craighead. 2020. Research at the Intersection of Entrepreneurship, Supply Chain Management, and Strategic Management: Opportunities Highlighted by COVID-19. *Journal of Management* 46: 1330–41. [CrossRef]

Krichen, Kamel, and Haithem Chaabouni. 2021. Entrepreneurial Intention of Academic Students in the Time of COVID-19 Pandemic. *Journal of Small Business and Enterprise Development* 29: 106–26. [CrossRef]

Kripa, Dorina, Edlira Luci, Klodiana Gorica, and Ermelinda Kordha. 2021. New Business Education Model for Entrepreneurial HEIs: University of Tirana Social Innovation and Internationalization. *Administrative Sciences* 11: 122. [CrossRef]

Krueger, Norris F., and Alan L. Carsrud. 1993. Entrepreneurial Intentions: Applying the Theory of Planned Behaviour. *Entrepreneurship & Regional Development* 5: 315–30. [CrossRef]

Krueger, Norris F., Michael D. Reilly, and Alan L. Carsrud. 2000. Competing Models of Entrepreneurial Intentions. *Journal of Business Venturing* 15: 411–32. [CrossRef]

Lehmann, Erik E., Jonah M. Otto, Laurenz Weiße, and Katharine Wirsching. 2022. Internationalization Meets Digitalization: Entrepreneurial Responses in Higher Education to the COVID-19 Pandemic. In *The COVID-19 Crisis and Entrepreneurship: Perspectives and Experiences of Researchers, Thought Leaders, and Policymakers*. Edited by David B. Audretsch and Iris A. M. Kunadt. International Studies in Entrepreneurship. Cham: Springer International Publishing, pp. 229–40. [CrossRef]

Li, Zonglong, Wenyi Zhang, Yanhui Zhou, Derong Kang, Biao Feng, Qing Zeng, Lingling Xu, and Minqiang Zhang. 2022. College Students' Entrepreneurial Intention and Alertness in the Context of the COVID-19 Pandemic. *Sustainability* 14: 7713. [CrossRef]

Lim, Dominic S.K., Chang Hoon Oh, and Dirk De Clercq. 2016. Engagement in Entrepreneurship in Emerging Economies: Interactive Effects of Individual-Level Factors and Institutional Conditions. *International Business Review* 25: 933–45. [CrossRef]

Liñán, Francisco, and Alain Fayolle. 2015. A Systematic Literature Review on Entrepreneurial Intentions: Citation, Thematic Analyses, and Research Agenda. *International Entrepreneurship and Management Journal* 11: 907–33. [CrossRef]

Liñán, Francisco, and Yi-Wen Chen. 2006. *Testing the Entrepreneurial Intention Model on a Two-Country Sample*. Working Papers 0607. Bellaterra: Departament Empresa, Universitat Autònoma de Barcelona.

Liñán, Francisco, and Yi-Wen Chen. 2009. Development and Cross–Cultural Application of a Specific Instrument to Measure Entrepreneurial Intentions. *Entrepreneurship Theory and Practice* 33: 593–617. [CrossRef]

Lopes, João M., Sofia Gomes, Tânia Santos, Márcio Oliveira, and José Oliveira. 2021. Entrepreneurial Intention before and during COVID-19—A Case Study on Portuguese University Students. *Education Sciences* 11: 273. [CrossRef]

Lucarelli, Caterina, Camilla Mazzoli, and Sabrina Severini. 2020. Applying the Theory of Planned Behavior to Examine Pro-Environmental Behavior: The Moderating Effect of COVID-19 Beliefs. *Sustainability* 12: 10556. [CrossRef]

Lungu, Anca Elena, Ioana Andreea Bogoslov, Eduard Alexandru Stoica, and Mircea Radu Georgescu. 2021. From Decision to Survival—Shifting the Paradigm in Entrepreneurship during the COVID-19 Pandemic. *Sustainability* 13: 7674. [CrossRef]

Lüthje, Christian, and Nikolaus Franke. 2003. The "making" of an Entrepreneur: Testing a Model of Entrepreneurial Intent among Engineering Students at MIT. *R and D Management* 33: 135–47. [CrossRef]

Maes, Johan, Hannes Leroy, and Luc Sels. 2014. Gender Differences in Entrepreneurial Intentions: A TPB Multi-Group Analysis at Factor and Indicator Level. *European Management Journal* 32: 784–94. [CrossRef]

Maheshwari, Greeni, Khanh Linh Kha, and Anantha Raj A. Arokiasamy. 2022. Factors Affecting Students' Entrepreneurial Intentions: A Systematic Review (2005–2022) for Future Directions in Theory and Practice. *Management Review Quarterly*. [CrossRef]

Maresch, Daniela, Rainer Harms, Norbert Kailer, and Birgit Wimmer-Wurm. 2016. The Impact of Entrepreneurship Education on the Entrepreneurial Intention of Students in Science and Engineering versus Business Studies University Programs. *Technological Forecasting and Social Change* 104: 172–79. [CrossRef]

McGee, Jeffrey E., and Ryan P. Terry. 2022. COVID-19 as an External Enabler: The Role of Entrepreneurial Self-Efficacy and Entrepreneurial Orientation. *Journal of Small Business Management*, 1–26. [CrossRef]

Meahjohn, Inshan, and Prakash Persad. 2020. The Impact of COVID-19 on Entrepreneurship Globally. SSRN Scholarly Paper. Rochester, NY. Available online: https://papers.ssrn.com/abstract=3687519 (accessed on 22 October 2022).

Mehrabian, Albert, and James A. Russell. 1974. *An Approach to Environmental Psychology*. Cambridge: The MIT Press. Available online: https://psycnet.apa.org/record/1974-22049-000 (accessed on 22 October 2022).

Menter, Matthias. 2022. Entrepreneurship and Economic Resilience in Times of Crisis: Insights from the COVID-19 Pandemic. In *The COVID-19 Crisis and Entrepreneurship: Perspectives and Experiences of Researchers, Thought Leaders, and Policymakers*. Edited by David B. Audretsch and Iris A. M. Kunadt. International Studies in Entrepreneurship. Cham: Springer International Publishing, pp. 97–104. [CrossRef]

Mirjana, Pejic Bach, Aleksic Ana, and Merkac-Skok Marjana. 2018. Examining Determinants of Entrepreneurial Intentions in Slovenia: Applying the Theory of Planned Behaviour and an Innovative Cognitive Style. *Economic Research-Ekonomska Istraživanja* 31: 1453–71. [CrossRef]

Misoska, Ana Tomovska, Makedonka Dimitrova, and Jadranka Mrsik. 2016. Drivers of Entrepreneurial Intentions among Business Students in Macedonia. *Economic Research-Ekonomska Istraživanja* 29: 1062–74. [CrossRef]

Moriano, Juan A., Marjan Gorgievski, Mariola Laguna, Ute Stephan, and Kiumars Zarafshani. 2012. A Cross-Cultural Approach to Understanding Entrepreneurial Intention. *Journal of Career Development* 39: 162–85. [CrossRef]

Mueller, Amber L., Maeve S. McNamara, and David A. Sinclair. 2020. Why Does COVID-19 Disproportionately Affect Older People? *Aging (Albany N. Y.)* 12: 9959–81. [CrossRef]

Munir, Hina, Cai Jianfeng, and Sidra Ramzan. 2019. Personality Traits and Theory of Planned Behavior Comparison of Entrepreneurial Intentions between an Emerging Economy and a Developing Country. *International Journal of Entrepreneurial Behaviour and Research* 25: 554–80. [CrossRef]

Murnieks, Charles Y., Anthony C. Klotz, and Dean A. Shepherd. 2020. Entrepreneurial Motivation: A Review of the Literature and an Agenda for Future Research. *Journal of Organizational Behavior* 41: 115–43. [CrossRef]

Mwasalwiba, Ernest Samwel. 2010. Entrepreneurship Education: A Review of Its Objectives, Teaching Methods, and Impact Indicators. *Education + Training* 52: 20–47. [CrossRef]

Ndou, Valentina. 2021. Social Entrepreneurship Education: A Combination of Knowledge Exploitation and Exploration Processes. *Administrative Sciences* 11: 112. [CrossRef]

Ndou, Valentina, Gioconda Mele, and Pasquale Del Vecchio. 2019. Entrepreneurship Education in Tourism: An Investigation among European Universities. *Journal of Hospitality, Leisure, Sport & Tourism Education* 25: 100175. [CrossRef]

Ndou, Valentina, Giustina Secundo, Giovanni Schiuma, and Giuseppina Passiante. 2018. Insights for Shaping Entrepreneurship Education: Evidence from the European Entrepreneurship Centers. *Sustainability* 10: 4323. [CrossRef]

Neneh, Brownhilder Ngek. 2019. From Entrepreneurial Intentions to Behavior: The Role of Anticipated Regret and Proactive Personality. *Journal of Vocational Behavior* 112: 311–24. [CrossRef]

Nguyen, Phuong Mai, Van Toan Dinh, Thi-Minh-Ngoc Luu, and Yongshik Choo. 2020. Sociological and Theory of Planned Behaviour Approach to Understanding Entrepreneurship: Comparison of Vietnam and South Korea. Edited by Maria Palazzo. *Cogent Business & Management* 7: 1815288. [CrossRef]

Noelia, Franco-Leal, and Diaz-Carrion Rosalia. 2020. A Dynamic Analysis of the Role of Entrepreneurial Ecosystems in Reducing Innovation Obstacles for Startups. *Journal of Business Venturing Insights* 14: e00192. [CrossRef]

Oo, Pyayt P., Arvin Sahaym, Sakdipon Juasrikul, and Sang-Youn Lee. 2018. The Interplay of Entrepreneurship Education and National Cultures in Entrepreneurial Activity: A Social Cognitive Perspective. *Journal of International Entrepreneurship* 16: 398–420. [CrossRef]

Oosterbeek, Hessel, Mirjam van Praag, and Auke Ijsselstein. 2010. The Impact of Entrepreneurship Education on Entrepreneurship Skills and Motivation. *European Economic Review* 54: 442–54. [CrossRef]

Our World in Data. n.d. COVID-19: Stringency Index. COVID-19 Data Explorer. Available online: https://ourworldindata.org/explorers/coronavirus-data-explorer (accessed on 21 October 2022).

Palalić, Ramo, Veland Ramadani, Arnela Đilović, Alina Dizdarević, and Vanessa Ratten. 2017. Entrepreneurial Intentions of University Students: A Case-Based Study. *Journal of Enterprising Communities* 11: 393–413. [CrossRef]

Papagiannis, George D. 2018. Entrepreneurship Education Programs: The Contribution of Courses, Seminars and Competitions to Entrepreneurial Activity Decision and to Entrepreneurial Spirit and Mindset of Young People in Greece. *Journal of Entrepreneurship Education* 21: 1–21.

Paray, Zahoor Ahmad, and Sumit Kumar. 2020. Does Entrepreneurship Education Influence Entrepreneurial Intention among Students in HEI's?: The Role of Age, Gender and Degree Background. *Journal of International Education in Business* 13: 55–72. [CrossRef]

Pedrini, Matteo, Valentina Langella, and Mario Molteni. 2017. Do Entrepreneurial Education Programs Impact the Antecedents of Entrepreneurial Intention?: An Analysis of an Entrepreneurship MBA in Ghana. *Journal of Enterprising Communities* 11: 373–92. [CrossRef]

Perez-Quintana, Anna, Esther Hormiga, Joan Carles Martori, and Rafa Madariaga. 2017. The Influence of Sex and Gender-Role Orientation in the Decision to Become an Entrepreneur. *International Journal of Gender and Entrepreneurship* 9: 8–30. [CrossRef]

Prasetyo, Yogi Tri, Allysa Mae Castillo, Louie John Salonga, John Allen Sia, and Joshua Adam Seneta. 2020. Factors Affecting Perceived Effectiveness of COVID-19 Prevention Measures among Filipinos during Enhanced Community Quarantine in Luzon, Philippines: Integrating Protection Motivation Theory and Extended Theory of Planned Behavior. *International Journal of Infectious Diseases* 99: 312–23. [CrossRef] [PubMed]

Premand, Patrick, Stefanie Brodmann, Rita Almeida, Rebekka Grun, and Mahdi Barouni. 2016. Entrepreneurship Education and Entry into Self-Employment Among University Graduates. *World Development* 77: 311–27. [CrossRef]

Rantanen, Teemu, and Timo Toikko. 2017. The Relationship between Individualism and Entrepreneurial Intention—A Finnish Perspective. *Journal of Enterprising Communities* 11: 289–306. [CrossRef]

Ratinho, Tiago, Alejandro Amezcua, Benson Honig, and Zhaocheng Zeng. 2020. Supporting Entrepreneurs: A Systematic Review of Literature and an Agenda for Research. *Technological Forecasting and Social Change* 154: 119956. [CrossRef]

Ratten, Vanessa. 2021. Coronavirus (COVID-19) and Entrepreneurship: Cultural, Lifestyle and Societal Changes. *Journal of Entrepreneurship in Emerging Economies* 13: 747–61. [CrossRef]

Rayburn, Steven W., Anthony McGeorge, Sidney Anderson, and Jeremy J. Sierra. 2022. Crisis-Induced Behavior: From Fear and Frugality to the Familiar. *International Journal of Consumer Studies* 46: 524–39. [CrossRef]

Ringle, C. M., S. Wende, and J.-M. Becker. 2015. *SmartPLS*. Boenningstedt: SmartPLS GmbH.

Ruiz-Rosa, Inés, Desiderio Gutiérrez-Taño, and Francisco J. García-Rodríguez. 2020. Social Entrepreneurial Intention and the Impact of COVID-19 Pandemic: A Structural Model. *Sustainability* 12: 6970. [CrossRef]

Schepers, Jelle, Pieter Vandekerkhof, and Yannick Dillen. 2021. The Impact of the COVID-19 Crisis on Growth-Oriented SMEs: Building Entrepreneurial Resilience. *Sustainability* 13: 9296. [CrossRef]

Shane, Scott, Edwin A. Locke, and Christopher J. Collins. 2003. Entrepreneurial Motivation. *Human Resource Management Review* 13: 257–79. [CrossRef]

Sharma, Sarika, and Sonica Rautela. 2021. Entrepreneurial Resilience and Self-Efficacy during Global Crisis: Study of Small Businesses in a Developing Economy. *Journal of Entrepreneurship in Emerging Economies. ahead-of-print*. [CrossRef]

Sinclair, Vaughn G., and Kenneth A. Wallston. 2004. The Development and Psychometric Evaluation of the Brief Resilient Coping Scale. *Assessment* 11: 94–101. [CrossRef] [PubMed]

Srisathan, Wutthiya A., and Phaninee Naruetharadhol. 2022. A COVID-19 Disruption: The Great Acceleration of Digitally Planned and Transformed Behaviors in Thailand. *Technology in Society* 68: 101912. [CrossRef] [PubMed]

Thompson, Edmund R. 2009. Individual Entrepreneurial Intent: Construct Clarification and Development of an Internationally Reliable Metric. *Entrepreneurship Theory and Practice* 33: 669–94. [CrossRef]

Traczyk, Jakub, and Tomasz Zaleskiewicz. 2016. Implicit Attitudes toward Risk: The Construction and Validation of the Measurement Method. *Journal of Risk Research* 19: 632–44. [CrossRef]

Trif, Simona Mihaela, Gratiela Georgiana Noja, Mirela Cristea, Cosmin Enache, and Otniel Didraga. 2022. Modelers of Students' Entrepreneurial Intention during the COVID-19 Pandemic and Post-Pandemic Times: The Role of Entrepreneurial University Environment. *Frontiers in Psychology* 13: 976675. [CrossRef]

Usman, Magaji Abdullahi, and Xinbo Sun. 2022. Global Pandemic and Entrepreneurial Intention: How Adversity Leads To Entrepreneurship. *SAGE Open* 12: 21582440221123420. [CrossRef]

van Gelderen, Marco, Maryse Brand, Mirjam van Praag, Wynand Bodewes, Erik Poutsma, and Anita van Gils. 2008. Explaining Entrepreneurial Intentions by Means of the Theory of Planned Behaviour. *Career Development International* 13: 538–59. [CrossRef]

Volkmann, Christine K., and Marc Grünhagen. 2022. The COVID-19 Pandemic as a Catalyst for Digital Entrepreneurship Education: Reflections on a Rapid Transformation of an Educational Ecosystem. In *The COVID-19 Crisis and Entrepreneurship: Perspectives and Experiences of Researchers, Thought Leaders, and Policymakers*. Edited by David B. Audretsch and Iris A. M. Kunadt. International Studies in Entrepreneurship. Cham: Springer International Publishing, pp. 253–73. [CrossRef]

Wismans, Annelot, Milco Lodder, and Roy Thurik. 2022. Entrepreneurial Intention of Dutch Students During the COVID-19 Pandemic: Are Today's Students Still Tomorrow's Entrepreneurs? In *The COVID-19 Crisis and Entrepreneurship: Perspectives and Experiences of Researchers, Thought Leaders, and Policymakers*. Edited by David B. Audretsch and Iris A. M. Kunadt. International Studies in Entrepreneurship. Cham: Springer International Publishing, pp. 187–207. [CrossRef]

Zampetakis, Leonidas A., Maria Bakatsaki, Charalambos Litos, Konstantinos G. Kafetsios, and Vassilis Moustakis. 2017. Gender-Based Differential Item Functioning in the Application of the Theory of Planned Behavior for the Study of Entrepreneurial Intentions. *Frontiers in Psychology* 8: 451. [CrossRef]

Zarnadze, Giorgi, Ines Dika, Gentjan Çera, and Humberto Nuno Rito Ribeiro. 2022. Personality Traits and Business Environment for Entrepreneurial Motivation. *Administrative Sciences* 12: 176. [CrossRef]

administrative sciences

MDPI

Article

Personality Traits and Business Environment for Entrepreneurial Motivation

Giorgi Zarnadze [1], Ines Dika [2], Gentjan Çera [3,*] and Humberto Nuno Rito Ribeiro [4]

1 Faculty of Management and Economics, Tomas Bata University in Zlin, 76001 Zlin, Czech Republic
2 Faculty of Economics, University of Tirana, 1001 Tirana, Albania
3 Faculty of Economics and Agribusiness, Agricultural University of Tirana, 1029 Tirana, Albania
4 ESTGA—Águeda School of Technology and Management, University of Aveiro, 3810-193 Aveiro, Portugal
* Correspondence: gcera@ubt.edu.al

Abstract: The influences of personality traits and business barriers on entrepreneurial behaviour have been studied by numerous researchers. However, the mechanisms by which changes in institutional contexts and personality traits affect individuals' entrepreneurial motivation have not received enough attention from researchers. This paper aims to fill this gap in the literature by examining the joint effects of personality traits and business barriers on one's motivation to engage in entrepreneurial activity. A structured questionnaire was developed to capture the above linkages. To measure the latent variables, factor and reliability analyses were utilised. To examine the relationships, a regression analysis was performed on a unique dataset of respondents from three counties: Czechia, Hungary, and Serbia. The results show that personality traits positively influence motivation towards entrepreneurship. Furthermore, external barriers (ecological and technological issues), and infrastructure factors positively affect entrepreneurial motivation. From an academic viewpoint, personality traits and motivation to engage in entrepreneurial activity can be nourished by education, highlighting the role of universities in this regard. Our paper's findings should trigger the interest of policymakers who aim to introduce new instruments or change the existing ones (designing policies) to boost entrepreneurship.

Keywords: entrepreneurship; entrepreneurial motivation; personality traits; business barriers; infrastructure; Czechia; Hungary; Serbia

Citation: Zarnadze, Giorgi, Ines Dika, Gentjan Çera, and Humberto Nuno Rito Ribeiro. 2022. Personality Traits and Business Environment for Entrepreneurial Motivation. *Administrative Sciences* 12: 176. https://doi.org/10.3390/admsci12040176

Received: 30 October 2022
Accepted: 22 November 2022
Published: 25 November 2022

1. Introduction

1.1. Research Background and Problem Statement

Entrepreneurship has been seen as a contributory factor and an economic engine across every country, since it helps to create new employment and boosts labour market productivity and competitiveness (Murnieks et al. 2020; Dvorský et al. 2019). Everyday entrepreneurship is being given more and more importance across developed and developing economies. Traditionally, entrepreneurship research has followed two distinct courses. The first course consists of studies that look at the personal characteristics of entrepreneurs to discover what sets them apart from the wider public. Gartner (2017) referred to this as the "traits" method, which discovered that some psychological characteristics had a strong influence on entrepreneurial motivation. The second path focuses on the external (environmental/business) conditions that seem to generate variations in the number of business start-ups over time (Taormina and Lao 2007; Çera et al. 2021a). In previous studies, researchers usually studied only personality traits' influence on motivation, or, on the other hand, only the business environment's impact on it. Shane et al. (2003) suggested that the one-sided approach of the research approached a "dead end" and it was important for scholars to consider collective impact on motivation. This paper follows the suggested path and analyses the joint effect of personality traits and business barriers on a person's motivation to engage in entrepreneurial activities. Rather than taking into account personal characteristics in isolation, the study is able to determine how specific personality traits and business barriers combine to achieve a high (or low) degree of entrepreneurial motivation.

Over the past twenty years, the personality debate in entrepreneurship has re-emerged, with many criticising the original trait research. The main point of the criticism was that personality traits are not adapted specifically to entrepreneurs and are more generic (Munir et al. 2019). The problem arises when there is a necessity of measuring many personality traits and linkages among the constructs appear to be weak (Şahin et al. 2019). As a result, scholars agree that personality traits are crucial in the process of entrepreneurial decisions or actions. This study considers personality traits including independence, education, locus of control, risk taking, creativity, need for achievement, and self-confidence as determinant factors influencing entrepreneur motivation. The decision to focus on these personal characteristics was inspired by prior studies (Frese 2009; Adler et al. 1992), which argue that these traits are human capital attributes affecting motivation. Despite their alleged significance, these personality traits have received very little attention from researchers in relation to their impact on entrepreneurial motivation among entrepreneurs across the globe. In order to assist the government's efforts to establish institutional arrangements and nurture entrepreneurship among residents, to help enable the launch of small businesses, it is important to identify which factors may motivate people to become entrepreneurs in the first place. This study contributes to a greater understanding of the key personality traits that influence entrepreneurial motivation.

As entrepreneurship has been a focus of academic research, several experts and academics have carried out research and concluded that business barriers have a substantial impact on entrepreneurial motivation. According to Ahmad and Xavier (2012), the business environment is a collection of elements that influence entrepreneurship motivation and activity. According to Bernhofer and Li (2014), business barriers include cultural, economic, and political conditions, and people have various incentives in different situations. The author argues that business barriers can be demotivational toward engagement in entrepreneurship. Turulja et al. (2020) defined business barriers as the total of the legal and institutional barriers, market barriers, financial barriers, and entrepreneurial infrastructure, among other things. He mentioned that different environments influence motivation and barriers can decrease the motivation to be involved in start-ups. Overall, the previous studies had less focus on elements such as legislative, ecological, and technological issues in the context of entrepreneurial motivation. Hence, we added these aspects and the other three elements of business barriers, political, economic, and social factors, to the dimensions of this study.

While there is a growing body of literature on barriers to entrepreneurs, less attention is paid to business infrastructure and how it affects entrepreneurial motivation. Gnyawali and Fogel (2017) divided the business environment into three parts: (a) general environmental conditions for entrepreneurship; (b) descriptive studies of the environmental conditions of a particular country or region; and (c) the role of public policy in shaping entrepreneurial environments. Authors considered business infrastructure as an element of general environmental conditions and the joint effect of all the above factors on entrepreneurial motivation was studied. There is a lack of research analysing the impact of business infrastructure, as a separate variable, on entrepreneur motivation. Hence, this paper is intended to fill this gap.

1.2. Aim and Structure of the Paper

The goal of this research is to enhance both theory and practice. More research is needed on person-adapted traits in the field of entrepreneurship, as well as the impact of business barriers and infrastructure, according to several studies. The joint impact on motivation is the focus of the current study, which expands on one-sided approach research in the field of entrepreneurship. Making such selections provides individuals with a novel perspective and new understanding. Additionally, this study may have practical applications, notably for entrepreneurship education. While some traits are challenging to alter, prior research shows examples of interventions that can quickly affect traits and, subsequently, motivation.

The remaining parts of this paper are as follows. After the introduction, it continues with a literature review of entrepreneurial intention, motivation, and behaviour, followed by personality traits and intention and finally business environment and intention. Then, it continues with the methods and procedures including unit of analysis and data collection, variable measurement, and methods. The next section consists of results, followed by a discussion. Finally, the conclusion is the last section, which discusses research findings and highlights recommendations and contributions of this paper.

2. Literature Review

2.1. Theoretical Background

Various models have been devised and utilized by various researchers to identify a person's entrepreneurial motivations. Bird (1988) developed a model, entrepreneurial intentionality, according to which a person's inclinations to entrepreneurship are based primarily on both environmental and personal factors. Environmental factors include political, economic, social, ecological, legislative, and technological elements that can influence one's mindset. The study is also based on the institutional theory, which states that individuals' and organizations' behaviours are shaped by the institutional context (North 1990). Regarding personal factors, they can influence a person's motivation about starting a new business and being involved in entrepreneurship. Boyd and Vozikis (1994) further developed the above-mentioned theory and added self-efficiency to the analyses as a factor impacting entrepreneurial intention, behaviour, and motivation.

Ajzen (1991) developed a model, the theory of planned behaviour (TPB), and he argues that a person's behaviour is based on voluntary control and explicit planning. TPB stresses that intentions are immediate provenances of action, and higher intentions can determine the possibility of accomplishing the behaviour (Ajzen 1991).

2.2. Entrepreneurial Intention, Motivation, and Behaviour

There is a need to distinguish three key concepts that at first sight seem to have close meaning, but actually differ in context: entrepreneurial intention, entrepreneurial motivation, and entrepreneurial behaviour. Entrepreneurial intention expresses a person's desire to pursue a profession as an entrepreneur (Dana et al. 2021; Çera and Çera 2020). People with entrepreneurial inclinations intend to take measured risks, accumulate necessary resources, and launch their businesses (Alshebami 2022). Entrepreneurial intent motivates entrepreneurial behaviour (Karabulut 2016). According to Bird and West (1998) intention is an attitude of mind that leads a person's intents and activities to entrepreneurship. Liñán et al. (2010) explain that intention is a group's endeavour to behave entrepreneurially. Hmieleski and Corbett (2006) argue that an entrepreneurial intention is an intention to achieve high growth in any type of business. The intention can be considered as originating in free will, and leads a person's perception, focus, experimentation with, and behaviour towards his/her objective (Bird and West 1998). People with a higher level of self-efficiency tend to challenge themselves with more complicated tasks and achieve higher goals; thus, self-efficiency can be found as a trigger for entrepreneurial intention (Cacciotti et al. 2020; García-Cabrera et al. 2020; Çera et al. 2021b; Şahin et al. 2019).

Similarly, it can be argued that people with high entrepreneurial motivation are more likely to become entrepreneurs (Estay et al. 2013). Entrepreneurial motivations are defined as urges or the proclivity to arrange, manage, and dominate institutions, people, or ideas as quickly and autonomously as feasible (Solesvik 2013). Different theories suggest different characteristics to be motivational to engage in entrepreneurship. According to Segal et al. (2005), motivation is defined as a combination of expectation, usefulness, and polarity. Economically based models argue that risk tolerance is the key factor; people with higher risk tolerance are willing to be self-employed in the future (Douglas and Shepherd 2000). Hessels et al. (2008) suggested that motivation is highly dependent on the country's specificity. Therefore, the latter study calls on researchers who try to better understand the determinants of an individual's motivation to become an entrepreneur to consider environmental factors (business barriers) in their analysis. This paper follows this suggestion in examining influencing factors of entrepreneurial motivation.

Different from entrepreneurial intention and motivation, entrepreneurial behaviour is defined as a planned action by different individuals (Ajzen 1985). As stated before, entrepreneurial intention or motivation can motivate behaviour (Alshebami et al. 2022). Other theories also suggest that these factors can be predictors of behaviour (O'Gorman 2019; Venesaar et al. 2021). According to Krueger (2017), three main constructs enable the individual to express identifiable entrepreneurial behaviour. It is also based on certain beliefs and requirements that will drive the intention of a person to acquire skills and experience in a certain field, so this person will be able to engage in entrepreneurial processes (Kirkley 2016).

2.3. Personality Traits and Entrepreneurial Motivation

Various studies indicate that entrepreneurs have personality differences compared to managers (Bazkiaei et al. 2020; Shane and Nicolaou 2013). This is why it is important to analyse the relationship between personality traits and motivation. Some behavioural traits might lead to individuals perceiving entrepreneurship as a more gratifying practice, and therefore they may persist in establishing a new firm and becoming entrepreneurs (Segal et al. 2005). Bird (1988) created the model, which states that motivations can be influenced by contextual and personal characteristics. The author mentions that personality traits are the factors that can trigger starting a new business. Other authors further developed the previous study by Bird (1988). The novelty brought by Boyd and Vozikis (1994) was that they added a self-efficiency factor to the analyses and demonstrated its importance in entrepreneurial motivation and behaviour. In this paper, we tried to cover as many dimensions of personality traits as possible: independence, being educated, internal locus of control, risk taking, creativity, need for achievement, and self-confidence. Below, each of them is described.

Independence. The desire for independence is critical to entrepreneurial ambition. Carter et al. (2003) define independence as a person's desire for freedom, autonomy, and flexibility in how they spend their time. Entrepreneurs are typically classified as self-starters and, accordingly, their level of independence is higher compared to other people (Raza et al. 2018).

Being educated. Entrepreneurs tend to be independent, risk-takers, creative, and confident, but none of these traits will work if they lack education (Mónico et al. 2021). According to Solesvik (2013), there are three types of education that entrepreneurs need to acquire. The first one is academic, where people acquire basic knowledge; the second is vocational education, which refers to skills on how to secure a job and earn money; and the third one is financial education, which represents knowledge of financial culture, understanding certain figures, and language. All three types of education are crucial for entrepreneurship and can be considered as a motivation for involvement in those processes.

Locus of control. Locus of control can be divided into two parts: internal and external. Locus of control measures what kind of control a person has over their life. Internal locus of control describes how a person controls their own life and believes in his/her decisions, while external shows how life can be affected by external factors such as luck, other people's behaviour, and so on (Karabulut 2016). Hisrich and Peters (1998) believe that locus of control is "an attribute indicating the sense of control that a person has over life". It is expected that people with an internal locus of control have motivations to become entrepreneurs and start their businesses.

Risk taking. Stewart and Roth (2001) believe that entrepreneurs are better risk takers compared to managers. They tolerate career, family, and financial risks and this is one of the major traits associated with achieving success (Alshebami and Seraj 2022). People with higher acceptance of risk tend to have more entrepreneurial motivations and may start a new business (Wu and Mao 2020).

Creativity. An individual's cognitive processing is aided by creativity, which constitutes the potential to develop new and useful ideas by combining existing and matching expertise and skills (Wei et al. 2020). Many scholars associate creativity with entrepreneurship as it is one of the most influential factors related to it and we can state that entrepreneurship itself is a type of creativity. People with a high level of creativity can maintain a positive disposition and identity when engaging in business activities (Mahto and McDowell 2018). Creativity can be one of the motivators for engaging in entrepreneurial activities.

Need for achievement. The need for achievement is described as possessing a strong motivation and intention to succeed. Several studies have found that the need for achievement has a considerable influence on entrepreneurial ambitions (Gürol and Atsan 2006; Shane et al. 2003). Researchers discovered that students tend to start new businesses and become entrepreneurs depending on their level of need for achievement, so they concluded that it influences intentions and further motivations as well (Gürol and Atsan 2006).

Self-confidence. Self-efficacy is described as a person's belief in his capacity to accomplish a task and his conviction that he will use this skill successfully to accomplish specific goals (Negara et al. 2019). According to Carsrud and Brännback (2011), self-efficacy influences entrepreneurship. It can contribute to the firm's performance (McGee and Terry 2022). Entrepreneurial motivation is impacted by self-efficacy via cognitive abilities, desire, and emotional responses. A person with strong self-efficacy has superior intellectual capacity, strategic adaptability, and is effective at environmental management (Bandura 1977). In other words, these individuals exercised greater control since they attempted to plan for the best- and worst-case scenarios, are capable of adapting to changes in plans and can manage environmental volatility.

A deeper knowledge of entrepreneurial motivation may be obtained by examining the combined effect of the above-discussed personality traits, as the literature study demonstrates that personality traits play a significant part in motivation to be an entrepreneur (Karabulut 2016). The widely used "Big Five" and narrow traits have been related to entrepreneurial motivation and success; many studies have researched this relationship, but in this paper, we narrowed down the traits and chose specific ones that can influence the motivation. Therefore, we form the first hypothesis:

Hypothesis 1 (H1). *Personality traits positively affect individuals' entrepreneurial motivation.*

2.4. Business Barriers and Entrepreneurial Motivation

Business barriers have a significant degree of influence on the entrepreneurial motivation of individuals. Exploring the business barriers factor is critical, since studying entrepreneurship determinants from the angle of personality traits alone would not be sufficient (Taormina and Lao 2007). Entrepreneurial motivation is a reaction to business environment stimuli. These elements have an impact on the formation of the brain and, as a result, on raising entrepreneurs (Pacut 2020). Entrepreneurs are bred by the business environment, and as a result, they acquire and apply what it has instilled in them (Shane et al. 2003).

Many academics who studied the aspects determining entrepreneurial motivation from the perspective of the entrepreneurial barriers saw the external environment as an objective condition for entrepreneurship (Munir et al. 2019). Suzuki et al. (2002) described entrepreneurial motivation as the result of environmental and individual factors; it is impacted by management abilities, managerial capabilities, market circumstances, corporate culture, and government backing. Alshebami and Seraj (2022) argue that high taxes, additional business laws, and limitations, particularly those pertaining to the labour force, deter people from opening small firms and have a negative impact on entrepreneurship. According to research by Gohmann (2012), economic barriers have a big impact on potential entrepreneurs and their motivation; individuals tend not to be engaged in entrepreneurship if their countries lack economic freedom. Some researchers studied the relationship between entrepreneurial motivation and socioeconomic barriers and highlighted that individual motivation was negatively affected by social and economic issues (Yao et al. 2016). According to Martínez-González et al. (2022), entrepreneurs' sentiments and motivations are greatly influenced by their impressions of the business barriers. Taormina and Kin-Mei Taormina and Lao (2007) completed a quantitative study on the relationship between entrepreneurial motivation and business barriers, discovering that the former is impacted by motivation towards accomplishment, optimism about life, and social networking. Previous studies mostly focused on the impact of political and socio-economic barriers on entrepreneurial motivations, while this study will take into consideration political, economic, social, legislative, ecological, and technological factors. We grouped barriers into two groups, and while previous research paid much attention on the first group of barriers (Pacut 2020), the influence of the second group has not been deeply studied yet. Based on the above viewpoints, we can form a second hypothesis:

Hypothesis 2 (H2). *Business barriers negatively affect an individual's entrepreneurial motivation.*

Certain infrastructure components appear to have a significant influence on the entrepreneurial climate (Gnyawali and Fogel 2017). These elements include an existing share of the market for products/services, existing resources for important raw materials, enough qualified workforce in the region, and the possibility to improve capacity, transportation systems, and the supply of electricity. Previously, various authors discussed the elements of infrastructure that influenced motivation, but in this study, we selected specific elements that we believe are crucial for motivation. Kontos (2010) argues that entrepreneurs make decisions about where to start their businesses based on these indicators. Global Entrepreneurship Monitor reported that a lack of transportation or a non-developed transport system is one of the challenge entrepreneurs face (Ahmad and Xavier 2012). The discussion shows that the existence of various infrastructure elements increases entrepreneurs' motivation, and therefore, increases the growth of businesses in a country. Though a single factor may have a less significant impact, the interaction of various factors may considerably increase the impact on entrepreneurial motivation (Suzuki et al. 2002). In addition, Carsrud and Brännback (2011) state that without existing market share for products or services, or the possibility to increase capacity, entrepreneurs will not be able to start a business and their motivation will decrease. Furthermore, Oosterbeek et al. (2010) argued that the qualified workforce directly impacts entrepreneurial motivation; the authors showed that entrepreneurs tend to start new businesses in regions where they see the possibility of having qualified employees for their new firms. On the other hand, other authors indicate that the data from their analysis highlight a connection between infrastructure and motivation, but the relation is negative (Çera et al. 2021a). Research conducted in central European studies showed that in Slovakia and Poland, infrastructure does not have a positive impact on motivation. Mixed results were given in the paper by (Walter and Block 2016). Other authors' efforts reveal discrepancies in the academic findings (Nowiński et al. 2020). They carried out study on individuals from the USA and Poland who wanted to start their own businesses. The direct impact of infrastructure on entrepreneurial motivation was shown to be modest, whereas the indirect impact was found to be considerable. This gives further motivation to conduct the current research and study the relationship between these two variables. As can be seen from the previous studies, infrastructure elements were studied separately along with their influence on motivation, but this study considers a specific group of elements and their joint effect on entrepreneurial motivation. Based on what was discussed above, we can form a third hypothesis:

Hypothesis 3 (H3). *Business infrastructure positively influences an individual's entrepreneurial motivation.*

2.5. Conceptual Model

Figure 1 shows the conceptual framework of the current research. As discussed in the literature review, entrepreneurial motivation can be influenced by three domains: personality traits, business barriers, and business infrastructure. They imply at least two theories, such as entrepreneurial intentionality (Bird 1988) and institutional theory (North 1990).

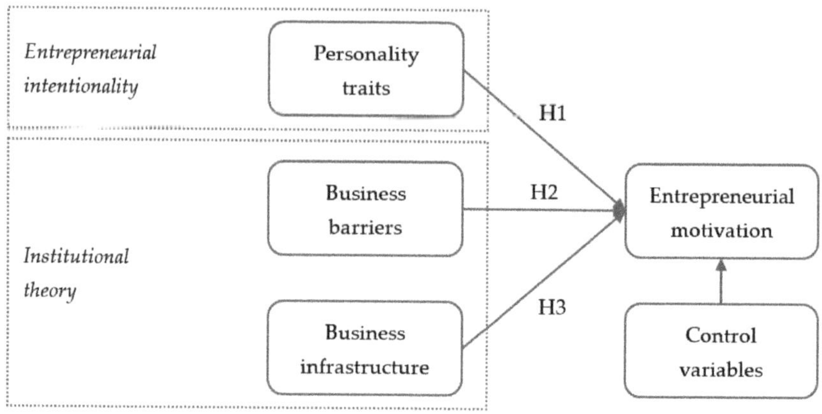

Figure 1. Conceptual framework.

3. Methods and Procedures

3.1. Data

In order to test the research model, primary data are needed for analysis. To collect primary data, a questionnaire should be developed. Therefore, to test the formulated hypotheses, a questionnaire is designed based on the literature in English. Then, it was translated into the Czech, Hungarian, and Serbian languages. It consists of two major parts: a demographic module and questions covering personality traits, entrepreneurial motivation, and business barriers.

As can be foreseen, the unit of analysis in this research consists of individuals. The respondents of the questionnaire were individuals from Czechia, Hungary, and Serbia. They were selected from business databases in their respective countries and reached via email. The respondents were asked to fill out an online form. Only the self-employed were considered as valid to fill in the form. They were targeted because they would offer a better understanding of the role of business barriers and enablers on motivation to start a business. The online form provided only one opportunity to fill it in.

After cleaning the collected data, the size of the sample consists of 329 valid responses. Regarding the distribution of the respondents by country, 28.9%, 30.4%, and 40.7% were from Czechia, Hungary, and Serbia, respectively. The majority of the respondents were males (61.7%), while less than two out of five of them were females. Regarding the highest level of completed education, the respondents are distributed as follows: 41.9% had high-school education or lower, 15.5% had professional education, 25.8% were undergraduates, and the rest were postgraduates (16.7%). Seven out of ten respondents said that are married, while 15.9% of them were single and less than 14% of the sample reported that they are divorced.

3.2. Variable Measurement

The measurement of the variables is shown in Table 1. The dependent variable is entrepreneurial motivation, measured on a Likert scale. This scale was used by scholars in prior studies (Mónico et al. 2021; Hessels et al. 2008; Segal et al. 2005; Jayawarna et al. 2013; Maheshwari et al. 2022). Personality traits are a scale used in the literature by numerous scholars, and are measured almost similarly (Karabulut 2016; Munir et al. 2019; Şahin et al.

2019; Premand et al. 2016; Lüthje and Franke 2003). In this paper, seven indicators are used for measurement, which are: independence, being educated, internal locus of control, risk taking, creativity, need for achievement, and self-confidence. Business barriers, known as well as institutions, are measured following the scale proposed by Çera et al. (2019a). Infrastructure measurement involves a Likert scale type of variable with six indicators as shown in Table 5. The measurement of this factor was inspired by prior studies (Gnyawali and Fogel 2017; Kontos 2010; Carsrud and Brännback 2011; Oosterbeek et al. 2010). The county, age of the respondent, his/her gender, being married or not, and working hours per week are used as control variables.

Table 1. Variable measurement.

Variable	Type	Measure
Country	Nominal	Where do you live? [1] Czechia; [2] Hungary; [3] Serbia
Age	Scale	What is your age?
Gender	Dummy	What is your gender? [1] Male, [2] Female
Married	Dummy	Are you married? [1] Yes, [2] No
Hours working	Scale	Please indicate how many hours do you spend on average at/or with work, weekly.
Entrepreneurial motivation (*dependent variable*)	Likert scale	Rate each of the following items (refer to Table 2) on how important they are for you. [1] 'Lowest' to [5] 'Highest'
Personality traits	Likert scale	Please rate the importance of the following factors for success (refer to Table 3). [1] 'Lowest' to [5] 'Highest'
Business barriers	Likert scale	To what extent the following factors (see Table 4) created difficulties for your business? [1] 'Lowest' to [5] 'Highest'
Infrastructure	Likert scale	Importance of the following infrastructure of the surrounding region where you live (see Table 5). [1] 'Not important to [5] 'Very important

Table 2. Component matrix: motivation for entrepreneurship.

	Loading	Mean	Standard Deviation
Access to additional financial resources	0.822	3.772	1.230
Good networks	0.779	3.681	1.168
Employment creation	0.769	3.264	1.332
Financial motives	0.512	4.277	0.8730
Eigenvalue	2.135		
Variance explained	0.534		
Cronbach's alpha	0.706		

Note: Extraction method, Principal Component Analysis. Rotation method, Varimax with Kaiser normalization. Only one component was extracted. The solution cannot be rotated. Kaiser–Meyer–Olkin Measure of sampling adequacy = 0.660. Sig. Bartlett's test < 0.001.

Table 3. Rotated component matrix: personality traits.

	Loading		Mean	Standard Deviation
	PersTr. 1	PersTr. 2		
Independence	0.806		3.979	1.060
Being educated	0.797		3.489	1.232
Internal locus of control	0.669		4.091	0.968
Risk taking		0.700	4.131	0.875
Creativity		0.689	4.198	0.982
Need for achievement		0.661	4.204	0.952
Self-confidence		0.576	4.401	0.839
Eigenvalue	2.792	1.070		
Variance explained	0.286	0.266		
Cronbach's alpha	0.792	0.736		

Note: Extraction method, Principal Component Analysis. Rotation method, Varimax with Kaiser normalization. Rotation converged in 3 iterations. Kaiser–Meyer–Olkin Measure of sampling adequacy = 0.765. Sig. Bartlett's test < 0.001. Coefficient loading displayed > |0.5|.

Table 4. Rotated component matrix: business barriers.

	Loading		Mean	Standard Deviation
	BusBar. 1	BusBar. 2		
Political issues	0.781		2.912	1.564
Economic issues	0.764		3.912	1.228
Social issues	0.590		3.167	1.244
Legislative issues	0.556		3.313	1.474
Ecological issues		0.844	2.100	1.160
Technological issues		0.830	2.757	1.312
Eigenvalue	2.182	1.286		
Variance explained	0.311	0.267		
Cronbach's alpha	0.730	0.733		

Note: Extraction method, Principal Component Analysis. Rotation method, Varimax with Kaiser normalization. Rotation converged in 3 iterations. Kaiser–Meyer–Olkin Measure of sampling adequacy = 0.652. Sig. Bartlett's test < 0.001. Coefficient loading displayed > |0.5|.

Table 5. Rotated component matrix: Infrastructure.

	Loading		Mean	Standard Deviation
	Infrast. 1	Infrast. 2		
Existing share of the market for products/services	0.697		3.910	1.148
Existing resources for important raw material	0.682		3.198	1.416
Sufficient qualified workforce in the region	0.677		3.723	1.373
Possibility to increase capacity	0.594		3.641	1.224
Transportation system		0.871	3.415	1.366
Supply of electricity		0.783	3.600	1.383
Eigenvalue	2.479	1.024		
Variance explained	0.300	0.283		
Cronbach's alpha	0.748	0.718		

Note: Extraction method, Principal Component Analysis. Rotation method, Varimax with Kaiser normalization. Rotation converged in 3 iterations. Kaiser–Meyer–Olkin Measure of sampling adequacy = 0.745. Sig. Bartlett's test < 0.001. Coefficient loading displayed > |0.5|.

3.3. Data Analysis

In this paper, four variables are measured using indicators. To reduce the number of these indicators, principal component analysis was performed per each set of indicators (Fabrigar and Wegener 2011). As the rotation method, Varimax with Kaiser normalization was selected. The output of the analyses is organized in table format per each set of indicators: entrepreneurial motivation (Table 2), personality traits (Table 3), business barriers (Table 4), and infrastructure (Table 5). Only factors reflecting eigenvalues higher than the value of one were kept in the analysis. The Kaiser–Meyer–Olkin value for each set of indicators was found to be greater than the standard threshold of 0.70 and Barlett's test of sphericity was reported to be statistically significant, indicating that the performed factor analysis is appropriate (Hair et al. 2010).

The factor analysis of indicators of entrepreneurial motivation emerged with only one factor explaining more than half of the variation in the data (refer to Table 2). Additionally, the factor loading was higher than the standard threshold of 0.40, meaning that the construct convergent validity is set (Stevens and Pituch 2015). Moreover, the scale results manifest good reliability, since the value of Cronbach's alpha was above the threshold of 0.70 (Hair et al. 2010). Besides the figures generated by factor and reliability analyses, in the table two descriptive statistics for each indicator are reported: mean and standard deviation.

The second performed factor analysis deals with personality traits indicators. Two factors emerged from the principal component analysis, explaining more than half of the variation in the data (refer to Table 3). In addition, the factor loading was higher than the value of 0.40, indicating that the construct convergent validity is not an issue (Stevens and Pituch 2015). Furthermore, the emerged factors reflected accepted reliability, since the value of Cronbach's alpha was not below the threshold of 0.70 (Hair et al. 2010). Apart from the figures provided by reliability and factor analyses, beside each indicator two additional statistics are shown: mean and standard deviation.

Table 4 summarizes the output of factor and reliability analyses about the indicators of business barriers. Similar to personality traits, two factors emerged from the principal component analysis. The explained variance by both emerged factors is almost 60% in the sample. Again, the factor loading was higher than the value of 0.40, providing evidence of construct convergent validity (Stevens and Pituch 2015). Like the personality traits case, both emerging factors showed acceptable scale reliability, since the value of Cronbach's alpha happened to be above 0.70 (Hair et al. 2010). To have a better view of the measured factors, next to each indicator are shown two additional statistics: mean and standard deviation.

Table 5 reports the output of the principal component and reliability analyses of the indicators that were used to represent infrastructure. Similarly to the case of business barriers, the performed principal component analysis resulted in two factors. Almost 60% of the variance in the sample is explained by these factors. The item reliability is set, since the factor loadings gave a value higher than 0.50, showing that construct convergent validity is set (Stevens and Pituch 2015). Additionally, both emerging factors showed acceptable scale reliability, since the value of Cronbach's alpha resulted above 0.70 (Hair et al. 2010). To have a better understanding of the indicators included in the analysis, the mean and standard deviation is shown for each of them.

3.4. Method

The research model is tested by regressing the emerging factors from principal component analysis against entrepreneurial motivation. Having the dependent variable characterized as a scale type of variable provides the possibility of testing the proposed relationships using standard regression (ordinary least square) (Harrell 2015). The assumption of normality of the interest variables is set since the emerged factors are generated by principal component analysis, which by default creates normally distributed variables. To better judge the influence of each factor on motivation for entrepreneurship, five regressions were performed. The first one includes only the control variables and the constant, while the last regression includes both the control variables and emerging factors from the factor analyses.

A general form of the fifth model can be seen below:

$$EntMot = \beta_0 + \beta_1\, PersTr_1 + \beta_2\, PersTr_2 + \beta_3\, BusBar_1 + \beta_4\, BusBar_2 + \beta_5\, Infrast_1 + \beta_6\, Infrast_2 + control\ variable + \varepsilon$$

where *EntMot*, *PersTr*, *BusBar*, *Infrast* stand for entrepreneurial motivation, personality traits, business barriers, and infrastructure; βs represent the standardized coefficients, and ε refers to the error term (unexplained variance).

All analyses in this paper are performed by utilizing SPSS, version 23, following the instructions outlined by Sarstedt and Mooi (2019) and Pallant (2016).

4. Results

To investigate the influences of personality traits, business barriers, and infrastructure on entrepreneurial motivation, five regression models are performed. The output of these regressions is shown in Table 6. The first model can be known as a baseline one, since it includes only control variables as regressors. The control variables in this paper are country, age, gender of the respondent, and the average number of hours worked per week. The model explains 26% of the variation in entrepreneurial motivation and is statistically significant (F = 17.8). Excluding gender (β = 0.068, t = 1.371, p > 0.10), all the other variables resulted in statistically significant determinants for entrepreneurial motivation.

Table 6. Regression results.

Effect	Variable	Model 1			Model 2			Model 3			Model 4			Model 5		
		Beta	*t*	Sig.	Beta	*t*	Sig.	Beta	*t*	Sig.	Beta	*t*	Sig.	Beta	*t*	Sig.
Control	Constant		−3.440	0.001		−2.877	0.004		−2.958	0.003		−2.883	0.004		−2.534	0.012
	SR	0.406	6.699	0.000	0.236	4.154	0.000	0.429	6.775	0.000	0.373	6.328	0.000	0.280	4.738	0.000
	HU	0.409	6.980	0.000	0.264	4.881	0.000	0.398	6.794	0.000	0.336	5.736	0.000	0.241	4.394	0.000
	Age	−0.109	−2.118	0.035	−0.038	−0.818	0.414	−0.142	−2.762	0.006	−0.099	−1.969	0.050	−0.046	−0.957	0.339
	Male	0.068	1.371	0.171	0.041	0.932	0.352	0.067	1.368	0.172	0.076	1.546	0.123	0.045	1.020	0.309
	Married	0.180	3.529	0.000	0.125	2.736	0.007	0.179	3.561	0.000	0.163	3.273	0.001	0.114	2.492	0.013
	Hours worked	0.192	3.740	0.000	0.111	2.400	0.017	0.182	3.593	0.000	0.139	2.720	0.007	0.090	1.936	0.054
H1	PersTr 1				0.359	7.522	0.000							0.331	6.839	0.000
	PersTr 2				0.286	6.267	0.000							0.272	5.498	0.000
H2	BusBar. 1							0.008	0.147	0.884				−0.074	−1.561	0.120
	BusBar. 2							−0.174	−3.459	0.001				−0.112	−2.384	0.018
H3	Infrast. 1										0.198	3.984	0.000	0.085	1.764	0.079
	Infrast. 2										0.159	3.193	0.002	0.061	1.296	0.196
Model fit	R square	0.259			0.426			0.287			0.316			0.453		
	F statistic	17.8			28.1			15.3			17.2			20.2		

Note: Dependent variable, motivation for entrepreneurship. Beta, standardized coefficients. PersTr, personality traits; BusBar, business barriers; Infrast, infrastructure. The reference country is Czechia. Male = 1, otherwise 0. Married = 1, otherwise 0.

In the second model, in addition to control variables, personality traits are regressed against motivation for entrepreneurship. The inclusion of the two factors that emerged from the factor analysis that covers personality traits almost doubled the explained variance compared to the baseline model (R^2 = 42.6%, F = 28.1). Both factors reflected a positive and statistically significant relationship with entrepreneurial motivation ($PersTr_1$: β = 0.359, t = 7.522, p < 0.001; $PersTr_2$: β = 0.286, t = 6.267, p < 0.001). Having these results, one can state that the data of this paper support H1, which claims that entrepreneurial motivation is positively influenced by personality traits.

The third model includes business barriers along with control variables. The two emerging factors of business barriers do not show the same importance for entrepreneurial motivation. Hence, the first factor of business barriers is found to be insignificant for entrepreneurial motivation (β = 0.008, t = 0.147, p > 0.10), while the second factor showed a negative and significant relationship (β = −0.174, t = −3.459, p < 0.01). The inclusion of the two factors that emerged from the factor analysis that covers business barriers did not improve the explained variance compared to the baseline model (ΔR^2 = 2.8%, F = 15.3). Since the result of the two factors is mixed regarding their influence on entrepreneurial motivation, H2 is partially supported.

Infrastructure is found to be an important determinator for individuals regarding the motivation towards entrepreneurship. The fourth model shown in Table 6 represents the inclusion of two factors that emerged from the principal component analysis about infrastructure. The addition of these two variables increased the explained variance by 6% compared to the baseline model (R^2 = 31.6%, F = 17.2). Thus, both factors reflected a positive and statistically significant influence on entrepreneurial motivation (*Infrast*$_1$: β = 0.198, t = 3.984, p < 0.001; *Infrast*$_2$: β = 0.159, t = 3.193, p < 0.01). Therefore, based on the analysis, it can be said that H3 is supported.

The above paragraphs correspond to the interpretation of models that do not include all variables at once in the regression. The last column of Table 6 is shown the output of the regression that includes all independent and control variables. The same results as in previous models are obtained even in this case. The difference here exists in the fact that the two emerged factors covering infrastructure do not manifest the same results as in model 4. To summarize, the data provide evidence in support of H1, since as the personality traits increase, so does individual entrepreneurial motivation. In addition, regarding business barriers, the results show mixed findings leading to the partial support of H2, which claims that entrepreneurial motivation is negatively affected by business constraints. Moreover, the data partially support H3, since only one factor of the infrastructure resulted in positive and statistical influences on motivation towards entrepreneurship. Nevertheless, these findings merit being discussed with reference to prior research as well. This discussion proceeds in the following section of the paper.

5. Discussion

The current article has provided new insights into the relationships between personality traits, business barriers, business infrastructure, and entrepreneurial motivation. Some findings were discordant with the current literature when contrasted with the study's purpose, which was to evaluate the influence of personality factors, business constraints, and business infrastructure on entrepreneurial motivation. According to the data, some business barriers have little effect on entrepreneurial motivation. These should be investigated further. Future studies need to focus on identifying the element of institutional constrains that do not affect entrepreneurial motivation. The primary findings of this study, as well as each carefully expressed hypothesis, are presented in the following paragraphs.

The research first explores the impact of personality traits on a person's entrepreneurial motivation. We performed a principal component analysis of the indicators that are linked to personality traits and two factors emerged: the first one includes independence, being educated, and internal locus of control, and the second one includes risk taking, creativity, need for achievement, and self-confidence. The evidence showed that personality traits (both factors) significantly impact entrepreneurial motivation. Findings showed that the more people display the above-discussed personality traits, the more they tend to be involved in entrepreneurship and starting a new business. Thus, a high level of traits positively affects motivation. More specifically, the results suggested that the complex influence of seven different configurations predicts a high level of entrepreneurial motivation. These findings are consistent with the existing literature (Bird 1988; Boyd and Vozikis 1994; Bird and West 1998; Karabulut 2016). Thus, we found evidence supporting the first hypothesis. It is important to stress that different personality traits can be increased by several boosting factors. Prior research indicates that being educated, the need for achievement and self-confidence can be increased by education (Negara et al. 2019; Secundo et al. 2021; Ndou et al. 2019; Ndou 2021). Some authors also suggest that role models can also improve the level of internal locus control and self-confidence (Murnieks et al. 2020) A role model helps individuals to learn, develop, and enhance their self-esteem by providing a picture of a realistic, good career choice experience.

The second examination studied the association between business barriers and entrepreneurship motivation. The importance of institutional barriers, regardless of their nature, in affecting the entrepreneurial behaviour of individuals, either negatively or positively, has been noted repeatedly in the existing literature (North 1990). Unlike many existing studies supporting business barriers that negatively impact motivation (Alshebami and Seraj 2022), the present study's results show a different outcome. The study findings reveal that the two emerging factors as business barriers do not attach the same importance to entrepreneurial motivation. As the first factor was found to be insignificant, it does not impact negatively on entrepreneurial motivation. This needs further investigation as to which element or set of elements does not influence motivation. It can be shown later in the studies that these elements do not have a direct impact, but may play a mediation role in the relationship. On the other hand, the second factor was found to be significant, and it indicates that the second factor negatively affects entrepreneurial motivation. This means that the greater the institutional requirements, the more likely motivation is to decrease. There are lots of studies that support the idea that business barriers negatively impact motivation (Martínez-González et al. 2022; Suzuki et al. 2002; Gohmann 2012; Taormina and Lao 2007; Munir et al. 2019), though some other studies show that barriers are not always significant for motivation, firm growth (Xheneti and Bartlett 2012; Ur Rehman et al. 2019), business climate (Çera et al. 2019b), and sales of under-reporting firms (Williams and Krasniqi 2018). Considering the results, we can partially support the third hypothesis and state that ecological and technological barriers decrease entrepreneurial motivation.

This paper also describes how business infrastructure impacts the motivation of a person toward entrepreneurship. It was expected that a positive relationship would be found between infrastructure and motivation. Results showed that both factors had significant, positive effects on motivation. This outcome is aligned with prior studies which also indicate a positive relationship between these variables (Gnyawali and Fogel 2017; Kontos 2010; Carsrud and Brännback 2011; Oosterbeek et al. 2010; Ahmad and Xavier 2012). On the other hand, this paper goes against the literature that previously suggested a negative relationship between infrastructure and motivation (Nowiński et al. 2020).

Furthermore, the uniqueness of this study is that it not only analyses the separate impact of each independent variable on entrepreneurial motivation, but the joint effect of all variables has also been researched. The results give a slightly different picture compared to what was shown before; specifically, infrastructure factors scored differently compared to when the independent relationship was explored relating to these constructs. The second factor was not significant, which means that transport systems and the supply of electricity may not have a positive effect on motivation. This could be caused by country specificity, and in further studies, more detailed analyses should be conducted.

6. Conclusions

The current study seeks to research the linkages between personality traits, business barriers, business infrastructure, and entrepreneurial motivation, all of which are now highly demanded. Furthermore, it was discovered that these factors are understudied in published research. The examined relationship adds to the current body of knowledge. It can assist policymakers and businesses interested in entrepreneurship in thinking from a variety of viewpoints and discovering innovative solutions. A deeper understanding of the factors that contribute to motivation may give the aforementioned players additional ideas to modify or develop new entrepreneurship-promoting strategies and regulations; accordingly, is important to investigate what triggers the motivation and what decreases it. The current research provides useful insights on entrepreneurial motivation determinants in the light of two theories that were tested in the study: entrepreneurial intentionality (Bird 1988) and institutional theory (North 1990).

6.1. Contribution of the Study

The article's significant addition, however, is that it examined business infrastructure, which is a present priority of the government and entrepreneurs. This research also proposed a model based on two theories (Bird 1988; North 1990), and after integrating them, proposed the conceptual framework. As a result, the study contributes to the current knowledge of which personality traits influence motivation. Hence, the study adds to the existing literature on how personality traits, institutional constraints, and business infrastructure affect entrepreneurial motivation. In this paper, personality traits were measured with seven indicators, which translates into covering a wider set of dimensions. The current study also shows that, contrary to previous studies, not all business constraints have a detrimental impact on entrepreneur motivation. Barriers should be classified into two types for future academics to examine their influence on entrepreneurship. However, barriers that have an indirect impact on motivation play a moderating role in the relationship. The outcomes reveal more about the entrepreneurs' thinking. Because entrepreneurial motivation is one of the most essential variables for the government and entrepreneurs, they seek higher-level knowledge and study to assist them in identifying the aspects that drive it.

The findings have some practical relevance for policymakers, who might utilize the study's findings to influence the important elements influencing motivation. The results of this investigation also help policymakers to learn how they might encourage the rise of entrepreneurship, which would increase national economic growth and decrease unemployment. It also assists aspiring entrepreneurs in understanding what influences entrepreneurial motivation, improving it, and becoming successful entrepreneurs. Policymakers should reduce business barriers to trigger entrepreneurial motivation; furthermore, they need to design more engaging business policies. This study helps to increase understanding of the key personality traits that influence entrepreneurial motivation, in order to better support the government's initiatives to build institutional structures and foster entrepreneurship among citizens, to better facilitate the launch of small businesses. As business infrastructure was found to influence motivation, it is important for the government to consider the findings and focus on improving infrastructure to create a friendly environment for businesses.

This study has brought contributions to educational institutions as well. They need to carefully adjust their curricula to increase personality trait levels among their students. Educational institutions play a significant role in increasing entrepreneurial thinking and developing an entrepreneurial mindset. These educational institutions could aim to foster and maintain these personality traits in youth, encouraging them to engage in entrepreneurial endeavours and activities. Personality traits level among university students can be increased by attending different educational programs. As it was found that traits positively affect motivation, this is a good area of focus for the students to become involved in start-ups.

6.2. Limitations and Future Research Avenues

Like every other study, this research also has some limitations that can be addressed in future research. The first and foremost limitation of the present examination is the limited sample size. It is believed that a bigger number of observations could lead to more robust results. This analysis was conducted in a limited number of countries; including other countries in the future might lead to a different result. Scholars can add more personality trait elements to the analyses in the future. As well as moderating and mediating variables in potential conceptual models, their joint effect on motivation would be an interesting topic to explore. In the same way, business barriers or business infrastructure elements can be added, and relationships can be studied. As discussed before, some barriers show an indirect influence on the motivation as they play a moderating role only; it would be interesting to explore this moderation effect as well. The research also may bring attention to the necessity of continuing to develop training programs and initiatives aimed at enhancing personal traits among prospective entrepreneurs, particularly those related to developing the above-discussed personality traits. In order to support society and culture and promote entrepreneurial activity, it is also necessary to create a suitable entrepreneurial ecosystem with the right institutional infrastructure.

Author Contributions: Conceptualization, I.D. and G.Z.; methodology, G.Ç. and H.N.R.R.; software, G.Ç.; validation, G.Z., G.Ç. and I.D.; formal analysis, G.Ç. and H.N.R.R.; investigation, G.Z.; resources, G.Ç. and G.Z.; data curation, G.Ç. and H.N.R.R.; writing—original draft preparation, G.Z. and I.D.;

writing—review and editing, G.Z., G.Ç. and I.D.; visualization, G.Ç.; supervision, I.D. and H.N.R.R.; project administration, I.D. and H.N.R.R.; funding acquisition, G.Ç. All authors have read and agreed to the published version of the manuscript.

Funding: This paper was supported by the International Visegrad Fund, project number 21820267, titled "How to prevent SMEs failure (Actions based on comparative analysis in Visegrad countries and Serbia)".

Institutional Review Board Statement: Not applicable.

Informed Consent Statement: Informed consent was obtained from all subjects involved in the study.

Data Availability Statement: The data presented in this study are available on request from the corresponding author. The data are not publicly available due to privacy and ethical reasons.

Conflicts of Interest: The authors declare no conflict of interest.

References

Adler, Nancy J., Richard Brahm, and John L. Graham. 1992. Strategy Implementation: A Comparison of Face-to-face Negotiations in the Peoples Republic of China and the United States. *Strategic Management Journal* 13: 449–66. [CrossRef]

Ahmad, Syed Zamberi, and Siri Roland Xavier. 2012. Entrepreneurial Environments and Growth: Evidence from Malaysia GEM Data. *Journal of Chinese Entrepreneurship* 4: 50–69. [CrossRef]

Ajzen, Icek. 1985. From Intentions to Actions: A Theory of Planned Behavior. In *Action Control*. Berlin and Heidelberg: Springer, pp. 11–39. [CrossRef]

Ajzen, Icek. 1991. The Theory of Planned Behavior. *Organizational Behavior and Human Decision Processes* 50: 179–211. [CrossRef]

Alshebami, Ali Saleh. 2022. Psychological Features and Entrepreneurial Intention among Saudi Small Entrepreneurs during Adverse Times. *Sustainability* 14: 7604. [CrossRef]

Alshebami, Ali Saleh, Abdullah Hamoud Ali Seraj, and Elham Alzain. 2022. Lecturers' Creativity and Students' Entrepreneurial Intention in Saudi Arabia. *Vision: The Journal of Business Perspective* 097226292210995. [CrossRef]

Alshebami, Ali Saleh, and Abdullah Hamoud Ali Seraj. 2022. Investigating the Impact of Institutions on Small Business Creation Among Saudi Entrepreneurs. *Frontiers in Psychology* 13: 897787. [CrossRef]

Bandura, Albert. 1977. Self-Efficacy: Toward a Unifying Theory of Behavioral Change. *Psychological Review* 84: 191–215. [CrossRef]

Bazkiaei, Hanieh Alipour, Low Hock Heng, Noor Ullah Khan, Roselina Binti Ahmad Saufi, and Raja Suzana Raja Kasim. 2020. Do Entrepreneurial Education and Big-Five Personality Traits Predict Entrepreneurial Intention among Universities Students? *Cogent Business & Management* 7: 7–11. [CrossRef]

Bernhofer, Barbara Lena, and Jun Li. 2014. Understanding the Entrepreneurial Intention of Chinese Students: The Preliminary Findings of the China Project of "Global University Entrepreneurial Spirits Students Survey" (GUESSS). *Journal of Entrepreneurship in Emerging Economies* 6: 21–37. [CrossRef]

Bird, Barbara. 1988. Implementing Entrepreneurial Ideas: The Case for Intention. *The Academy of Management Review* 13: 442–53. [CrossRef]

Bird, Barbara, and Page West. 1998. Time and Entrepreneurship. *Entrepreneurship Theory and Practice* 22: 5–9. [CrossRef]

Boyd, Nancy G., and George S. Vozikis. 1994. The Influence of Self-Efficacy on the Development of Entrepreneurial Intentions and Actions. *Entrepreneurship Theory and Practice* 18: 63–77. [CrossRef]

Cacciotti, Gabriella, James C. Hayton, J. Robert Mitchell, and David G. Allen. 2020. Entrepreneurial Fear of Failure: Scale Development and Validation. *Journal of Business Venturing* 35: 106041. [CrossRef]

Carsrud, Alan, and Malin Brännback. 2011. Entrepreneurial Motivations: What Do We Still Need to Know? *Journal of Small Business Management* 49: 9–26. [CrossRef]

Carter, Nancy M., William B. Gartner, Kelly G. Shaver, Elizabeth J. Gatewood, Nancy M. Carter, William B. Gartner, Kelly G. Shaver, and Elizabeth J. Gatewood. 2003. The Career Reasons of Nascent Entrepreneurs. *Journal of Business Venturing* 18: 13–39. [CrossRef]

Çera, Gentjan, and Edmond Çera. 2020. Intention to Start a Business and Entrepreneurship Education Programme: A Pre- and Post-Programme Research Design. *Journal of Enterprising Communities: People and Places in the Global Economy* 14: 603–19. [CrossRef]

Çera, Gentjan, Edmond Çera, Humberto Nuno Rito Ribeiro, and Sadik Maloku. 2021a. The Role of Trust in Government and Institutional Environment in Future Business Climate. *International Journal of Entrepreneurial Venturing* 13: 425–42. [CrossRef]

Çera, Gentjan, Edmond Çera, Zoltan Rozsa, and Svitlana Bilan. 2021b. Entrepreneurial Intention as a Function of University Atmosphere, Macroeconomic Environment and Business Support: A Multi-Group Analysis. *European Journal of Training and Development* 45: 706–24. [CrossRef]

Çera, Gentjan, Jaroslav Belas, and Eliska Zapletalikova. 2019a. Explaining Business Failure through Determinist and Voluntarist Perspectives. *Serbian Journal of Management* 14: 257–75. [CrossRef]

Çera, Gentjan, Pavla Breckova, Edmond Çera, and Zoltan Rozsa. 2019b. The Effect of Business Enabling Policies, Tax Treatment, Corruption and Political Connections on Business Climate. *Acta Polytechnica Hungarica* 16: 113–32. [CrossRef]

Dana, Leo Paul, Mehdi Tajpour, Aidin Salamzadeh, Elahe Hosseini, and Mahnaz Zolfaghari. 2021. The Impact of Entrepreneurial Education on Technology-Based Enterprises Development: The Mediating Role of Motivation. *Administrative Sciences* 11: 105. [CrossRef]

Douglas, Evan J., and Dean A. Shepherd. 2000. Entrepreneurship as a Utility Maximizing Response. *Journal of Business Venturing* 15: 231–51. [CrossRef]

Dvorský, Ján, Zora Petráková, Gentjan Çera, and Andrea Folvarčna. 2019. Important Factors for the Entrepreneurship in Central Europe. *Innovative Marketing* 15: 71–83. [CrossRef]

Estay, Christophe, François Durrieu, and Manzoom Akhter. 2013. Entrepreneurship: From Motivation to Start-Up. *Journal of International Entrepreneurship* 11: 243–67. [CrossRef]

Fabrigar, Leandre R., and Duane T. Wegener. 2011. *Exploratory Factor Analysis*. New York: Oxford University Press.

Frese, Michael. 2009. Towards a Psychology of Entrepreneurship—An Action Theory Perspective. *Foundations and Trends® in Entrepreneurship* 5: 437–96. [CrossRef]

García-Cabrera, Antonia Mercedes, Ana Maria Lucía-Casademunt, and Laura Padilla-Angulo. 2020. Immigrants' Entrepreneurial Motivation in Europe: Liabilities and Assets. *International Journal of Entrepreneurial Behaviour and Research* 26: 1707–37. [CrossRef]

Gartner, William B. 2017. "Who Is an Entrepreneur?" Is the Wrong Question. *Entrepreneurship Theory and Practice* 13: 47–68. [CrossRef]

Gnyawali, Devi R., and Daniel S. Fogel. 2017. Environments for Entrepreneurship Development: Key Dimensions and Research Implications. *Entrepreneurship Theory and Practice* 18: 43–62. [CrossRef]

Gohmann, Stephan F. 2012. Institutions, Latent Entrepreneurship, and Self–Employment: An International Comparison. *Entrepreneurship Theory and Practice* 36: 295–321. [CrossRef]

Gürol, Yonca, and Nuray Atsan. 2006. Entrepreneurial Characteristics amongst University Students: Some Insights for Entrepreneurship Education and Training in Turkey. *Education and Training* 48: 25–38. [CrossRef]

Hair, Joseph F., William C. Black, Barry J. Babin, and Rolph E. Anderson. 2010. *Multivariate Data Analysis*, 7th ed. Harlow: Pearson Education Limited.

Harrell, Frank E. 2015. *Regression Modeling Strategies: With Applications to Linear Models, Logistic and Ordinal Regression, and Survival Analysis*. Berlin and Heidelberg: Springer. [CrossRef]

Hessels, Jolanda, Marco Van Gelderen, and Roy Thurik. 2008. Entrepreneurial Aspirations, Motivations, and Their Drivers. *Small Business Economics* 31: 323–39. [CrossRef]

Hisrich, Robert D., and Michael P. Peters. 1998. *Entrepreneurship*. New York: Irwin/McGraw-Hill.

Hmieleski, Keith M., and Andrew C. Corbett. 2006. Proclivity for Improvisation as a Predictor of Entrepreneurial Intentions. *Journal of Small Business Management* 44: 45–63. [CrossRef]

Jayawarna, Dilani, Julia Rouse, and John Kitching. 2013. Entrepreneur Motivations and Life Course. *International Small Business Journal: Researching Entrepreneurship* 31: 34–56. [CrossRef]

Karabulut, Ahu Tuğba. 2016. Personality Traits on Entrepreneurial Intention. *Procedia-Social and Behavioral Sciences* 229: 12–21. [CrossRef]

Kirkley, William Walton. 2016. Entrepreneurial Behaviour: The Role of Values. *International Journal of Entrepreneurial Behaviour and Research* 22: 290–328. [CrossRef]

Kontos, Maria. 2010. Considering the Concept of Entrepreneurial Resources in Ethnic Business: Motivation as a Biographical Resource? *International Review of Sociology* 13: 183–204. [CrossRef]

Krueger, Norris F. 2017. What Lies Beneath? The Experiential Essence of Entrepreneurial Thinking. *Entrepreneurship Theory and Practice* 31: 123–38. [CrossRef]

Liñán, Francisco, Juan Carlos Rodríguez-Cohard, and José M. Rueda-Cantuche. 2010. Factors Affecting Entrepreneurial Intention Levels: A Role for Education. *International Entrepreneurship and Management Journal* 7: 195–218. [CrossRef]

Lüthje, Christian, and Nikolaus Franke. 2003. The "making" of an Entrepreneur: Testing a Model of Entrepreneurial Intent among Engineering Students at MIT. *R and D Management* 33: 135–47. [CrossRef]

Maheshwari, Greeni, Khanh Linh Kha, and Anantha Raj A. Arokiasamy. 2022. Factors Affecting Students' Entrepreneurial Intentions: A Systematic Review (2005–2022) for Future Directions in Theory and Practice. *Management Review Quarterly*. [CrossRef]

Mahto, Raj V., and William C. McDowell. 2018. Entrepreneurial Motivation: A Non-Entrepreneur's Journey to Become an Entrepreneur. *International Entrepreneurship and Management Journal* 14: 513–26. [CrossRef]

Martínez-González, José Alberto, Carmen Dolores Álvarez-Albelo, Javier Mendoza-Jiménez, and Urszula Kobylinska. 2022. Predicting the Entrepreneurial Behaviour of Starting Up a New Company: A Regional Study Using PLS-SEM and Data from the Global Entrepreneurship Monitor. *Mathematics* 10: 704. [CrossRef]

McGee, Jeffrey E., and Ryan P. Terry. 2022. COVID-19 as an external enabler: The role of entrepreneurial self-efficacy and entrepreneurial orientation. *Journal of Small Business Management: 1–26*. [CrossRef]

Mónico, Lisete, Carla Carvalho, Samuel Nejati, Marco Arraya, Pedro Parreira, Carlo Gabriel, and Porto Bellini. 2021. Entrepreneurship Education and Its Influence on Higher Education Students' Entrepreneurial Intentions and Motivation in Portugal. *BAR—Brazilian Administration Review* 18: 6–8. [CrossRef]

Munir, Hina, Cai Jianfeng, and Sidra Ramzan. 2019. Personality Traits and Theory of Planned Behavior Comparison of Entrepreneurial Intentions between an Emerging Economy and a Developing Country. *International Journal of Entrepreneurial Behaviour and Research* 25: 554–80. [CrossRef]

Murnieks, Charles Y., Anthony C. Klotz, and Dean A. Shepherd. 2020. Entrepreneurial Motivation: A Review of the Literature and an Agenda for Future Research. *Journal of Organizational Behavior* 41: 115–43. [CrossRef]

Ndou, Valentina. 2021. Social Entrepreneurship Education: A Combination of Knowledge Exploitation and Exploration Processes. *Administrative Sciences* 11: 112. [CrossRef]

Ndou, Valentina, Gioconda Mele, and Pasquale Del Vecchio. 2019. Entrepreneurship Education in Tourism: An Investigation among European Universities. *Journal of Hospitality, Leisure, Sport & Tourism Education* 25: 100175. [CrossRef]

Negara, Stie Trisna, Helisia Margahana, and Sumatera Selatan. 2019. Self Efficacy, Self Personality and Self Confidence on Entrepreneurial Intention: Study on Young Enterprises. *Journal of Entrepreneurship Education* 22: 2–4. Available online: https://www.bernas.id/64070-anak-muda-jogja-berpotensi- (accessed on 26 October 2022).

North, Douglass C. 1990. *Institutions, Institutional Change, and Economic Performance*. Cambridge: Cambridge University Press.

Nowiński, Witold, Mohamed Yacine Haddoud, Krzysztof Wach, and Renata Schaefer. 2020. Perceived Public Support and Entrepreneurship Attitudes: A Little Reciprocity Can Go a Long Way! *Journal of Vocational Behavior* 121: 103474. [CrossRef]

O'Gorman, Colm. 2019. Entrepreneurial Intentions and Entrepreneurial Behaviour. *Entrepreneurial Behaviour: Individual, Contextual and Microfoundational Perspectives 17–37*. [CrossRef]

Oosterbeek, Hessel, Mirjam van Praag, and Auke Ijsselstein. 2010. The Impact of Entrepreneurship Education on Entrepreneurship Skills and Motivation. *European Economic Review* 54: 442–54. [CrossRef]

Pacut, Agnieszka. 2020. Drivers toward Social Entrepreneurs Engagement in Poland: An Institutional Approach. *Administrative Sciences* 10: 5. [CrossRef]

Pallant, Julie. 2016. *SPSS Survival Manual: A Step by Step Guide to Data Analysis Using IBM SPSS*. New York: McGraw-Hill Education.

Premand, Patrick, Stefanie Brodmann, Rita Almeida, Rebekka Grun, and Mahdi Barouni. 2016. Entrepreneurship Education and Entry into Self-Employment Among University Graduates. *World Development* 77: 311–27. [CrossRef]

Raza, Syed Ali, Wasim Qazi, and Nida Shah. 2018. Factors Affecting the Motivation and Intention to Become an Entrepreneur among Business University Students. *International Journal of Knowledge and Learning* 12: 221–41. [CrossRef]

Şahin, Faruk, Hande Karadağ, and Büşra Tuncer. 2019. Big Five Personality Traits, Entrepreneurial Self-Efficacy and Entrepreneurial Intention: A Configurational Approach. *International Journal of Entrepreneurial Behaviour and Research* 25: 1188–211. [CrossRef]

Sarstedt, Marko, and Erik Mooi. 2019. *A Concise Guide to Market Research: The Process, Data, and Methods Using IBM SPSS Statistics*. Springer Texts in Business and Economics. Berlin and Heidelberg: Springer. [CrossRef]

Secundo, Giustina, Gioconda Mele, Pasquale Del Vecchio, Gianluca Elia, Alessandro Margherita, and Valentina Ndou. 2021. Threat or Opportunity? A Case Study of Digital-Enabled Redesign of Entrepreneurship Education in the COVID-19 Emergency. *Technological Forecasting and Social Change* 166: 120565. [CrossRef]

Segal, Gerry, Dan Borgia, and Jerry Schoenfeld. 2005. The Motivation to Become an Entrepreneur. *International Journal of Entrepreneurial Behaviour and Research* 11: 42–57. [CrossRef]

Shane, Scott, and Nicos Nicolaou. 2013. The Genetics of Entrepreneurial Performance. *International Small Business Journal* 31: 473–95. [CrossRef]

Shane, Scott, Edwin A. Locke, and Christopher J. Collins. 2003. Entrepreneurial Motivation. *Human Resource Management Review* 13: 257–79. [CrossRef]

Solesvik, Marina Z. 2013. Entrepreneurial Motivations and Intentions: Investigating the Role of Education Major. *Education + Training* 55: 253–71. [CrossRef]

Stevens, James P., and Keenan A. Pituch. 2015. *Applied Multivariate Statistics for the Social Sciences: Analyses with SAS and IBM's SPSS*, 6th ed. London: Routledge.

Stewart, Wayne H., and Philip L. Roth. 2001. Risk Propensity Differences between Entrepreneurs and Managers: A Meta-Analytic Review. *The Journal of Applied Psychology* 86: 145–53. [CrossRef]

Suzuki, Kan-ichiro, Sang-Hoon Kim, and Zong-Tae Bae. 2002. Entrepreneurship in Japan and Silicon Valley: A Comparative Study. *Technovation* 22: 595–606. [CrossRef]

Taormina, Robert J., and Sammi Kin-Mei Lao. 2007. Measuring Chinese Entrepreneurial Motivation. *International Journal of Entrepreneurial Behavior & Research* 13: 200–21. [CrossRef]

Turulja, Lejla, Emir Agic, and Ljiljan Veselinovic. 2020. Motivation of the Youth of Bosnia and Herzegovina to Start a Business: Examining Aspects of Education and Social and Political Engagement. *Economic Research-Ekonomska Istrazivanja* 33: 2692–712. [CrossRef]

Ur Rehman, Naqeeb, Arjona Çela, Fatbardha Morina, and Kriselda Sulçaj Gura. 2019. Barriers to Growth of SMEs in Western Balkan Countries. *Journal of Management Development* 38: 2–24. [CrossRef]

Venesaar, Urve, Elina Malleus, Grete Arro, and Martin Toding. 2021. Entrepreneurship Competence Model for Supporting Learners Development at All Educational Levels. *Administrative Sciences* 12: 2. [CrossRef]

Walter, Sascha G., and Jörn H. Block. 2016. Outcomes of entrepreneurship education: An institutional perspective. *Journal of Business Venturing* 31: 216–33. [CrossRef]

Wei, Jiangru, Yuting Chen, Yamin Zhang, and Jing Zhang. 2020. How Does Entrepreneurial Self-Efficacy Influence Innovation Behavior? Exploring the Mechanism of Job Satisfaction and Zhongyong Thinking. *Frontiers in Psychology* 11: 708. [CrossRef]

Williams, Colin C., and Besnik Krasniqi. 2018. Explaining Informal Sector Entrepreneurship in Kosovo: An Institutionalist Perspective. *Journal of Developmental Entrepreneurship* 23: 1850011. [CrossRef]

Wu, Feixia, and Chuanyu Mao. 2020. Business Environment and Entrepreneurial Motivations of Urban Students. *Frontiers in Psychology* 11: 1483. [CrossRef]

Xheneti, Mirela, and Will Bartlett. 2012. Institutional Constraints and SME Growth in Post-communist Albania. *Journal of Small Business and Enterprise Development* 19: 607–26. [CrossRef]

Yao, Xiaofang, Xiyue Wu, and Dan Long. 2016. University Students' Entrepreneurial Tendency in China: Effect of Students' Perceived Entrepreneurial Environment. *Journal of Entrepreneurship in Emerging Economies* 8: 60–81. [CrossRef]

The following aspects ensure the originality of the study. Firstly, it focuses on an essential research topic regarding PHEIs' performance reporting, considering the fact that the improvement and modernization of the public system (Grossi et al. 2020) have traditionally received international support (e.g., rules, funds, assistance) regarding entrepreneurial behavior (Capella-Peris et al. 2020). More than that, performance measurement has been a major path for preserving public trust and securing continued funding and resources (Lee 2021). Exploring the factors that influence the credibility of the performance measurement system is a serious step in understanding and improving the effectiveness of performance in the public sector (Ghosh and Wu 2012). In this context, we want to fill the gap in the literature regarding the entrepreneurial behavior of HEIs, in terms of performance and its measurement. The connection between financial, non-financial reporting/information role (Grossi et al. 2020), the influence of internal/external factors (Ghosh and Wu 2012), and the request for a realistic reporting system (Caputo et al. 2021) are justified by global reform theories. Secondly, in the empirical sphere, we create a unique performance-reporting model, grounded on an Eastern European ex-communist country's particularities, which at the same time is a member of the European Union. The paper succeeded in enriching the scientific literature by taking a holistic approach to performance and its reporting in universities, including in comprehensive puzzle elements related to realities and perspectives of evolution, influencing factors, and reporting models. The hybrid-view achievement in public universities' performance reporting may be more significant as a behavior and effect than in a country where the tradition and experience of a well-established reporting model has created stability and coherence.

Moreover, the study addresses a broad range of users. First, there are theoreticians to understand the evolution of public performance reporting. Second, practitioners familiarize themselves with the implications of the concept analyzed through the eyes of a specialist. Third, for professional bodies/legislators, the study offers a concrete, precise basis for future analysis to improve the mandatory items in national HEIs' performance-reporting regulations.

The remainder of the paper is structured as follows. The next section presents the debate concerning the reporting performance of PHEIs and then reporting as viewed in the Romanian context, followed by this study's theoretical framework. The methodological section describes the research design, research hypotheses' development, sample selection and data collection, and variable description. The results and discussion section begins with the sample description and then discusses each of the analyzed dimensions: teaching, research, interaction with the external environment, and funding. Results for the entire university are presented, and the hypothesis validation is discussed. The conclusion section highlights the theoretical and practical contributions of the study, limitations, and further developments.

2. The Context of the Research

2.1. Debates in the Literature Concerning the Reporting Performance in PHEIs

Reporting the performance of organizations may be a critical element of overall performance. Performance reports involve information about how effectively the organization is fulfilling its mission, expressed in specified goals and objectives (Grossi et al. 2020). Hatry (2013) claims that reporting performance measures to external stakeholders allows citizens, elected officials, and interested parties to understand what public organizations are doing with their allocated resources. External reporting may encourage the organization to perform better on the measures it reports. Further, comparisons made of similar reported measures allow one organization to measure or benchmark success against another.

As has happened in many public service organizations, the last decade has seen unprecedented pressure to reform universities. The most relevant of these reforms is a shift from an elite to a mass higher education system; cuts in state funding and resulting difficulty financing the institutions exclusively with public funds; the emergence of new approaches, such as NPM; and greater competition between universities (Siegel and Wright 2015). Those changes to universities' hybrid behaviors use private sector mechanisms and tools within the public sector (Grossi et al. 2020).

Growing demands to become more competitive, efficient, effective, and accountable have led to an increased interest in introducing control mechanisms to assess organizational performance. Consequently, performance management systems have been implemented in some universities, and the measurement and reporting of research and teaching performance have become increasingly common within universities. Wide varieties of performance indicators have been developed in different jurisdictions, mostly by government initiatives, to monitor the quantitative aspects of performance (Ter Bogt and Scapens 2012). Here, we mention some examples in the United States (Gordon et al. 2002), Finland (Orr et al. 2014), Italy (Bonollo and Merli 2018), and a few in Eastern European former communist contexts (Scott 2007; Dobija et al. 2018).

Concerning performance reporting at the international level, the IPSASB (2015) advances a principle-based recommended practice guideline (RPG3) regarding reporting the service performance information that may be considered a useful reference for a harmonized performance measurement and reporting system across EU member states (Aversano et al. 2018). RPG3 defines effectiveness, efficiency, inputs, output, outcome, performance indicators, and service performance objectives. Moreover, the implementation examples that accompany RPG3 illustrate the terms defined above. RPG3 mentions that the reporting of service performance information should be annual, and it should cover the same period of reference as the financial reporting covers. IPSASB encourages the disclosure of all additional information relevant to the users.

Performance has long been a concern in higher education, especially in PHEIs, as it is connected with accountability and quality assessment, and international rankings. In this context, PHEIs' voluntary or mandatory established performance measurements are useful in assessing the progress towards established goals (Kyrillidou 2002). Moreover, there is an increased demand for PHEIs to disclose their contributions to society as part of the third mission regarding teaching and research (Maingot and Zeghal 2008).

In Europe, the United States, Canada, and Australia, central governments have been involved directly in developing 'indicators'. Thus, the managers of public organizations may not have complete freedom to choose their performance measures. They may have to pay attention to the measures chosen by the government. Even when they must respond to outsiders' measures, however, the managers of a public organizations are responsible for establishing an internal performance management system, including measurement and reporting performance information that will allow it to manage the organization (Behn 2003). In the case of universities, the performance reporting needs to be connected to the stated mission by revealing whether the goals were achieved (Kauppila et al. 2015). According to this, the performance reporting might include a section dedicated to teaching, a section dedicated to research, and another one dedicated to the relationship with the external environment (Bonollo and Merli 2018).

The performance-reporting initiatives in PHEIs represent a set of changes in the relationship between governments and PHEIs. The state entrusts universities to meet the needs of the national economy in a dynamic global marketplace. However, in the meantime, the state carefully monitors universities' overall progress and performance following national needs and objectives using financial incentives and disincentives. On the other hand, universities are trying to attract other financial resources from the state, different funders, or private partners. Tensions are possible between the state funding based on the state's pre-established performance indicators and the university in this context. To obtain state funds, the universities must adjust this behavior (Capella-Peris et al. 2020). In this utilitarian environment, governments will inevitably seek greater accountability and performance (Alexander 1998). States are not driven in this direction because of an authoritative desire to control and regulate (Levine 1997). States attempt to better monitor and assess PHEIs because they are responsible for acquiring more value for resources.

In the particular case of HEIs, some studies have demonstrated that voluntary disclosure results from multiple factors interacting with each other, including regulatory oversight, market forces, costs of disclosure, and organizational structure and governance (Hyndman and Eden 2001). An organization's mandatory reporting and voluntary disclosure are influenced by the coercion of legislation, reporting and funding regulations, and other stakeholder requirements for information.

By adopting entrepreneurial behaviors and operating in a competitive market, universities have become aware that it is not enough to focus only on short-term economic and financial results to gain credibility and become competitive in the medium and long term. Non-financial information on governance, their social and environmental impact, and universities' strategies could provide a better approach to addressing stakeholder concerns.

However, the role of financial information and its interdependence with non-financial information should not be overlooked. Favorable performance on non-financial measures seems to be irrelevant when performance on financial measures is unfavorable (Ghosh and Wu 2012). At the same time, financial results indicate the consequences of decisions made in the past and are sometimes unable to show the causal factors that could lead to the given outcomes. In this dispute, universities would no longer have to publish their activities through disconnected financial and non-financial reports but would produce a single integrated report (Caputo et al. 2021; Lee 2021). The harmonization of financial and non-financial standards and research on universities' IR are steps toward sustainability reports.

Moreover, in public organizations, the state can exercise its coercive powers directly or indirectly. For PHEIs, the state can enact laws and regulations that guide performance disclosure (Bonollo and Merli 2018). The government can also indirectly exercise its coercive powers through institutions or establish specific governmental policies. Another significant fact refers to market forces. For universities, the rankings represent an essential indicator of their position in the national, regional, or worldwide educational market (Urdari et al. 2017). Moreover, performance reporting may be used as an inexpensive marketing tool because the managers may voluntarily disclose that information on performance that gives the organization advantages compared with other educational competitors in the market. Other factors influencing the HEIs' performance reporting are the ones that are organizational in nature, such as structure, implemented performance measurements and reporting procedures, internal politics (Aversano et al. 2018), and governance in line with its declared mission.

2.2. Romanian PHEIs Main Characteristics

Since 1990, Romania has undergone a dramatic transformation, from a highly centralized totalitarian regime to democratic governance, following a radical change in its political and governing system. These changes affect the education sector as much as the economic sectors (Curaj et al. 2015). The first 15 years (1990–2006) of the new higher education system were characterized by the new elements and changes imposed by the new legislative framework, including the universities' autonomy, public financing mechanisms and a performance approach (Education Law 84 1995). University autonomy offers HEIs the right to establish and implement their developmental strategies and policies. However, the autonomy was limited in certain aspects, such as the personnel and financial policies that remained under state control. In this period, the number of public HEIs and specialized programs increased. Moreover, Romania became an exciting destination for international students (Pricopie et al. 2009), but Romanian authorities were not prepared to accommodate them. The Bologna process alignment started in 2005, leading to an increase in the European comparability of the Romanian higher education system and the position of Romanian HEIs in the international higher education market (Pricopie et al. 2009).

A change in the national educational strategy intervened in 2007, focusing on promoting excellence and scientific production. All universities were classified according to their mission into three main groups: advanced research and education, education and science, and education-centered (Coste and Tiron-Tudor 2015). Another critical change refers to the ranking of study programs' and universities' classifications. This process aimed to provide information to potential beneficiaries regarding the quality of teaching, research, student services, community services, and internationalization, delineated into five categories (A > B > C > D > E, where > means 'better results than'). Stakeholders argued against the process because the data processing methodology considered in the evaluation process was not made public.

Regarding the public funding of universities, the law introduced different types of financing, depending on their objectives, such as principal financing, complementary financing, institutional development financing, and so forth. The principal financing is allotted according to a per capita cost-differentiation formula, as the main part of the overall universities' public funding. Complementary financing is based on a qualitative component (i.e., calculated by considering qualitative indicators, which were updated regularly).

The principal financing of universities, according to Education Law 2011, considers the results of the national classification exercise and the different ranking processes. Unfortunately, the implementation of this issue was unsuccessful, as the link between these instruments and the funding methodology was not well maintained. Moreover, even if the law stipulated other forms of disbursing higher education funding, such as institutional development financing that was not influenced by the classification, the subsequent methodologies were developed and implemented only after 2015.

To increase university autonomy and public responsibility, the law proposed that universities establish their mission, institutional development strategy, curricula design and implementation, quality-assurance mechanisms, and financial and human resources management protocols. All these elements must be operationalized using performance measurements and must be periodically reported.

According to the National Education Law No. 84 adopted in 1995, reporting Romanian universities' performance is mandatory. The rector is responsible for preparing a report describing the institutional state. Over time, the law has undergone several changes, and Law No. 1/2011 repealed it. The new education law brought significant changes, one of which was the rectors' obligation to publish the annual report on the university's website (Education Law 2011, art. 130). The report presents the state of the university. The annual report must be publicly available to all interested parties. The minimum information to be provided in the report refers to the following issues: financial situation of the university by funding sources and types of expenses; study programs; staff structure and evolution; results of the research activity; quality assurance internal system; the degree of ethics in all university activities; and the situation of the professional insertion of the graduates from the previous promotions.

The information presented in the report on HEI status is based on relevant performance indicators for each category, demonstrating how financial and human resources have been used to fulfil the missions of teaching and learning, research and community-based services, or the impact on the economic and social environments (Coste and Tiron-Tudor 2015). In addition to demonstrating the performance obtained, the rector's report is an element of public responsibility and a primary condition for public funding (Education Law 2011, art. 150, paragraph 3, p. 32).

The rector's report represents the document in which the essential information that defines the HEI is presented. Law does not establish the specific performance indicators for each group of information; the rector can opt for specific indicators. Additionally, many universities provide additional voluntary information concerning significant elements for the previous year's performance activity to the mandatory requirements. As an example, for 2019, the UBB Rector Report includes voluntary information concerning the following issues: non-traditional education; the University Publishing House (Cluj University Press); university extensions; relationships with the business environment, student and alumni practices; international cooperation; computerization and data communications; communication and public relations; relationships with students; administration and patrimony; and the UBB Centenary.

2.3. Theoretical Framework

The literature supports the implementation of performance reporting from several other theoretical frameworks, such as agency theory, public choice theory, and institutional theory (Grossi et al. 2019). For our purpose, we use the institutional theory, as it explains why organizational structures and practices become entrenched and how and why changes occur (Greve and Argote 2015). According to this theory, institutions impose norms or social coherence on human activity by producing and reproducing an environment for thinking and acting (Burns and Scapens 2000). Research focuses on extra-organizational (social, economic, and political) influences on organizational practices (Fligstein 1998). The theory describes the organization's ability to change, following how institutionalized norms and values affect assumptions (Liguori and Steccolini 2012) and espouses that the process of change finally generates an isomorphic equilibrium (Dumitru et al. 2014).

The PHEI system is pressed for greater performance and quality. Efforts are made in order to adopt business-like attitudes that will keep them sustainable in a competitive economy and turn them into hybrid entities (Grossi et al. 2020). This global reform movement is sustained by new public management principles (NPMs). According to this theory, public sector organizations can be managed and evaluated in the same way as private organizations—namely, through demands for accountability, transparency, efficiency, and responsiveness (Gomes and Yasin 2017).

Performance indicators intended to measure progress towards established national/international goals (Kyrillidou 2002) are asked to describe their contribution to society, often related to the quality of the university's teaching and research process (Maingot and Zeghal 2008). They are the cornerstone of adequate governance mechanisms in the universities, based on performance management measurement and reporting.

Romanian higher education reforms are the results of both global pressures and local demands. As a European Union member, we consider that the primary higher education system reforms' main conditions are justified by global reform ideologies. As part of this, the factors that influence performance reporting are subject to global isomorphic pressures concerning transforming the process of governance models in higher education.

DiMaggio and Powell (1983) defined isomorphism as the factor that encourages the similitude by which institutions tend to adopt the same structure and practices, resulting in their homogeneity (Dobija et al. 2018). There are three processes of institutional isomorphism: coercive, mimetic, and normative. Coercive isomorphism refers to the influence of political and governmental regulations on organizations. In our case, the universities are likely to implement changes within their policies to adjust to the government's requirements due to coercive pressures (Najeeb 2013). Mimetic isomorphism occurs when actors face uncertainty and try to emulate successful organizations as a solution. With an increasing level of competition and internationalization in the higher education context, universities have tried to model themselves on other prototypes in similar contexts through mimetic processes. Normative isomorphism (Paauwe and Boselie 2003) arises primarily from professionalization. Within the higher education context, professionalization involves two aspects: one is the homogenizing influence of established norms (regulatory bodies), and the other is the professional organizations (e.g., accreditation agencies).

Under Chen et al. (2010) and Dobija et al. (2018), the study identifies the following theoretical decision-making mechanisms related to the isomorphism of Romanian national policy (factors) regarding performance reporting: level of implementation of performance-reporting systems; the size of the institution/type of university (coercive and normative); mobile/immobile resource; financing (coercive); and personnel recruitment/staff remuneration (coercive and normative). In all these items, we can find mimetic learning (Cai 2010), as long as national higher education systems attempt to imitate prosperous nations when they face uncertainties or ambiguous development goals (e.g., Bologna process, European/international universities' ranking).

However, some studies (e.g., Gonzales 2012) show that institutional analysis in higher education research focuses mainly on policy and management issues. Given that fund allocation (financing) for higher education institutions depends on the quality of human resources evaluated through the research (number/type/indexation of the article) and teaching outputs (Agyemang and Broadbent 2015), we also consider that employees need to have the potential to impact the performance that is measured (Bouckaert 1995). Performance evaluation is a mechanism by which individual goals and behaviors are aligned with organizational objectives, a process that helps employees understand and accept organizational norms (Ayers 2015).

In this sense, our research combines the institutional theory with the operant theory. The reporting systems for which the operant theory may be applied include performance evaluations, whether they be quality control reports, personnel evaluations, or variance reports (Lovata 1992). Many studies conclude that specific behavior results from its consequences (Ulrich et al. 1974). The theory is suitable for our context—the universities' funding levels depend on the employers' results. Additionally, the individual income level is influenced by this individual factor. Thus, rewards such as money are considered positive reinforcement if their presence increases the likelihood that the behavior will recur. At the same time, the behavior is most easily modified when it produces a negative consequence. If the expected reward is not satisfactory, the motivation to exercise an increased quality in the developed activity will upturn. If, on the other hand, the system ignores or criticizes the employee's results (e.g., changing the framing articles, excluding journals/databases, diminishing the article/journal ranking), this consequence is likely to make them avoid working hard in the future. This mix between the institutional and operant theories is meant to justify and analyze the main factors that influence the form and evolution of performance reporting used for Romanian higher education.

2.4. Research Hypothesis Development

The hypotheses developed further refer to the factors likely to influence the reporting of information on the performance of PHEIs in Romania. The main arguments considered for the factors' inclusion refer to scientific references from the literature and their relevance in the Romanian context.

Hypothesis 1: *Performance reporting is associated with the financial resources attracted: the more resources a PHEI attracts/obtains, the more information is reported.*

The literature often mentions the level of financial resources as having a positive significance on reporting information in the public sector (Maingot and Zeghal 2008; Gallego et al. 2010; Gallego-Álvarez et al. 2011). The link between the financial resources attracted by a PHEI and the level of information presented is a topic that, at first glance, seems to have no significance in the public sector. However, the decrease in public resources' capabilities to satisfy the level, quality, and extent of public needs of the population is an indisputable reality (Manolescu 2009). Thus, the fundamental nature of the traditional relationship between the government and higher education is in the process of undergoing significant changes to sustain more students in attempting to maximize economic returns (Alexander 2000; Garde Sanchez et al. 2020). This unequal ratio can be rebalanced by identifying solutions needed to supplement public funding. State reporting and funding mechanisms for HEIs are in the midst of a significant transformation from an input-based system to a more competitive outcomes-based approach (Aversano et al. 2018). Through mimetic isomorphism, PHIE borrows solutions from private entities' behaviors (e.g., marketing and promotion policies, a public–private partnership between universities and community, tuition fee strategies, other forms of collaborations with the private environment, and use of bank loans granted to the university/student), which, through coercive and normative isomorphism, are implementing legal rules in universities.

In this context, PHEI might be concerned with finding the proper balance between institutional autonomy and performance-based assessments to become competitive in the market. With greater expectations being placed on it, higher education is obliged to examine itself or be examined by others (Alexander 1998). Accordingly, the growing societal necessity dictates that universities must become more responsive to economic needs and governmental demands for increased performance (Sangiorgi and Siboni 2017; Brusca et al.

2019; Garde Sanchez et al. 2020). In Romania, higher education is funded from the state budget according to the size and the university category, under granted basic and additional financing forms; for the second factor, the higher education institution's performance is an essential criterion. For primary funding, the government typically determines the value of resources for various students in their fields of study at a centralized level.

Hypothesis 2: *Performance reporting is associated with salary costs.*

Universities are 'communities', where individuals gather to invest in their human capital (Alexander 2000). Overall, higher education systems' regulatory framework has become more complex and expensive to sustain, particularly regarding, on the one hand, employee salaries and rights, and on the other hand, employer obligations and contributions related to employees. Moreover, equity issues concerning disability, race, and gender are entered into force (Gordon and Whitchurch 2007).

The reality also shows an increasing diversification of academic tasks (teaching, scholarship, research, consultancy, community service, and administration). Kogan et al. (1994) note that the range of roles that an academic may be expected to undertake can include 'teacher, scholar, practitioner, demonstrator, writer, model, discoverer, inventor, investigator, designer, architect, explorer, expert, learner, developer, collaborator, transformer, facilitator, enabler, evaluator, critic, assessor, setter, guide, colleague, supervisor, mentor, listener, adviser, coach, counselor, negotiator, mediator, juggler'. Therefore, the historical trilogy (teaching, research, and administration) of academic work (Garde Sanchez et al. 2020) would appear to have been enlarged. Since the public institution's financial resources did not increase directly proportional to the work's complexity, the effect was to increase the level of quality/involvement required to correspond to a different reward based on performance criteria.

Thus, performance disclosure appears as a mediating, bidirectional positioned factor. Through coercive isomorphism, the educational institution has conditioned the financing of human capital's performance, which in turn, according to the operant theory, can be motivated/demotivated by the level of remuneration received. A favorable expectation/reward will determine a motivated behavior in the future, with a favorable effect on the increase in the financing sources, while a negative one will demoralize the human factor, with adverse effects on the future performance indicators. Through normative isomorphism, these human capital politics and rules are put in a particular view to sustain the institutional or personal interests, based on academic particularities.

In too many cases, the primary performance-reporting quantifiable item is considered to be the quality or quantity of research (Dunkin 2005; Siegel and Wright 2015). This approach can decrease the importance of other duties, roles, and functions, especially teaching, serving, and displaying good academic citizenship. As a general remark, Alexander (2000) notes that the tension between the numbers and quality dominates higher education debates in most advanced countries.

Each university is responsible for having the capacity and intelligence to stimulate human capital and build on this capacity, both academic and professional. As Dunkin (2005, p. 8) notes, 'The capacity to develop business/earn one's salary/manage 'client' relationships, once missing from academics, is now part of the skills repertoire of our next generation of academics'.

In the empirical study, we considered the ratio between costs with salaries and the number of students as a factor that can influence the degree of reporting of information on public service performance. The scientific literature uses this item to a lesser extent (Suryadi 2007), but in this research, we want to show that costs with salaries in the Romanian public sector positively influence the reporting of performance information.

Salary costs were included in the case study because, after studying the annual reports on universities' states, we came across several documents claiming that the university's performance was encouraged by providing financial benefits to teachers. Through mimetic behavior, also in some public entities, salary costs are an element that can influence the reporting of information on the performance of services because, in the private sector, the entity's staff can obtain bonuses from the performance achieved; their payment correlates to the success achieved. Thus, the main arguments supporting the inclusion of variable costs with salaries in the empirical study are given by the university employees' financial advantage and by the university's performance objectives, specified in the annual report on the university's state.

The universities' management was often interested in the result. The pecuniary benefits granted to the teachers encouraged the performance, thus obtaining at least two advantages for the two parties involved: one for the institution and the other for the academic staff. The publication of scientific research in internationally renowned journals is encouraged by the management of higher education institutions, because, on the one hand, it receives international recognition for research activity, and on the other hand, several performance indicators are met, and more information on scientific research will be published. First, accessing research projects is a method of attracting financial resources, an essential activity for the higher education institution and those involved in developing the projects. Secondly, with the help of research projects, academic staff publish scientific research in international databases and participate in international conferences, thus receiving international recognition for their studies; they are highlighted by the number of citations. The advantage of universities, in this case, is to meet performance indicators.

Hypothesis 3: *Performance reporting is associated with the size of the higher education institution.*

One of the most used factors in empirical research, identified as having a significant effect on the level of information presentation, is the institution's size. The role of this factor in the private sector has a positive impact on information reporting (Glaum et al. 2013), and one of the reasons is the need to inform shareholders about the position and the obtained performance so that the investments are made in the best conditions (Gallego-Álvarez et al. 2011).

In private sector research, the size of a company has been determined by total assets, a form also found in empirical studies in the public sector, in which the positive link between the two has been demonstrated (Gordon et al. 2002; Gordon and Fischer 2018). Researching the university environment has led some studies to choose a specific factor for quantifying the size of a university, and in the literature (Maingot and Zeghal 2008; Gallego et al. 2010; Suhaiza and Nur 2011; Gallego-Álvarez et al. 2011), the number of students is an indicator of the size and has a positive influence on information reporting.

Regardless of how the size of public sector institutions, total assets, or the number of students was determined, the studies above proved a positive link between the institution's size and a high degree of information reporting. The arguments presented above led us to set the first hypothesis for this study, the total assets being the size indicator of higher education institutions.

Hypothesis 4: *Performance reporting is associated with the quality category of the higher education institution.*

Performance disclosure can be used as a tool for HIE to compete in the international university arena and attract students and researchers (Brusca et al. 2019). This target depends on the university's profile: teaching or research. Thus, the research universities have a more substantial opportunity to obtain more research development funds through national or European competitions. For teaching universities, the theoretical and practical implications of the future profession—collaborations with employers—are just some of the specific information that increases this category's market credibility. In this context, in line with Maingot and Zeghal (2008), we consider that these entities are stimulated to disclose performance information following their mission (type).

Based on specific legal rules and professionalization implications (coercive and normative isomorphism), university performance disclosure in two directions (research or teaching) becomes a critical factor in stakeholders' visions. More specifically, Maingot and Zeghal (2008) state that students 'outcomes are the results of universities' developed educational offering and activities rather than the results of inputs (selection effects) or exogenous influences, such as economic conditions.

3. Methodology

3.1. Research Design

The investigation performed consists of the following steps. Firstly, in the previous section, based on the literature, we argued that PHEIs report information concerning their performance to demonstrate their accountability, responsibility, and transparency, and different factors influence the level of performance-reporting disclosure. Then, we formulated the hypotheses according to which factors, such as size, resources attracted per student, costs with salaries and category, influence the reporting of performance measurement information. Secondly, a performance-reporting disclosure index (DI) is proposed, based on existing national legislation that includes mandatory reporting elements and international rankings indicators and recommended practices issued by the International Public Sector Accounting Standards Board (IPSASB). The DI captured the general and specific elements required for the preparation of performance measurement reporting. The DI may be a useful tool for Romanian universities in terms of reporting performance information. Thirdly, the data were collected using a sample of PHEIs from economic sciences. Fourthly, we test the hypotheses, running a linear regression. We then correlate the results with the literature, and we discuss and contextualize the results.

3.2. Sample Selection and Data Collection

Of the total, 23 Romanian PHEIs are included in the sample, all of them with at least one specialization belonging to the field of economic sciences. In the case of each university included in our sample, we connected to their official website and identified the reporting section, where we found and downloaded the reports for the last five years (2015–2019). We conducted this analysis between September and November of 2020. We acknowledge that online information is continuously updated, and we marked the date of verification (within the formerly mentioned period of analysis). The cached versions of the websites in the specified interval provide the validity of our data. Afterward, we proceeded to analyze the content of the documents and identify the disclosure index items. Ultimately, we performed the corresponding calculations to attain the disclosure levels concerning each of the dimensions analyzed for each university and each year.

The universities' distribution, according to the classification made by the Romanian Ministry of Education and Scientific Research (MENCS), is as follows:

- Universities of advanced research and education: 4;
- Universities of education and scientific research: 6;
- Universities focused on education: 13.

3.3. Description of Variables

3.3.1. Dependent Variable

The first step to achieve the research objective was to determine the performance-related disclosure index (DI), acting as a dependent variable. DI might be an acceptable tool for measuring a series of elements in documents published by institutions (Banks and Nelson 1994; Gallego et al. 2010). The information index model used in this study measures the appearance of information in the report and is marked with '1' and '0' if the information is not presented, and the calculation formula of the index is as follows:

$$DI = \frac{\sum_{i=1}^{m} d_i}{\sum_{i=1}^{n} d_i}$$

where:

DI = performance-reporting disclosure index;
d – the element of form i presented, performance indicator;
m = number of items submitted or disclosed;
n = number of items expected to be presented.

In order to refine our study, we divided this disclosure index into components:

$$DI_{Total} = DI_{Teach} + DI_{Research} + DI_{Ext} + DI_{Fin}$$

where:

DI_{Total} = performance-reporting disclosure index (DI);
DI_{Teach} = performance-reporting disclosure index regarding teaching activity;
$DI_{Research}$ = performance-reporting disclosure index regarding research activity;
DI_{Ext} = performance-reporting disclosure index regarding external environment;
DI_{Fin} = performance-reporting disclosure index regarding financials;

In the construction of *DI* (Table A1), performance information was selected based on the following arguments:

- Used in previous studies (Guthrie and Neumann 2007; Suryadi 2007; Lukman et al. 2010; Ramos-Vielba et al. 2010; Shin 2010; Al-Ashaab et al. 2011; Perkmann et al. 2011; Ter Bogt and Scapens 2012; Seppo and Lilles 2012; Rossi and Rosli 2015; Asif and Searcy 2013; Hegarty 2014; Chan 2015; Albats et al. 2018; Bonollo and Merli 2018; Francesconi and Guarini 2018);
- Required by the Romanian National Education Law (Law 1/2011);
- Recommended by the IPSASB in the RPG3;
- Included in important world rankings (Academic Ranking of World Universities or Shanghai Ranking, Times Higher Education World University Rankings, Quacquarelli Symonds World University Rankings and U-Multirank).

In the literature, there are a series of studies using different DI constructs. For this study, among them, the ones that are relevant are those that: used the indicators requested by legislation in force (Gordon and Fischer 2018; Maingot and Zeghal 2008; Suhaiza and Nur 2011; Montesinos et al. 2013; Gomes and Yasin 2017) or are proposed by certain empirical studies (Gordon et al. 2002; Gallego et al. 2010; Gallego-Álvarez et al. 2011; Dobija et al. 2018). Based on these reasons set out above, the performance indicators included in the disclosure index include 46 elements (listed in Table A1) grouped into four categories as follows:

- Teaching process: 15 performance indicators;
- Research: 13 performance indicators;
- External environment: 11 performance indicators;
- Financial resources: 7 performance indicators.

The information on the performance included in the report on the state of the university has the central role of demonstrating how the material, human, financial, and informational resources were managed but also the degree of fulfillment of the objectives proposed by the rector in the management contract and the commitment to fulfilling them once taking office for four years. Thus, we analyzed to what extent the rector's annual report covers the mandatory disclosure requirements stipulated by law (Education Law 2011, no. 1, art. 130, paragraph 2), and those included in addition to our index based on IPSAS recommended good practices concerning service performance reporting and international educational rankings.

3.3.2. Independent Variables

Based on the elements used in the studies in the literature, the hypotheses were defined. To determine the level of reporting information on the performance of services of higher education institutions in Romania, we used the following factor variables: financial resources attracted per student, personnel costs, and size and university category. An element of novelty consists of testing the influence of salary costs on reporting information on Romanian universities' performance.

To determine the financial resources attracted, the personnel costs, and the size of the higher education institution, we collected the data from the financial statements, documents that are required to be published on each university's website. In order to eliminate size effect of the financial information, we reported the incomes, costs with salaries, and total assets to the number of students, and data collected from the annual reports. The university's categorization in advanced research and education, education and scientific research, and education centered was extracted from the national classification (Education Law 2011), and three control variables were considered. The first two control variables were introduced in regression to interpret the third data, which were not included. The choice of the first two variables is motivated by higher education institutions' membership in the category that includes scientific research activity. We show the dependent and factorial variables in the statistical model to test the hypotheses (Table 1).

Table 1. Description of the variables used in the statistical model.

Symbol	Variable	Computation	Hypothesis	Authors
DI	Disclosure index	Disclosure index for information related to performance	Dependentvariable	Gallego-Álvarez et al. 2011; Suhaiza and Nur 2011
RES	Financial resources attracted/per student	Ratio between total revenue and total number of students	H1	Guthrie and Neumann 2007; Chan 2015
SAL	Costs with salaries	Ratio between total costs with salaries and total number of students	H2	Gordon and Whitchurch 2007
SIZE	University's size	Ratio between total assets and total number of students	H3	Ter Bogt and Scapens 2012; Chan 2015
	University's category		H4	
CAT_1		1. Control variable with value '1' for universities from the category advanced research and education; value '0' for other categories;	Control	Law 1/2011
CAT_2		2. Control variable with value '1' for universities from the category teaching and scientific research; value '0' for other categories;		Law 1/2011
CAT_3		3. Control variable with value '1' for universities from the category education centered; value '0' for other categories.		Law 1/2011

3.4. Development of the Linear Regression

In order to determine the influences on the DI in Romania and to test the hypotheses, we formulated a hierarchical linear model, through which the dependent variable (DI) is linked to its factors (explanatory variables) as discussed in the literature and the design of the hypotheses. Hierarchical linear regression is the statistical tool that reorganizes information and examines the consistency of the underlying theory (Lindenberger and Pötter 1998), similar to the least squares method (Osborne 2000). The analysis that can be performed—namely, modeling both within and between individual variations (Terracciano et al. 2005)—motivates the hierarchical linear model's choice. We used the SPSS 2.0 program to develop the model and explore statistical data.

The variables included later were selected to improve the model and explain the model's variation. The following factor variables were included in the first level: financial resources attracted per student (RES) and costs with salaries (SAL). These explanatory variables are indicators important for measuring the institution's financial performance, showing the economic and financial aspects of the operational activity. In the second level, the variable size of the university (SIZE) was introduced. We chose to include the university category factor variable in the third level, used as the control variable. The advantage of using hierarchical linear regression is the gradual highlighting of the influence of factors on reporting information on service performance.

The hierarchical linear regression used in determining econometric models has the following form:

$$DI = \propto_0 + \propto_1 RES + \propto_2 SAL + \propto_3 SIZE + \propto_4 CAT\_1 + \propto_5 CAT\_2 + \propto_5 CAT\_3 + \varepsilon$$

3.5. Sources of Data

We studied the annual reports of HEIs for five years, and hand collected data for each year. Since this study does not follow the evolution of performance information reporting, but there is a causal relationship between the dependent and factor variables in terms of direction, level, and significance, we merged these years into one singe period because the conditions of analysis are identical throughout the years, without fundamental legislative changes and no organizational and accounting perspective changes. This way, our pooled sample increased in observations and provided opportunity for better data analysis.

Pooled data is a modified cross-sectional method to analyze fixed and random effects (Johnson 1995). The reason for choosing this technique is argued by increasing the statistical significance of the model or comparing institutions' effects (Johnson 1995). The purpose of applying this analysis is to determine the random effects, which allows the estimations of the differences (Jesilow and Ohlander 2010) between higher education institutions. Thus, each university will have four records in the study. Fixed effects, resulting from the model's application, exhibit the characteristic of not showing variations over time, regardless of the measured and unmeasured effects (Johnson 1995).

Based on the content analysis of 115 reports published by the 23 public universities in Romania, we obtained a statistically significant model over 5 years.

4. Results

The empirical research highlighted the extent to which the information on service performance is presented in the annual report prepared by the HEIs. Linear regressions, composed of factors whose influence has been demonstrated in various studies found in the literature, were tested, and from a statistical point of view, their significance on the presentation of information on service performance was reconfirmed.

4.1. Sample Description

The sample of PHEIs consisted of 23 public universities. For each university, the rector's reports for the previous five years were analyzed in detail. The average level of reporting information on the Romanian PHEI environment's performance is 54%, meaning that the PHEI presents only half of the relevant information that reflects the performance obtained. The lack of a standardized model for reporting information on performance leaves the assessment of the report's content and format on the institution's state to the management. Some of the reports studied were rich in content and described several procedures and events, but the information considered essential by international classifications, literature in the field, and national and international bodies was not presented. Reports belonging to the teaching process and the financial dimension, presented 60% and 71%, respectively, of the indicators followed in this study. Scientific-research activity is described in the annual reports, but the presentation level that reflects the performance is 49%.

Regarding the interaction with the external environment, the information is presented in a proportion of 42%, most often providing data for the number of partnerships with private environments and other higher education institutions. In contrast, the number of international students or the number of study programs in other international circulation languages was rarely presented. The variables' descriptive statistics (Table 2) reflect the results obtained and described above.

Table 2. Variables' descriptive statistics.

Variable	Average	The Mean Square Deviation
DI or DI_{Total}	0.54	0.14
DI_{Teach}	0.60	0.21
$DI_{Research}$	0.49	0.14
DI_{Ext}	0.42	0.24
DI_{Fin}	0.71	0.15
RES	8201.38	3237.99
SAL	4429.14	923.99
SIZE	23,645.61	7904.41
CAT_1	0.25	0.44
CAT_2	0.28	0.45
CAT_3	0.47	0.50

Firstly, the hierarchical linear regression was tested by using as a dependent variable the PRDI on the teaching process (DI_{Teach}). Secondly, we used the PRDI of the research activity ($DI_{Research}$) and, thirdly, the PRDI for the relationship with the external environment (DI_{Ext}). Fourthly, we tested the financial dimension index (DI_{Fin}). Thus, the regression was run for the four dimensions in which the initially established performance indicators were included, motivating this choice by the homogeneity of the indicators in each dimension. Subsequently, the hierarchical linear regression was tested by using as a dependent variable the presentation index that contained the performance indicators set for each dimension (DI_{Total}) to test the variation of the information presented according to the selected factor variables.

4.2. The Level of Disclosure Concerning the Entire University Performance

The university DI for the four dimensions and factor variables—resources attracted per student, salary costs, size of university and category of universities classified as advanced research and education institutions, according to the Pearson correlation coefficient—shows a strong correlation. The correlation matrix for the information presentation index is shown in Table 3.

Table 3. Correlation matrix for RPDI.

	DI_{Total}	RES	SAL	SIZE	CAT_1	CAT_2
DI_{Total}	1					
RES	0.349 ***	1				
SAL	0.541 ***	0.583 ***	1			
SIZE	0.236 **	0.302 ***	0.255 **	1		
CAT_1	0.727 ***	0.246 **	0.429 ***	0.276 **	1	
CAT_2	0.072	0.099	0.209 **	−0.271 **	−0.361 ***	1

Significance test: ** significant at level 0.05; *** significant at level 0.01.

The Enter method was used to enter the statistical model of all the factor variables to explain the connection between them and the dependent variable. Level 1 included factorial variables: (1) financial resources attracted per student because in empirical studies in the literature the importance of the variable was demonstrated, and (2) salary costs because, on the one hand, the variable was not found in public sector studies (the novelty of this study), and on the other hand, we wanted to demonstrate the implications of the variable on the performance reporting. In levels 2 and 3, new factorial variables were introduced: size, respectively, and the university category (Table 4). The factorial variables listed above have a positive impact on the level of reporting of information on higher education institutions' performance. The collinearity between the factorial variables is accepted because the VIF value is smaller than 3.

Table 4. Performance-reporting model characteristics.

Variable \ Model	Model 1		Model 2		Model 3	
	Coefficient	t	Coefficient	T	Coefficient	t
RES	0.22×10^{-5}	0.391	0.12×10^{-5}	0.21	0.21×10^{-5}	0.539
SAL	0.75×10^{-4}	3.3858 ***	0.73×10^{-4}	3.741 ***	0.12×10^{-4}	0.724
SIZE	-	-	0.17×10^{-5}	0.880	0.14×10^{-5}	0.952
CAT_1	-	-	-	-	0.246	8.203 ***
CAT_2	-	-	-	-	0.107	3.846 ***
R^2	0.294		0.303		0.677	
F	12.711 ***		8.701 ***		24.365 ***	

Significance test: *** significant at level 0.01.

The factor variables included in the statistical model explain the level of reporting of the information on the performance of higher education institutions in Romania in a proportion of 68%; the value R^2 is 0.677. In this case, too, the model's representativeness gradually improved, increasing significantly in model 3. The value recorded by F is 24,365, with a significance of 0.000. In model 2, the variation of DI_{Total} is influenced by wage costs, as in model 1. In model 3, the significant increase in the index of information on service performance is influenced by category 1 and category 2 universities, compared with education universities (category 3). Compared with higher education institutions classified as universities of education, those in the category of advanced research and education and universities of education and scientific research present more performance information.

The robustness of the model was tested by two methods: the homoscedasticity test and the Shapiro–Wilk test. The regression model has robustness because all the hypotheses regarding the normal distribution of residues were accepted.

4.3. The Level of Disclosure Concerning Each of the Four Components of University Performance Disclosure

4.3.1. Teaching Dimension of PHEIs' Performance Reporting

The correlation between the variables was tested using the Pearson coefficient (Table 5). The test results reveal a strong correlation between DI_{Teach} and the factor variables RES, SAL, and CAT_1. Thus, we can conclude that the resources attracted per student, salary costs, and the inclusion of universities in the advanced research category and education, positively impact the level of reporting of information on higher education institutions' performance. The correlation between DI_{Teach} and university size (SIZE) is average. DI_{Teach} is correlated at level 0.32 with the financial resources attracted per student; in other words, the higher the financial resources per student, the higher the level of reporting the information on the teaching process's performance, and vice versa. The collinearity between the factorial variables is an accepted one because the value of the variance inflation factor or VIF is below 3, with the collinearity being within the tolerated limits.

In model 1, the variation of the dependent variable DI_{Teach} is influenced by salary costs, so a high level of salary costs leads to more information on the teaching process's performance. In model 2, the disclosure index is significantly influenced by wage costs. In model 3, the university category influences DI_{Teach}. In all three models, the size of the university does not influence the variation of DI_{Teach}. The model's robustness was verified with the Shapiro–Wilk test and the homoscedasticity test, with all hypotheses regarding the normal distribution of residues being accepted. The tested model can be considered an average because the value recorded by the coefficient of determination is 0.43, meaning that the selected factorial variables influence the degree of reporting of the information on the teaching process's performance in a proportion of 43%. This shows how much of the variation in the performance index is measured by the analyzed factors. The significance test has a limit of 0.05, and the result obtained falls within this limit (the value F being 8.804, with a significance of 0.002), thus obtaining a significant regression model. In this case, too, the model's robustness was tested, accepting all the hypotheses regarding the normal distribution of residues.

Table 5. Correlation matrix for teaching dimension.

	DI_{Teach}	RES	SAL	SIZE	CAT_1	CAT_2
DI_{Teach}	1					
RES	0.32 ***	1				
SAL	0.38 ***	0.583 ***	1			
SIZE	0.26 **	0.302 ***	0.255 **	1		
CAT_1	0.54 ***	0.246 **	0.429 ***	0.276 **	1	
CAT_2	0.10	0.099	0.209 **	−0.271 **	−0.361 ***	1

Significance test: ** significant at 0.05; *** significant at 0.01.

4.3.2. Research Dimension of PHEIs' Performance Reporting

Regarding the level of research activity performance reporting, salary costs and advanced research and education universities give positive influences on $DI_{Research}$. The classification of universities in education and scientific research negatively influences the level of reporting of information on research activity performance. The correlation of the variables is shown in Table 6. The value recorded by the VIF indicator is below 3, which allows us to accept the model because the collinearity falls within the tolerated limits.

Table 6. Correlation matrix for research dimension.

	$DI_{Research}$	RES	SAL	SIZE	CAT_1	CAT_2
$DI_{Research}$	1					
RES	0.144	1				
SAL	0.413 ***	0.583 ***	1			
SIZE	−0.015	0.302 ***	0.255 **	1		
CAT_1	0.623 ***	0.246 **	0.429 ***	0.429 ***	1	
CAT_2	−0.215 **	0.099	0.209 **	−0.271 **	−0.361 ***	1

Significance test: ** significant at level 0.05; *** significant at level 0.01.

The level of reporting of information on research activity performance is influenced by the costs with salaries, as demonstrated by all three models. The result is explained by teachers' financial benefits for publishing scientific research in internationally renowned journals. Thus, the first benefit resulting from the publication of articles is related to teachers. The second benefit is for the university, which, in addition to receiving international recognition for its research activity, meets several performance indicators and will publish more information on scientific research. The research projects accessed attracted financial resources, and the people involved were remunerated for the activity carried out within the project. However, at the same time, the project's performance indicators bring the university the advantage to excel in research activity by publishing articles and participating in various conferences, indicators demonstrating the performance of higher education institutions for the research mission. The higher the salary costs, the more performance the university will publish on the research activity In model 3, the variation of $DI_{Research}$ is influenced by the salary costs, the financial resources attracted per student, and by the classification of the universities in the category of those of advanced research and education. R^2 is 0.487, meaning that the selected factorial variables influence the variation of the degree of reporting the information on the performance of the research activity in a proportion of 49%. The value shows a significant jump from model 2 to model 3, thus demonstrating the gradual improvement of the model's representativeness; however, the significance of coefficients is more important for our study since these indicate the direction and level of influence of factor variables.

The hierarchical linear regression model is robust because all assumptions regarding the normal distribution of residues have been accepted.

4.3.3. Interaction with External Environment Dimension of PHEIs' Performance Reporting

The correlation between the variables and the index of presentation of information on university performance concerning the external environment is positively influenced by the factorial variables: resources attracted per student, salary costs, and university size and category (Table 7) The correlation between the factorial variables is strong or average, but they are accepted because the VIF indicator's value is below 3%, with the collinearity falling within the tolerated limits.

Table 7. Correlation matrix for external environment.

	DI_{Ext}	RES	SAL	SIZE	CAT_1	CAT_2
DI_{Ext}	1					
RES	0.252 **	1				
SAL	0.342 ***	0.583 ***	1			
SIZE	0.184 *	0.302 ***	0.255 **	1		
CAT_1	0.407 ***	0.246 **	0.429 ***	0.276 ***	1	
CAT_2	0.228 **	0.099	0.209 **	−0.271 **	−0.361 ***	1

Significance test: * significant at level 0.1; ** significant at level 0.05; *** significant at level 0.01.

Following the same procedure as in the previous cases, all three modes were tested. In model 3, the value of R^2 is 0.354; the factor variables influence the degree of reporting of the information regarding the university performance for the interaction with the external environment in a proportion of 35%. The model's representativeness gradually improved, and the value R^2 increased significantly from model 2 to model 3. The value of F is 6.366, with a significance of 0.000. In model 1, the variation of DI_{Ext} was demonstrated, depending on the salary costs. Model 2 retains the same influence. In model 3, the category of universities classified as advanced research and education institutions and those of education and research represent factorial variables that influence the variation of DI_{Ext}.

The PRDI concerning financial dimension is positively correlated with the resources attracted per student, salary costs, the university's size, and the category of universities assigned to advanced research and education institutions; the results are presented in Table 8. There is a negative correlation between the factorial variables university size and university category classified as educational and research institutions. Moreover, in this case, the VIF indicator's value is below 3, so the variables included in the model are accepted because they fall within the tolerance limit.

Table 8. Correlation matrix for financial dimension.

	DI_{Fin}	RES	SAL	SIZE	CAT_1	CAT_2
DI_{Fin}	1					
RES	0.261 **	1				
SAL	0.549 ***	0.583 ***	1			
SIZE	0.187 *	0.302 ***	0.255 **	1		
CAT_1	0.640 ***	0.246 **	0.429 ***	2.76 **	1	
CAT_2	−0.34	0.09	0.209 **	−0.271 **	−0.361 ***	1

Significance test: * significant at level 0.1; ** significant at level 0.05; *** significant at level 0.01.

Model 3 was formed for testing the variation of DI_{Fin} according to the factor variables, which are explained in a proportion of 51%; the representativeness of the regression gradually improved from one model to another. Costs with salaries are statistically high in all three models. The university's category influences the presentation of financial results, and compared with universities focused on education, higher education institutions in the higher category will present more information on funding activity. This result is explained by universities' abilities of advanced research and education to attract financial resources from various activities. In this case, too, the model's robustness was tested, accepting all the hypotheses regarding the normal distribution of residues.

5. Validation of Hypotheses and Discussion

The resources attracted per student is a controversial variable for the public sector, being included in a small number of empirical studies, but its positive influence on the degree of information reporting has often been demonstrated (Maingot and Zeghal 2008; Gallego-Álvarez et al. 2011) using the Pearson coefficient. The reporting of information on service performance is positively associated with the level of financial resources attracted by public universities in Romania. The results of the Pearson coefficient from regression testing on research are shown. Otherwise, the financial resources attracted per student do not significantly influence the reporting of information on the research activity performance.

The novel factor variable 'costs with salaries' positively influences the level of reporting information on the performance of the teaching process, research activity, the relationship with the external environment, financial activity, and activities listed above as a whole.

The size of higher education institutions in Romania, determined by the total assets related to the number of students, is positively associated with reporting performance information. As in the other empirical studies found (Maingot and Zeghal 2008; Gordon and Fischer 2018; Gallego-Álvarez et al. 2011), we demonstrated that in Romania, too, the size of the university positively influences information reporting. This result is interpreted as follows: higher education institutions with several assets report performance information at a significantly higher level, and vice versa, with the effects and cause being determined by logical reasoning. The exception to the rule was encountered in reporting information on the research activity's performance, where the university's size does not influence the reporting process.

In most cases, the university category is a factor with a significant influence on the presentation of information on service performance, as has been demonstrated in the literature (Gallego-Álvarez et al. 2011). The variable was used with the classifications made by MENCS, and higher education institutions included in the category of advanced research and education universities present more performance information, which positively influences the reporting in all situations analyzed.

The reporting of information on the teaching process's performance is positively influenced by the financial resources attracted per student, salary costs, and size and category of higher education institutions. The larger the educational institutions, and the more they are part of advanced research and education universities, the more information they present. This result indicates that universities that have gained a reputation over time are also those that have developed because they met users' expectations and maintained their position in academia, bringing improvements and diversifying their educational offerings. Reporting information on performance in scientific research is influenced by salary costs and the category of the university. The stability of higher education institutions has attracted experience in scientific approaches and in proposing new research projects. The presentation of information attesting to scientific research performance would not be possible if several indicators were not met. The publication of research in renowned journals, indexed in databases, is encouraged by the university, and in some cases, financial benefits are offered. Firstly, the financial benefits obtained by the academic staff in cases of exceptional results were confirmed through the information presented in the annual report on the state of the university. Secondly, they were demonstrated by the empirical research conducted.

The resources attracted per student, salary costs, and size and category of the university outline the relationship with the external environment and the positive influence on information reporting. The results obtained in this case depend on the number of partnerships concluded with universities in different states. The number of international students who choose to study at a partner university in Romania increases, thus validating teaching and research performances and attracting financial resources. Partnerships with private entities and attracting resources from this environment also show the trust of private entities, building a strong and lasting relationship with the external environment while validating education and research conducted at the university.

In Romania, financing from the state budget is made according to the university's size and category, being granted basic and additional financing, the latter based on the higher education institution's performance, aspects supported by MENCS. From the empirical research, we can conclude that the presentation of information on financial performance is positively influenced by the resources attracted per student, salary costs, and size and category of higher education institutions, classified as universities of advanced research and education.

The results obtained by applying the model with all four dimensions, taken as a whole, demonstrate that the level of reporting information on the performance of the higher education institution is positively influenced by financial resources attracted per student, salary costs, and size and category of a university in the best category—namely, universities of advanced research and education (Table 9).

Table 9. Variables' influence over the level of performance-reporting disclosure index.

Variable	DI_{Teach}	$DI_{Research}$	DI_{Ext}	DI_{Fin}	DI_{Total}
RES	+	0	+	+	+
SAL	+	+	+	+	+
SIZE	+	0	+	+	+
Cat_1	+	+	+	+	+
Cat_2	0	−	+	0	0

The variation in the level of reporting of information on higher education institutions' performance has different characteristics, depending on each dimension considered in the study. The formulated hypotheses were partially validated and are briefly presented in Table 10.

Table 10. Hypotheses validation.

Hypotheses	DI_{Teach}	$DI_{Research}$	DI_{Ext}	DI_{Fin}	DI_{Total}
Hypothesis 1: *Performance reporting is associated with the financial resources attracted.*	Reject	Reject	Reject	Reject	Reject
Hypothesis 2: *Performance reporting is associated with salary costs.*	Partially accept	Accept	Partially accept	Accept	Partially accept
Hypothesis 3: *Performance reporting is associated with the size of the higher education institution.*	Reject	Partially accept	Reject	Reject	Reject
Hypothesis 4: *Performance reporting is associated with the quality category of the higher education institution.*	Accept	Accept	Accept	Accept	Accept

Hypothesis 1 has not been validated in any model, so the resources attracted per student using our data sample is not associated with reporting of performance information. The results obtained are contrary to expectations, but we can say that in Romania, the financial resources attracted per student do not influence the reporting of information on service performance.

Most of the time, hypothesis 2 was validated, but, contrary to expectations, in the regression in which the variable 'university category' was introduced, it was shown that salary costs do not influence performance information reporting. Instead, the variation in the level of information presented on research and funding performance is influenced by wage costs.

The assertion presented by hypothesis 3 was validated in a single version of the models ($DI_{Research}$). Thus, the reporting of information on the performance of research is influenced by the university's size. In the other models, the hypothesis was not confirmed. As we described before, in Romania, higher education is funded from the state budget according to the size and the university category. Therefore, especially for the research domain, the visibility of the results helps not only in the international recognition of the university (rankings and databases) but can also bring in counterpart national and international financial sources.

Hypothesis 4 was validated in all five cases, with the university category being thus an essential factor in determining the level of presentation of performance information and influencing the variation of reporting.

The regression efficiency test is an important element underlying the model and the conclusions drawn from the data analysis, and the homoscedasticity test was performed to validate the results. The homoscedasticity test involves scattering the dependent variables' values without influencing the factorial variables' values (Bai et al. 2016; Balkin et al. 2016). Based on the data obtained from the homoscedasticity verification, we can state that the tested model is an efficient one, and the conclusions formulated based on the obtained results are validated. The Shapiro–Wilk test was also performed to determine whether the residue had a normal distribution. This test involves validating the hypothesis that the residues have a normal distribution; otherwise, the hypothesis is rejected (Shapiro and Wilk 1965). The hypothesis is accepted if the value of Sig. > 0.005. We obtained a normal residue distribution for all models run, so the normal distribution hypothesis was accepted.

In the university context, the follower of entrepreneurial behavior, the (mainly) voluntary content, and involuntary information are adapted to the 'stakeholders' requirements to increase the university's attractiveness and credibility on the market.

6. Conclusions

A university is a remarkable framework for the study. During the past few decades, universities' behavioral contexts have changed from an exclusive public principle to a hybrid view. These actions define adjustments at the national level that have had several effects at the organizational and individual levels. Our study poses an essential argument regarding performance measurement reporting in HEIs, analyzed from a theoretical and empirical point of view through the factors that influence it. Customization on Romania's HEI particularity increases the final value of research because it fulfils an essential gap in the literature regarding this subject approach in former communist countries of Eastern Europe.

Our results sustain that universities' performance management systems must be designed to respond to the various external and internal stakeholders' information requirements. In universities with hybrid behavior, the content of voluntary and mandatory information is adapted to the 'stakeholders' requirements to increase its attractiveness and credibility in the market.

Hence, it is directed towards improving universities' quality (i.e., in teaching and scientific research). Under the pressure of rationality and efficiency of public financial resources, we noticed that establishing a performance-driven system for higher education is determined by linking universities' governmental funding with an assessment of the results.

Our study reveals that financial resources attracted per student, salary costs, size, and category influence performance reporting in Romanian HEIs. To increase their accountability, responsibility, and transparency, public universities are more exposed to coercive isomorphism linked to legislation—primary funding, salary costs, and size established according to the Ministry's rules. The category of universities' or employees' behavior (which affects the salary cost in the research component) also has a significant normative isomorphism—starting, for example, from the influence of international rankings or accreditations. This is why we also insist on examining employees' attitudes through the operant theory. The link between the institutional and operant theories contributes to a better understanding of dependence amongst HEIs and their employees. In general, HEIs are less likely affected by mimetic isomorphism, although we can find them assuming behaviors specific to the private sphere, given their evolution towards a hybrid one. In this approach, based on mimetic isomorphism, we found that size is significant in explaining the total extent of disclosure, in accordance with most previous research that has found that public university size (such as a corporate one) positively influences the amount of information disclosed on websites and webpage navigability. Therefore, institutions may experience all three types of isomorphism to legitimize themselves, although in different degrees.

Based on previous approaches, Romanian higher education is subject to global isomorphic processes in the context of internationalization. The isomorphism pressure is felt on Romanian HEIs, mainly from the moment of accession in the European Union or several international organizations in the field. The coercive forces lead to the homogenization of behaviors and internal structures of organizations, as they impose by law a single set of criteria, standards, and indicators. The reference to European models and the adoption of their characteristics (mimetic) creates similar behaviors at the international level (e.g., Bologna system, articles ranking, subject categories, or teaching plans). Human resources, through employees' attitudes (normative isomorphism), is requested to sustain the institutional or personal interests, based on academic particularities.

Based on these factors, by empirical research, our study proposes a performance-reporting model based on Romanian HEIs' analysis, with applicability to all public HEIs. Our performance reporting aims to fix responsibility for performance results by suggesting a limited list of common indicators for use in the institutional reports on performance. They are in accordance with IPSASB principles; thus, they also have international approval.

The paper succeeds to add value to scientific literature from several points of view. Firstly, it contributes to the development of knowledge by covering a gap in the literature concerning Eastern Europe PHEIs' performance-reporting literature. From a theoretical viewpoint, our paper manages in an original way to combine institutional theory with operant theory to highlight and explain the factors that influence the HEIs' performance-reporting system. In HEIs, the particularity of the involvement of the human factor in quantifying performance level generates effects for both the behavior of the institution and the academic staff. This is why we combine the two theories we consider useful for the elaborated research. Secondly, from an empirical viewpoint, the study proposes a performance-reporting model based on mandatory performance-reporting issues, international rankings indicators, and international recommended practices. The reporting model is developed on four universities' dimensions: teaching, research, interaction with the external environment, and financial resources.

The practical implications of the study are more important as performance is connected with factors that influence it and these factors become key elements in the development and assessment objectives of any public institution. HEIs are no exception, linking to a competitive process not only at national level but especially international. The conception of a performance reporting model based on the particularities of the Romanian system and the insertion of Romanian universities in international systems (e.g., Bologna) or rankings increases its applicability to other HEIs.

Nevertheless, this research has certain limitations. Despite its advantages (uniqueness and low visibility), an Eastern European country's analysis may contain some specific particularities due to history or political, economic, or social influences that can affect comparability/generalization with HEIs with long traditions of democratic regimes. From an empirical point of view, the restriction related to the number of public universities and the ability to collect other material can be considered a limitation. For future research, we consider the obvious potential for cross-country and longitudinal studies on performance-reporting systems.

Author Contributions: Conceptualization, A.T.-T.; methodology, S.F.; validation, S.F.; formal analysis, S.F.; investigation, C.S.N.; resources, A.A.; data curation, A.A.; writing—original draft, C.S.N.; writing—review & editing, A.A.; supervision, A.T.-T., project administration, C.S.N.; funding acquisition, A.T.-T. All authors have read and agreed to the published version of the manuscript.

Funding: This research received no external funding.

Institutional Review Board Statement: Not applicable.

Informed Consent Statement: Not applicable.

Conflicts of Interest: The authors declare no conflict of interest.

Appendix A

Table A1. Performance-reporting disclosure index.

Dimension	Performance Information	ARWU	THE	QS	U-Multirank	Authors
Teaching dimension	5. Number of students at bachelor level			X		Lukman et al. (2010); Ter Bogt and Scapens (2012); Asif and Searcy (2013)
	2. Number of students at master level			X		
	3. Number of students at doctoral level			X		
	4. Number of graduates at bachelor level		X		X	Lukman et al. (2010); Shin (2010); Ter Bogt and Scapens (2012); Asif and Searcy (2013)
	5. Number of graduates at master level		X		X	
	6. Number of graduates at doctoral level		X			
	7. Number of graduates at bachelor and master level me				X	
	8. Number of graduates with prestigious prizes	X				
	9. Number of PhD obtained abroad or in joint supervision				X	Al-Ashaab et al. (2011); Ramos-Vielba et al. (2010); Albats et al. (2018)
	10. Number of PhD with prestigious prizes		X			
	11. Insertion in the labor market			X	X	Asif and Searcy (2013); Lukman et al. (2010)
	12. Teachers with prestigious prizes	X				
	13. Number of programs of studies					Suryadi (2007)
	14. Didactic staff periodical assessment			X		Suryadi (2007); Asif and Searcy (2013); Ter Bogt and Scapens (2012); Lukman et al. (2010)
	15. University's reputation in teaching		X			
Research dimension	16. University's reputation in research		X			
	17. Number of publications				X	Suryadi (2007); Guthrie and Neumann (2007); Lukman et al. (2010); Al-Ashaab et al. (2011); Ter Bogt and Scapens (2012); Asif and Searcy (2013) Kauppila et al. 2015; Albats et al. 2018; Bonollo and Merli (2018)
	18. Number of publications with coauthors from abroad				X	
	19. Ratio between publications number and teaching staff		X			
	20. Number of publications with teachers form the region				X	
	21. Number of publications in Nature and Science	X				
	22. Number of interdisciplinary publications				X	
	23. Number of publications indexed in Science Citation Index and Social Science Citation Index				X	
	24. Number of citations		X	X	X	Suryadi (2007); Guthrie and Neumann (2007); Lukman et al. (2010); Al-Ashaab et al. (2011); Perkmann et al. 2011; Rossi and Rosli 2015; Asif and Searcy (2013)
	25. Number of research projects				X	
	26. Number of patent cited publications				X	
	27. Number of inventions certificates and patents				X	
	28. Number of post-doctoral places				X	Seppo and Lilles (2012)
External environment dimension	29. Number of international students			X		Lukman et al. (2010); Hegarty (2014)
	30. Ratio between number of students with Romanian residence and international students		X			
	31. Number of students from mobilities				X	Lukman et al. (2010); Hegarty (2014)
	32. Number of international professors			X	X	Bonollo and Merli (2018)
	33. Ratio between the didactic staff with Romanian residence and from abroad		X			
	34. Number of study programs la bachelor and master level in foreign languages				X	Asif and Searcy (2013)
	35. Number of partnerships with other universities				X	Albats et al. (2018)
	36. Number of partnerships with companies and not for profit entities				X	Seppo and Lilles 2012; Asif and Searcy (2013); Rossi and Rosli 2015
	37. Number of students in internships				X	Lukman et al. (2010); Al-Ashaab et al. 2011; Bonollo and Merli (2018)
	38. Number of spin-offs				X	Seppo and Lilles 2012; Rossi and Rosli 2015; Asif and Searcy (2013); Bonollo and Merli (2018)
	39. Number of start-ups				X	
Financial dimension	40. Financial resources obtained from the state budget					Seppo and Lilles (2012); Francesconi and Guarini (2018)
	41. Financial resources obtained from research activities		X		X	Guthrie and Neumann (2007); Shin (2010)
	42. Ratio between financial and research resources		X			
	43. Financial resources obtained from private sector				X	Seppo and Lilles 2012; Asif and Searcy (2013); Rossi and Rosli 2015
	44. Financial resources obtained from lifelong learning trainings and professional programs				X	Guthrie and Neumann (2007); Hegarty (2014)
	45. Financial resources obtained from local partners				X	Asif and Searcy (2013)
	46. Ration between financial resources and didactic staff		X			Guthrie and Neumann (2007); Hegarty (2014)

References

Agyemang, Gloria, and Jane Broadbent. 2015. Management control systems and research management in universities: An empirical and conceptual exploration. *Accounting, Auditing and Accountability Journal* 29: 1018–46. [CrossRef]

Alach, Zhivan. 2017. The use of performance measurement in universities. *International Journal of Public Sector Management* 30: 102–17. [CrossRef]

Al-Ashaab, Ahmed, Myrna Flores, Athanasia Doultsinou, and Andrea Magyar. 2011. A balanced scorecard for measuring the impact of industry–university collaboration. *Production Planning and Control* 22: 554–70. [CrossRef]

Albats, Ekaterina, Irina Fiegenbaum, and James A. Cunningham. 2018. A micro level study of university industry collaborative lifecycle key performance indicators. *Journal Technological Transformation* 43: 389–431. [CrossRef]

Alexander, F. King. 1998. The Endless Pursuit of Efficiency: The International Movement to Increase Accountability and Performance in Higher Education. Paper presented at the ASHE Annual Meeting Paper, Miami, FL, USA, November 5–8.

Alexander, F. King. 2000. The Changing Face of Accountability: Monitoring and Assessing Institutional Performance in Higher Education. *The Journal of Higher Education* 71: 411–31. [CrossRef]

Asif, Muhammad, and Cory Searcy. 2013. Determining the key capabilities required for performance excellence in higher education. *Total Quality Management and Business Excellence* 25: 22–35. [CrossRef]

Aversano, Natalia, Francesca Manes-Rossi, and Paolo Tartaglia-Polcini. 2018. Performance Measurement Systems in Universities: A Critical Review of the Italian System. In *Outcome-Based Performance Management in the Public Sector*. Edited by Borgonovi Elio, Anessi-Pessina Eugenio and Bianchi Carmine. Berlin: Springer, pp. 269–88.

Ayers, Rebecca S. 2015. Aligning Individual and Organizational Performance: Goal Alignment in Federal Government Agency Performance Appraisal Programs. *Public Personnel Management* 44: 169–91. [CrossRef]

Bai, Zhidong, Guangming Pan, and Yanqing Yin. 2016. Homoscedasticity tests for both low and high-dimensional fixed design regressions. *arXiv* arXiv:1603.03830.

Balkin, Richard S., Katelyn M. Richey Gosnell, Andrew Holmgren, and Jason W. Osborne. 2016. Nonlinear Analysis in Counseling Research. *Measurement and Evaluation in Counseling and Development* 50: 109–15. [CrossRef]

Banks, William, and Morton Nelson. 1994. Financial disclosures by Ontario universities: 1988–1993. *Journal of International Accounting Auditing and Taxation* 3: 287–305. [CrossRef]

Behn, Robert D. 2003. Why measure performance? Different purposes require different measures. *Public Administration Review* 63: 586–606. [CrossRef]

Bonollo, Elisa, and Mara Zuccardi Merli. 2018. Performance Reporting in Italian Public Universities: Activities in Support of Research, Teaching and the "Third Mission". In *Outcome-Based Performance Management in the Public Sector*. Edited by Borgonovi Elio, Anessi-Pessina Eugenio and Bianchi Carmine. Berlin: Springer, pp. 307–30.

Bouckaert, Geert. 1993. Measurement and meaningful management. *Public Productivity and Management Review* 17: 31–43. [CrossRef]

Brusca, Isabel, Sandra Cohen, Francesca Manes-Rossi, and Giuseppe Nicolò. 2019. Intellectual capital disclosure and academic rankings in European universities: Do they go hand in hand? *Meditari Accountancy Research* 28: 51–71. [CrossRef]

Burke, Joseph C., Henrik Minassians, and Po Yang. 2002. State Performance Reporting Indicators: What Do They Indicate? *Planning for Higher Education* 31: 15–30.

Burns, John, and Robert W. Scapens. 2000. Conceptualizing management accounting change: An institutional framework. *Management Accounting Research* 11: 3–25. [CrossRef]

Cai, Yuzhuo. 2010. Global isomorphism and governance reform in Chinese higher education. *Tertiary Education and Management* 16: 229–41. [CrossRef]

Capella-Peris, Carlos, Jesús Gil-Gómez, Manuel Martí-Puig, and Paola Ruíz-Bernardo. 2020. Development and validation of a scale to assess social entrepreneurship competency in higher education. *Journal of Social Entrepreneurship* 11: 23–39. [CrossRef]

Caputo, Fabio, Lorenzo Ligorio, and Simone Pizzi. 2021. The contribution of higher education institutions to the SDGs—An evaluation of sustainability reporting practices. *Administrative Sciences* 11: 97. [CrossRef]

Chan, Vivian. 2015. Implications of key performance indicator issues in Ontario universities explored. *Journal of Higher Education Policy and Management* 37: 41–51. [CrossRef]

Chen, Huifa, Qingliang Tang, Yihong Jiang, and Zhijun Lin. 2010. The role of international financial reporting standards in accounting quality: Evidence from the European Union. *Journal of International Financial Management and Accounting* 21: 220–78. [CrossRef]

Choong, Kwee Keong. 2013. Understanding the features of performance measurement system: A literature review. *Measuring Business Excellence* 17: 102–21. [CrossRef]

Coste, Andreea Ioana, and Adriana Tiron-Tudor. 2015. Performance indicators in Romanian Higher Education. *SEA: Practical Application of Science* 3: 177–80.

Curaj, Adrian, Liviu Matei, Remus Pricopie, Jamil Salmi, and Peter Scott. 2015. *The European Higher Education Area: Between Critical Reflections and Future Policies*. Berlin: Springer Nature, p. 898.

DiMaggio, Paul J., and Walter W. Powell. 1983. The iron cage revisited: Institutional isomorphism and collective rationality in organizational fields. *American Sociological Review* 48: 147–60. [CrossRef]

Dobija, Dorota, Anna Maria Górska, and Anna Pikos. 2018. The impact of accreditation agencies and other powerful stakeholders on the performance measurement in Polish universities. *Baltic Journal of Management* 14: 84–102. [CrossRef]

Dumitru, Valentin Florentin, Andrei Stanciu, Madalina Dumitru, and Liliana Feleaga. 2014. Pressure and Isomorphism in business education. *Amfiteatru Economic Journal* 16: 784–99.
Dunkin, Ruth. 2005. The HR challenge: Some more thoughts. Paper presented at the OECD Conference on Trends in the Management of Human Resources, Paris, France, August 25–26.
Education Law. 2011. Education Law No 1 of 2011, Romanian Official Gazette, XXIII, No 18, 10 January 2011. Available online: https://www.edu.ro/sites/default/files/legea-educatiei_actualizata%20august%202018.pdf (accessed on 8 November 2022).
Fligstein, Neil. 1998. The politics of quantification. *Accounting Organizations and Society* 23: 325–33. [CrossRef]
Francesconi, Andrea, and Enrico Guarini. 2018. Performance-Based Funding and Internal Resource Allocation: The Case of Italian Universities. In *Outcome-Based Performance Management in the Public Sector*. Edited by Borgonovi Elio, Anessi-Pessina Eugenio and Bianchi Carmine. Berlin: Springer, pp. 289–306.
Gallego, Isabel, Isabel-María García, and Luis Rodríguez. 2010. Universities' websites: Disclosure practices and the revelation of financial information. *The International Journal of Digital Accounting Research* 9: 153–92. [CrossRef]
Gallego-Álvarez, Isabel, José Manuel Prado-Lorenzo, and Isabel-María García-Sánchez. 2011. Corporate social responsibility and innovation: A resource-based theory. *Management Decission* 49: 1709–27. [CrossRef]
Garde Sanchez, Raquel, Jesús Mauricio Flórez-Parra, María Victoria López-Pérez, and Antonio Manuel López-Hernández. 2020. Corporate governance and disclosure of information on corporate social responsibility: An analysis of the top 200 universities in the shanghai ranking. *Sustainability* 12: 1549. [CrossRef]
Ghosh, Dipankar, and Anne Wu. 2012. The Effect of Positive and Negative Financial and Nonfinancial Performance Measures on Analysts' Recommendations. *Behavioral Research in Accounting* 24: 47–64. [CrossRef]
Glaum, Martin, Peter Schmidt, Donna L. Street, and Silvia Vogel. 2013. Compliance with IFRS 3-and IAS 36-required disclosures across 17 European countries: Company-and country-level determinants. *Accounting and Business Research* 43: 163–204. [CrossRef]
Gomes, Carlos F., and Mahmoud M. Yasin. 2017. Toward promoting effective strategic performance: The relevance of the alignment of performance measurement and competitive strategic choices. *International Journal of Business Excellence* 12: 329–50. [CrossRef]
Gonzales, Leslie D. 2012. Responding to mission creep: Faculty members as cosmopolitan agents. *Higher Education* 64: 337–53. [CrossRef]
Gordon, George, and Celia Whitchurch. 2007. Managing human resources in higher education: The implications of a diversifying workforce. *Higher Education Management and Policy* 19: 1–21. [CrossRef]
Gordon, Gus, and Mary Fischer. 2018. The need for performance management systems in public higher education. *Journal of Higher Education Theory and Practice* 18: 79.
Gordon, Teresa, Mary Fischer, David Malone, and Greg Tower. 2002. A Comparative Empirical Examination of Extent of Disclosure by Private versus Public Colleges and Universities in the United States. *Journal of Accounting and Public Policy* 21: 235–75. [CrossRef]
Greve, Henrich R., and Linda Argote. 2015. Behavioral theories of organization. In *International Encyclopedia of the Social and Behavioral Sciences*, 2nd ed. Amsterdam: Elsevier, pp. 481–86.
Grossi, Giuseppe, Dorota Dobija, and Wojciech Strzelczyk. 2020. The Impact of Competing Institutional Pressures and Logics on the Use of Performance Measurement in Hybrid Universities. *Public Performance and Management Review* 43: 818–44. [CrossRef]
Grossi, Giuseppe, Kirsi-Mari Kallio, Massimo Sargiacomo, and Matti Skoog. 2019. Accounting, performance management systems, and accountability changes in knowledge-intensive public organizations: A literature review and research agenda. *Accounting, Auditing and Accountability Journal* 33: 256–80. [CrossRef]
Guthrie, James, and Ruth Neumann. 2007. Economic and non-financial performance indicators in universities: The establishment of a performance-driven system for Australian higher education. *Public Management Review* 9: 231–52. [CrossRef]
Hatry, Harry P. 2013. Sorting the relationships among performance measurement, program evaluation, and performance management. *New Directions for Evaluation* 137: 19–32. [CrossRef]
Hegarty, Niall. 2014. Where We Are Now-The Presence and Importance of International Students to Universities in the United States. *Journal of International Students* 4: 223–35. [CrossRef]
Hyndman, Noel, and Ron Eden. 2001. Rational management, performance targets and executive agencies: Views from agency chief executives in Northern Ireland. *Public Administration* 79: 579–98. [CrossRef]
IPSASB. 2015. Recommended Practice Guideline (RPG3) on Reporting Service Performance Information. IFAC. Available online: http://www.ifac.org/system/files/publications/files/IPSASBRPG-3-Reporting-Service-Performance-Information.pdf (accessed on 1 November 2022).
Jesilow, Paul, and Julianne Ohlander. 2010. The impact of the National Practitioner Data Bank on licensing actions by state medical licensing boards. *Journal of Health and Human Services Administration* 33: 94–126. [PubMed]
Johnson, David R. 1995. Alternative methods for the quantitative analysis of panel data in family research: Pooled time-series models. *Journal of Marriage and the Family* 77: 1065–77. [CrossRef]
Kauppila, Osmo, Anu Mursula, Janne Harkonen, and Jaakko Kujala. 2015. Evaluating university–industry collaboration: The European foundation of quality management excellence model-based evaluation of university– industry collaboration. *Tertiary Education and Management* 21: 229–44. [CrossRef]
Kogan, M., I. Moses, and E. El-Khawas. 1994. The changing framework for the academic profession. In *Staffing Higher Education Meeting New Challenges. Report on the IMHE Project on Policies for Academic Staffing in Higher Education*. Bristol: Jessica Kingsley Publishers Ltd., pp. 12–32.

Kyrillidou, M. 2002. *Current Context for Performance Indicators in Higher Education*. Washington, DC: ARL Statistics and Measurement Program Home, Association of Research Libraries.

Lee, Chongmyoung. 2021. Factors influencing the credibility of performance measurement in nonprofits. *International Review of Public Administration* 26: 156–74. [CrossRef]

Levine, Arthur. 1997. *Higher Education's New Status as a Mature Industry*. Washington, DC: Chronicle of Higher Education.

Liguori, Mariannunziata, and Ileana Steccolini. 2012. Accounting change: Explaining the outcomes, interpreting the process. *Accounting, Auditing and Accountability Journal* 25: 27–70. [CrossRef]

Lindenberger, Ulman, and Ulrich Pötter. 1998. The complex nature of unique and shared effects in hierarchical linear regression: Implications for developmental psychology. *Psychological Methods* 3: 218. [CrossRef]

Lovata, Linda M. 1992. The use of operant theory in the design of performance reporting systems. *Management Accounting Research* 3: 273–89. [CrossRef]

Lukman, Rebeka, Damjan Krajnc, and Peter Glavič. 2010. University ranking using research, educational and environmental indicators. *Journal of Cleaner Production* 18: 619–28. [CrossRef]

Maingot, Michael, and Daniel Zeghal. 2008. An Analysis of Voluntary Disclosure of Performance Indicators by Canadian Universities. *Tertiary Education and Management* 14: 269–83. [CrossRef]

Manolescu, G. 2009. *Uncertainty in Financing Higher Education*. Working Papers. Bucharest: Spiru Haret University, Faculty of Finance and Banking, Center for Advanced Economic and Financial Research, vol. 3.

Montesinos, Vicente, Isabel Brusca, Francesca Manes Rossi, and Natalia Aversano. 2013. The usefulness of performance reporting in local government: Comparing Italy and Spain. *Public Money and Management* 33: 171–76. [CrossRef]

Najeeb, Ali. 2013. Institutional perspectives in HRM and MNC research: A review of key concepts. *International Journal of Employment Studies* 21: 79–100.

Orr, Dominic, Alex Usher, and Johannes Wespel. 2014. *Do Changes in Cost-Sharing Have an Impact on the Behaviour of Students and Higher Education Institutions? Evidence from Nine Case Studies, Volume II: National Reports*. Luxembourg: Publications Office of the European Union.

Osborne, Jason W. 2000. Advantages of hierarchical linear modeling. *Practical Assessment, Research and Evaluation* 7: 1–3.

Paauwe, Jaap, and Paul Boselie. 2003. Challenging 'strategic HRM' and the relevance of the institutional setting. *Human Resource Management Journal* 13: 56–70. [CrossRef]

Perkmann, Markus, Andy Neely, and Kathryn Walsh. 2011. How should firms evaluate success in university–industry alliances? A performance measurement system. *R&D Management* 41: 202–16.

Pollitt, Christopher, and Sorin Dan. 2013. Searching for impacts in performance-oriented management reform. *Public Performance and Management Review* 37: 7–32. [CrossRef]

Pricopie, Remus, Luminiţa Nicolescu, Zeno Reinhardt, and Oana Almăşan. 2009. Dynamics in the Internationalization of Higher Education at the Global Level and Specific Trends in Romania. *Revista Romana de Comunicare Si Relatii Publice* 11: 103–9.

Ramos-Vielba, Irene, Manuel Fernández-Esquinas, and Elena Espinosa-de-los-Monteros. 2010. Measuring university–industry collaboration in a regional innovation system. *Scientometrics* 84: 649–67. [CrossRef]

Rossi, Federica, and Ainurul Rosli. 2015. Indicators of university–industry knowledge transfer performance and their implications for universities: Evidence from the United Kingdom. *Studies in Higher Education* 40: 1970–91. [CrossRef]

Rossi, Francesca Manes, and Natalia Aversano. 2015. Advancing performance measurement Italian local government vis-à-vis the IPSASB project. *The International Journal of Productivity and Performance Management* 64: 76–93. [CrossRef]

Sangiorgi, Daniela, and Benedetta Siboni. 2017. The disclosure of intellectual capital in Italian universities: What has been done and what should be done. *Journal of Intellectual Capital* 18: 354–72. [CrossRef]

Scott, Peter. 2007. Higher education in Central and Eastern Europe. In *International Handbook of Higher Education*. Edited by James J. F. Forest and Philip G. Altbach. Berlin: Springer, pp. 423–42.

Seppo, Marge, and Alo Lilles. 2012. Indicators measuring university-industry cooperation. *Discussions on Estonian Economic Policy: Theory and Practice of Economic Policy* 20: 204–25.

Shapiro, Samuel Sanford, and Martin B. Wilk. 1965. An analysis of variance test for normality (complete samples). *Biometrika* 52: 591–611. [CrossRef]

Shin, Jung Cheol. 2010. Impacts of performance-based accountability on institutional performance in the US. *Higher Education* 60: 47–68. [CrossRef]

Siegel, Donald S., and Mike Wright. 2015. Academic entrepreneurship: Time for a rethink? *British Journal of Management* 26: 582–95. [CrossRef]

Suhaiza, Ismail, and Barizah Abu Bakar Nur. 2011. Reporting practices of Malaysian public universities: The extent of accountability disclosure. *African Journal of Business Management* 5: 6366.

Suryadi, Kadarsah. 2007. Framework of measuring key performance indicators for decision support in higher education institution. *Journal of Applied Sciences Research* 3: 1689–95.

Ter Bogt, Henk J., and Robert W. Scapens. 2012. Performance management in universities: Effects of the transition to more quantitative measurement systems. *European Accounting Review* 21: 451–97. [CrossRef]

Terracciano, Antonio, Robert R. McCrae, Larry J. Brant, and Paul T. Costa Jr. 2005. Hierarchical linear modeling analyses of the NEO-PI-R scales in the Baltimore Longitudinal Study of Aging. *Psychology and Aging* 20: 493. [CrossRef]

Turner, D. 2011. *Quality in Higher Education*. Rotterdam: Sense.

Ulrich, R., T. Stachnik, and J. Mabry, eds. 1974. *Control of Human Behavior: Behavior Modification in Education*. Glenview: Scott, Foresman, vol. 3.

Urdari, Claudia, Teodora Viorica Farcas, and Adriana Tiron-Tudor. 2017. Assessing the legitimacy of HEIs' contributions to society: The perspective of international rankings. *Sustainability Accounting, Management and Policy Journal* 8: 191–215. [CrossRef]

administrative sciences

MDPI

Article

Economic Growth through the Lenses of Education, Entrepreneurship, and Innovation

Simona Andreea Apostu [1,2,*], Lindita Mukli [3], Mirela Panait [2,4,*], Iza Gigauri [5] and Eglantina Hysa [6]

1 Department of Statistics and Econometrics, Bucharest University of Economic Studies, 010552 Bucharest, Romania
2 Institute of National Economy, 050711 Bucharest, Romania
3 Department of Mathematics, Aleksander Moisiu University, 2001 Durres, Albania; linditamukli@gmail.com
4 Department of Cybernetics, Economic Informatics, Finance and Accounting, Petroleum-Gas University of Ploiești, 100680 Ploiești, Romania
5 School of Business, Computing and Social Sciences, St. Andrew the First-Called Georgian University, 0179 Tbilisi, Georgia; i.gigauri@sangu.edu.ge
6 Department of Economics, Epoka University, 1032 Tirana, Albania; ehysa@epoka.edu.al
* Correspondence: simona.apostu@csie.ase.ro (S.A.A.); mirela.matei@upg-ploiesti.ro (M.P.)

Abstract: Economic growth is a major goal pursued by public authorities but can be achieved with the involvement of several categories of stakeholders given the complexity of the phenomenon and the many influencing factors. In this research paper, the authors analyze specific current issues that are representative as influencers of economic growth. This study brings into focus the importance of education, particularly tertiary education, entrepreneurship skills, and innovation capacities of businesses. The objectives are (1) to find out if tertiary education leads to economic growth; (2) to examine if innovation is one of the promotors of economic growth; and (3) to discuss the impact of the dynamic of businesses (enterprise birth) on economic growth. The methodology used in this research is panel regression (static model) for a sample consisting of 30 European countries for the period 2003–2020. The main findings are associated with a positive influence of tertiary education on economic growth, whereas the two other variables, that of entrepreneurship and innovation, are found to be insignificant for this time period.

Keywords: economic growth; education; entrepreneurship; innovation

Citation: Apostu, Simona Andreea, Lindita Mukli, Mirela Panait, Iza Gigauri, and Eglantina Hysa. 2022. Economic Growth through the Lenses of Education, Entrepreneurship, and Innovation. *Administrative Sciences* 12: 74. https://doi.org/10.3390/admsci12030074

Received: 24 May 2022
Accepted: 24 June 2022
Published: 29 June 2022

1. Introduction

There are many studies that have focused their research on finding the reasons for the factors pushing economic growth (Vasile et al. 2007; Enache et al. 2013; Yusuf and Nabeshima 2007; Klofsten et al. 2019; Hysa et al. 2020; Morina et al. 2020; Xu et al. 2020; Anghelache et al. 2021; Ur Rehman and Hysa 2021; Panait et al. 2022). One of the channels addressing economic growth is the education level. Intuitively, we can say that because of education, the living standard increases. This is related to the capability to function and the development of society (Hysa 2014). The higher the amount of skilled labor force, the higher the productivity, and the higher the technological advancements. The second intuition behind this positive relation is the idea that the higher the education level is, the higher the wage level, implying high spending. This can be seen as increased demand, pushing the supply side, thus there is a generation in the market.

Throughout time, European countries have placed a heavy emphasis on education. Strengthening education results in a more skillful and knowledgeable population, which in turn puts these acquired skills and information to use in the market (Ndou et al. 2019; Secundo et al. 2019; Ndou 2021). The results of many studies throughout the years align with the fact that education has a positive influence on the overall improvement of the economy (Vasile et al. 2007; Blessinger and Cozza 2016; Blessinger et al. 2019). Naturally, the type, level, and place where the education is received affect the overall capital gain in a person's library of skills and knowledge, and this starts with the primary education, which, according to Hanushek and Woessmann (2010), is emphasized as an important influencer in the overall construction education system as the very basis and fundamentals of the same system. Primary education functions as a solid, strong foundation, while secondary, tertiary, and education beyond continue to further the assimilation of knowledge and skills in individuals. Cognitive skills value quality over quantity of education, which is stated by Hanushek and Woessmann (2010), "when quality of education is introduced as a variable, quantity becomes nearly insignificant". Thus, receiving quality education and starting at an early stage with strong foundations proved to be the most useful in creating skilled and knowledgeable individuals.

Furthermore, in addition to primary and secondary education, several researchers such as Chatterji (1998), Pillay (2011), and Hanushek (2016) have investigated the contribution of tertiary education to economic growth, as a fundamental study level that supports the skilled labor force. Again, however, the results of these works are still conflicting with each other and do not properly identify a sole relation type. In addition to education, innovation and entrepreneurial incentives are seen as important factor to economic growth (Hysa and Mansi 2020; Cozza and Blessinger 2017). Hence, based on the above-mentioned debates, this study aimed to identify if some selected factors such as education level, innovation, and entrepreneurship are determinants in the economic growth of European countries. The main data employed in our model were retrieved from Eurostat and World Intellectual Property Organization for the time period 2003–2020. The paper used descriptive statistics and panel regression analysis to address the following research questions:

RQ1. As the literature supports, does education lead to economic growth while considering European countries?

RQ2. Is innovation one of the promotors of economic growth in the case of the European countries?

RQ3. What is the impact of the dynamic of businesses (enterprise birth) on economic growth?

These research questions are crucial to be investigated given that governments often have to make investment choices and target concrete determinants that foster economic growth. Good choices would help to create the appropriate frameworks to get through development and growth by implementing the right policies.

Furthermore, this study contributes to the literature on the contribution of educational level, innovation aspects in the economy, and the business dynamics to enter markets to push economic growth in European countries. What this study implies and intends to emphasize is a comprehensive understanding of economic growth though the integration of (1) tertiary education, as a main contributor to the skilled labor force, with (2) the innovation capacities, and (3) entrepreneurial incentives that assist in enterprises birth. Lastly, this paper is organized as follows: the first section represents an introduction of the issues, the second offers an overview of the main findings of the literature in terms of economic growth in relation to other determinants mentioned above, the third part of the paper presents the data and methodology applied to verify the research hypotheses, the fourth section describes the results of the empirical analysis, and the last two sections give some important insights in form of discussions and conclusions.

2. Literature Review

While previous studies have largely focused on the impact of primary and secondary education on economic growth, Chatterji (1998) includes tertiary education and finds an important role of this variable in economic growth. Furthermore, Pillay (2011) states that "tertiary education is a major driver of economic competitiveness, especially in the knowledge-driven global economy". Contrary to that, Hanushek (2016) performs an empirical analysis, finding that adding more years of schooling, when not having an increase in cognitive skills, does not have a significant impact on economic growth. Meanwhile, Benos and Zotou (2014) took the discussion to another level. In their research, they reviewed a total of 57 studies that measured the impact of education on economic growth. The result of their work was again ambiguous, and not homogenous across countries. These divergences were mainly driven because of differences in the type of data used. In recent years, more and more studies have focused on the impact that universities have on promoting the principles of sustainable development both through the transmission of knowledge and the formation of specific competencies among students and through the power of example, by initiating various CSR programs that seek to protect the environment, the development of local communities, and the fight against corruption (Matei 2013; Panait et al. 2016; Blessinger et al. 2018; Sengupta et al. 2020; Gigauri et al. 2021).

Highly educated individuals possess skills and display attitudes that have various positive effects on the labor force. The first and most significant contribution of these individuals is their higher predisposition to involvement in entrepreneurship (Chaganti and Greene 2002). The latter has a direct positive effect on economic welfare and growth. However, a classic debate is the debate about whether the entrepreneur is born, equipped with talent and emotional intelligence that endorses empathy, social skills driving to problem solving, smartly getting though proper networking, or whether the entrepreneur is formed. Some studies argue that the entrepreneur is born, and they support their opinions based on some examples of people that could build their successful business involving new ideas (Purwatiningsih et al. 2018). Some other studies support the idea that entrepreneurs can be taught and formed through education and training (Garavan and Barra 1994).

Entrepreneurs, especially educated entrepreneurs, contribute in a few critical ways. They introduce the spilling of knowledge directly or indirectly, that is introducing innovative ideas and improvements to current practices in the market. This way they indirectly inspire change and an increase in the overall knowledge of individuals working with or for them. This also proves to be a very effective way of stimulating competition and rivalry (Wong et al. 2005). By introducing new and improved ideas to the market, the new investors force the existing ones to make respective changes in order to maintain their current place in the market. Consequently, this brings another major influence of entrepreneurship in the economy: the introduction of new technology and innovation (Avram and Hysa 2022; Baumol and Strom 2007). Technology is created to facilitate different aspects of life; however, in an economy, it can either directly improve the speed and accuracy of production or take over some more laborious and time-consuming duties. This allows the individuals previously engaged with these tasks to invest and engage with different sectors of the economy.

In addition to the improvement of current instances, entrepreneurs are more likely to recognize new potential takes on investment risks (Baumol and Strom 2007; Berhani and Hysa 2014). These investments can result in more innovative services offered and most importantly more employment opportunities. The individuals employed under these successful investments are more likely to expand the range of their knowledge and skillset, which in turn increases the number of qualified and prepared individuals in the market. Another important point to consider is the types of entrepreneurship. The types of entrepreneurs depend on the level of education. Jiménez et al. (2015) observed in their study that secondary and tertiary education have a positive effect on the increase of formal entrepreneurship. Tertiary education in particular had a negative effect on informal entrepreneurship, while secondary education did not have such an effect (Jiménez et al. 2015). A formal type of entrepreneurship is overall more reliable for the workers and the customers since it includes businesses that are regulated by the law and are registered legally. Thus, a higher level of education produces individuals who are more compliant with the regulations and are more prepared to understand and successfully navigate the more complicated legal and financial aspects of a business.

Likewise, further studies were looking at the relationship between entrepreneurs and innovation. According to Hysa and Mansi (2020), entrepreneurs only are not sufficient enough to achieve a large-scaled innovation, being defined as the collection of four dimensions, namely (1) entrepreneurial capacities; (2) productive opportunities; (3) small-i innovation; and (4) organizational structure. Comparing developing and developed countries, developing countries highlight the existence of entrepreneurial capacities and small-i innovations, but again, being insufficient to make big-I innovations happen. Moreover, in the study by Audretsch et al. (2006), they argued that innovation is not considered the sole factor, but it has a critical role to play economically. Both knowledge creation and competition are important as the useful accelerators of economic growth (Audretsch et al. 2006). Thus, as emphasized in the study by Alfaro et al. (2019), it is important to understand how the combination of innovation and enterprises tasks are to be implemented.

Having gone through all the above discussions, we constructed an integrated framework that demonstrates the synergy of factors affecting economic growth. This framework is shown in Figure 1.

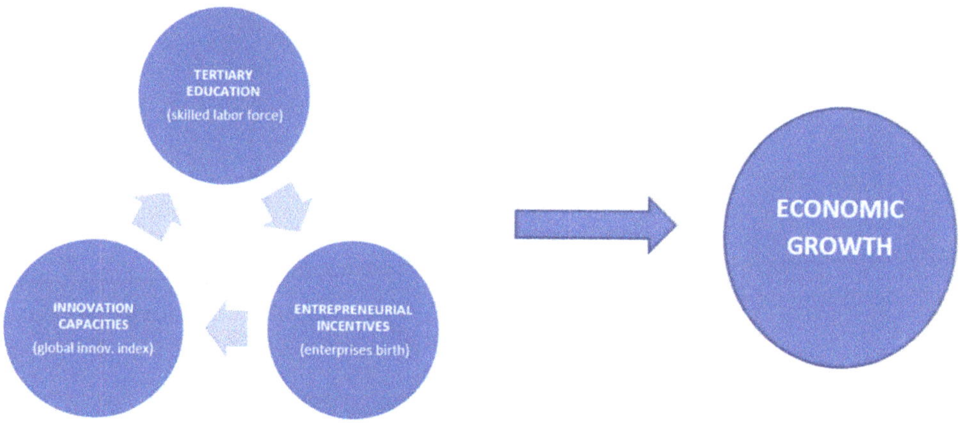

Figure 1. Integrated framework of economic growth. Source: Adapted by authors.

3. Data and Methodology

The main objective of our research is to identify if economic growth is influenced by education level, innovation, and entrepreneurship in European countries. Thus, we constructed our research considering four indicators: economic growth, births of enterprise, Global Innovation Index, and tertiary educational attainment. Our sample consists of 30 European countries due to data availability for the period 2003–2020. The countries included in the analysis are Switzerland, Sweden, the UK, Netherlands, Finland, Denmark, Germany, France, Austria, Ireland, Norway, Estonia, Belgium, Luxembourg, Czech Republic, Malta, Cyprus, Italy, Spain, Portugal, Slovenia, Hungary, Bulgaria, Slovakia, Latvia, Lithuania, Poland, Turkey, Croatia, and Romania. The data were provided by Eurostat, the variables' descriptions are presented in Table 1.

To highlight the variables influencing economic growth in terms of education, innovations, and entrepreneurship in the European countries, a panel regression (static model) was used.

For this, the following specification representing the static nature of model (Saini and Singhania 2018) can be used:

$$EG_{it} = c + \sum_{j=1}^{J} \beta_i X_{it}^{j} + \sum_{k=1}^{K} \beta_k Y_{it}^{k} + \sum_{l=1}^{L} \beta_l Z_{it}^{l} + e_{it} \tag{1}$$

$$e_{it} = v_i + u_{it} \tag{2}$$

where X, Y, and Z are different vectors of pull and push determinants. Economic growth (EG) is the dependent variable.

Table 1. Description of the variables and source of data.

Variable	Description	Source of Data
Births of enterprise	Net business population growth percentage, except activities of holding companies.	http://appsso.eurostat.ec.europa.eu/nui/submitViewTableAction.do (accessed on 12 March 2022)
Economic growth	The ratio of real GDP to the average population. GDP is considered at market prices, chain linked volumes (2010).	https://ec.europa.eu/eurostat/web/products-datasets/-/sdg_08_10 (accessed on 12 March 2022)
Global Innovation Index	The innovation ecosystem performance of economies. It comprises around 80 indicators, including measures on the political environment, education, infrastructure, and knowledge creation of each economy.	https://www.wipo.int/global_innovation_index/en/ (accessed on 12 March 2022)
Tertiary educational attainment	The share of the population aged 25–34 who have successfully completed tertiary studies (e.g., university, higher technical institution, etc.).	http://appsso.eurostat.ec.europa.eu/nui/show.do?dataset=sdg_04_20&lang=en (accessed on 12 March 2022)

In order to test the variables' stationarity, we used the Levin, Lin, and Chu—LLC (Levin et al. 2002), Im, Pesaran, and Shin W-Stat—IPS (Im et al. 2003), ADF-Fisher Chi-Square, and PP-Fisher Chi-Square tests. According to the panel unit root test, all variables except Global Innovation Index rejected the null hypothesis of the common unit root. The Global Innovation Index became stationary after the first difference.

In our analysis, we considered four unit root tests: IMP, LLC, ADF, and Phillips–Perron test (PP). To select between random and fixed effects, the Hausmann test was used (Hausman 1978). To check the robustness, we used the Wooldridge autocorrelation test (Wooldridge 2002), Wald test (heteroskedasticity of residues), Pesaran test (dependence of residues between the panels), Greene heteroscedasticity test (Greene 2003), and LM test (autocorrelation of residues).

We used Eviews 13 Student version to estimate the analysis models.

4. Results

To answer the research objective related to the determinant factors influencing economic growth in the European countries, we used the panel data equation model as follows:

$$EGI_{it} = \beta_{it} + \beta_1 BE_{it} + \beta_2 GI_{it} + \beta_3 TEI_{it} + \varepsilon_{it} \quad (3)$$

The dependent variable is represented by economic growth (EG). The explanatory variables included in the regression equations are births of enterprise (BE), Global Innovation Index (GI), and tertiary educational attainment (TE). In order to examine the characteristics of the countries included in the sample, descriptive analyses of the data were conducted (Table 2). The average births of enterprise in the sample is 3.12%, varying from −7.41% to 36.2%, with a standard deviation 6.24%. The economic growth registers a medium value of 28,093.14 Euros, ranging between 5390 Euros and 85,030.14 Euros, the standard deviation being 18,951.34 Euros. The Global Innovation Index registers an average of 50.07 with standard deviation of 7.95, the minimum value is 34.90 and maximum value is 68.40. The tertiary educational attainment presents values between 21.50% and 60.30%, the average 40.06%, and standard deviation of 8.64%. All the variables are normally distributed.

with the variables, we used the Arrelano Bond test (Arellano and Bond 1991), the result indicating to accept the non-autocorrelation.

The results presented in this paper indicate that in the case of the static model, tertiary educational attainment and birth of enterprise significantly influence economic growth. An increase in tertiary attainment generates an increase in economic growth, the link being a direct one, and an increase in birth of enterprise generates a decrease in economic growth, the link being inverse.

The findings are similar to those in the literature (Gyimah-Brempong 2011; Tsai et al. 2010; Nowak and Dahal 2016; Babatunde and Adefabi 2005; Hanushek 2013; Permani 2009), according to which educational level significantly influences economic growth. School attainment rates are adopted to approximate human capital accumulation, stimulating economic growth (Iamsiraroj 2016).

Education sustains economic growth based on three paradigms: (1) human capital theory; (2) catch-up models; (3) the interactions between education and technological innovation and change (Wolff 2000). The catch-up reflects the diffusion of technical knowledge from leading economies to the more backward ones (Gerschenkron 1952).

According to Reynolds et al. (1999) and Audretsch and Fritsch (1996), the relationship between enterprises and economic growth is inverse, especially in the case of Germany during the 1980s. There are also some studies according to which the relationship is positive: increasing the number of entrepreneurs leads to increasing economic growth (Wong et al. 2005). Holtz-Eakin and Kao (2003) concluded that entrepreneurship has a significant influence on productivity growth, at least for the United States.

5. Discussion

Economic growth is measured by the gross domestic product (GDP) and GDP per capita of a country. Economic development is characterized by less unemployment, a percentage of the population above the poverty line, and human development and wellbeing. Not only large companies but also enterprises are considerably contributed to by GDP, thus suggesting that governments should devote attention to creating an entrepreneurship-friendly environment, stimulating education and innovativeness.

The impact of entrepreneurial activities through newly founded firms on economic growth is widely recognized. Entrepreneurs facilitate economic development through labor, technologies, and capital. Numerous studies confirmed entrepreneurship as the main driver of economic growth and argue its contribution to employment opportunities (e.g., Naudé 2010; Chavis et al. 2011; Marcotte 2013; Fairlee and Chatterji 2013; Fritsch 2013; Hodges et al. 2015; Karimi et al. 2017). Entrepreneurs setting up new firms positively influence economic growth, when there are fewer legal, institutional, or cultural barriers. Cumming et al. (2014) empirically analyzed a sample of all countries available between the years 2004 and 2011 from three datasets from the World Bank, OECD, and Compendia, and concluded that entrepreneurship has a considerably positive impact on GDP per capita, exports per GDP, and patents per population, and has a negative effect on unemployment. It is noteworthy that these conclusions are not supported only by the OECD data, the reason for which could be the incomplete data in contrast to the World Bank's accurate dataset (Cumming et al. 2014).

Another important point to consider is the impact of innovation on economic development. An empirical study in CEE countries—Poland, Czech Republic, and Hungary—demonstrated long-term economic growth through innovation (Pece et al. 2015). Innovation leads to increased productivity which in turn enables the production of more goods and services resulting in economic growth. Innovative technologies serve the same mission to increase productivity which also induces wage growth. A study of 19 European countries from 1989 to 2014 asserted the long-run reciprocal correlation between innovation and per capita economic growth (Maradana et al. 2017). Economic growth induces innovation and innovation leads to per capita economic growth. Innovation contributes to economic growth through competitiveness, trade, financial systems, infrastructure development, and employment, which ultimately leads to improved quality of life and economic development (Maradana et al. 2017; Thurik 2009).

However, these study results differ from those described in the previous studies that confirm the significance of innovation for economic growth. This gap could have resulted from the time period, country context, and data available. In fact, many countries consider innovation as an important factor for economic growth and introduce supportive programs aiming at stimulating innovation in the countries. Li et al. (2018) underline the role of government in improving the innovation level of educational institutions by providing research funding (Li et al. 2018).

Moreover, the study results have not confirmed the statistically significant influence on economic growth for innovativeness and birth of enterprise in contrast to the previous studies highlighted in the literature, which can be explained by several reasons. First, the variable birth of enterprise can have no statistically significant effect on economic growth because a majority of newly established enterprises cannot survive due to their vulnerability. Moreover, the reason that innovativeness has less statistical significance for economic growth according to our study results could lie in the dataset by the Global Innovation Index. There could be missing data for some indicators for some EU countries in the selected time period. In addition, the indicators and measurements used by the Global Innovation Index could have limitations. The entrepreneurial process is cyclical—enterprises are born and disappear from the market. Thus, birth of enterprise might not be directly linked to economic growth but contributes to a country's development. Moreover, the effect of entrepreneurship and innovation on economic growth varies based on the development of a country. People in developed countries are less entrepreneurial compared to the number of entrepreneurs and self-employed people in developing countries. Consequently, more enterprises are set up in developing economies while citizens of developed countries prefer to work for big companies (Chang 2010). Furthermore, the reason for an individual to establish an enterprise (necessity or opportunity) impacts entrepreneurial outcomes (Rusu and Roman 2017; Stoica et al. 2020). Intention to become an entrepreneur emerges mostly from necessity and therefore, necessity-driven entrepreneurship has a negative correlation to economic growth in EU countries (Szabo and Herman 2012; Stoica et al. 2020). Thus, more empirical studies in this direction are needed, which will take into account other variables as well. Bosma et al. (2018) note that restaurants and retail stores also do not show a significant effect on economic growth, but they confirm the contribution of entrepreneurial activities to economic growth. The correlation between GDP per capita and enterprises introducing product or process innovations is averagely positive as there are significant differences among EU countries (Szabo and Herman 2012). This relationship is stronger in north-western Europe than in central, eastern, and southern European countries (Szabo and Herman 2012).

It is noteworthy that industries that grow are employing highly educated people. Skilled workers contribute to company success. Moreover, new enterprises foster employment, especially in regions. Universities have the ability to develop entrepreneurial skills of students by special programs which give participants motivation to start new enterprises (Cooper and Lucas 2007). Leadership style is positively associated to entrapreneurial behavior of universities (Stefani and Blessinger 2017; Farrukh et al. 2019). The curriculum, course content, pedagogical technics, theory, and practice can affect students' beliefs, attitudes, and intentions towards entrepreneurship, and therefore, can develop entrepreneurial skills, competences, and confidence necessary for entrepreneurial activities (Cooper and Lucas 2007; Sengupta and Blessinger 2019). In this regard, the significance of a multidisciplinary environment in entrepreneurial programs is also highlighted (Fiore et al. 2019). In addition, business incubators created by researchers help entrepreneurs and start-ups to generate and evaluate business ideas, set up teams and receive suitable training, establish an enterprise, and operate independently (Finardi 2013). Thus, educational institutions can deliberately inspire entrepreneurship intention and in this way stimulate economic growth.

Audretsch (2014) suggests that the role of universities is broad in an entrepreneurial society where organizations are established to encourage entrepreneurial activities and hence, drive economic growth. In the context of an entrepreneurial society, knowledge-based entrepreneurship is a driving engine for providing employment and ensuring economic growth (Guerrero and Urbano 2012). Investments in knowledge lead to commercialization of innovation as well as technology transfer from the university to for- and non-profit organizations producing economic growth (Audretsch 2014).

Furthermore, universities not only encourage students to launch enterprises but also build necessary skills to grow companies with innovativeness (Lewrick et al. 2010). The further development of a start-up is of paramount importance for a company to survive and consequently, entrepreneurship education must also encompass this topic (Lewrick et al. 2010). Educational programs impact on graduates' decision to start their own business while entrepreneurial behavior has an influence economic growth (Lewrick et al. 2010). Consequently, entrepreneurial university models strive to become change agents for economic and social development (Klofsten et al. 2019). For example, in the Netherlands, universities try to improve the entrepreneurial behavior of students so that they start new enterprises (Harkema and Schout 2008).

Personality traits of students also play an important role in education. As shown by the work of Qazi et al. (2020), personality traits are positively connected with entrepreneurial intention. It resonates with the study conducted by Räty et al. (2019) emphasizing the perception of innovative and competitive abilities, which are connected with entrepreneurial intention. By the same token, students' entrepreneurial intention depends on several factors and can be, for instance, determined by entrepreneurial education, the need for achievement, and locus of control (Vodă and Florea 2019). In general, education, both formal and informal, contributes to economic and social development as educational systems can improve the business and innovative potential of a country (Xu et al. 2020; Tvaronavičienė et al. 2018; Yusuf and Nabeshima 2007). Thus, education leads to economic growth. Moreover, universities should teach creativity for innovation, entrepreneurship, and encourage graduates to set up their firms since newly established companies contribute to economic and social development. Educational institutions can contribute to the development through teaching and research of entrepreneurship, innovation, and business with special courses and educational programs devoted specifically to developing necessary abilities among students.

6. Conclusions

This paper analyzed the relations between education, innovation, birth of enterprise, and economic growth. It explored that tertiary education positively correlates to economic growth while innovation and enterprise birth have less statistically significant effects. A contribution of this paper is to demonstrate how birth of enterprise, innovation, and education level impact economic growth in EU countries.

European universities are in a complex process of metamorphosis, their role is not only changing in society but also improving substantially. Universities are not only trainers of specialists in various fields but are also creators of regional and even national/international partnerships and networks that bring together companies, NGOs, and associations based on scientific relationships. In this way, the innovation activity is not only supported, but is nurtured, and the partnerships create synergies between the participating stakeholders. The involvement of students in research activity creates the premises for their professional development but also benefits for economic agents who thus rely on the energy specific to the young generation. In addition to the didactic and research function, the new entrepreneurial function of the universities supports the students in acquiring competencies that will allow them to set up start-ups and small companies through which to implement their innovative ideas.

The activity of the universities is more and more important considering the extension of the functions they have in the society. The increasing complexity of the university's activity and the existence of more and more sophisticated ecosystems have generated the gradual transition from the double helix model to the quintuple helix. The university–industry–government–public–environment interactions are a reality currently, and public policies in the field of education are being reconfigured taking into account the contribution that universities can have on different categories of stakeholders. In addition, the importance of education for the process of economic growth generates the growing interest of public authorities in the proper financing of this sector.

This study contributes to the literature by emphasizing that different variables affect the economic growth of a country and education is of significant importance in this regard. It also enhances theory by finding innovation and birth of enterprise as having a less significant influence on economic growth and highlighting the need for evaluating other variables such as individual country differences and the level of economic development. Government policies should focus on education strategies that support teaching and research as well as encouraging citizens to graduate from tertiary education. In addition, newly established enterprises need more support to survive and begin contributing to economic growth.

The authors are aware of the limitations of their research, generated by the choice of the sample of countries, the indicators used, and the selected analysis period. This gap can be explained by differences among the economic development of analyzed EU countries.

Further study can generate more findings to the nexus between education level, entrepreneurship intention, birth of enterprise, innovation, and economic growth in different country contexts. One direction is to identify the innovative and entrepreneurial potential of universities in the former communist countries of central and eastern Europe. The closed nature of communist economies has led to a lack of promotion of entrepreneurship.

Author Contributions: Conceptualization, S.A.A., L.M., M.P., I.G. and E.H.; methodology, S.A.A., L.M., M.P., I.G. and E.H.; software, S.A.A., L.M., M.P., I.G. and E.H.; validation, S.A.A., L.M., M.P., I.G. and E.H.; formal analysis, S.A.A., L.M., M.P., I.G. and E.H.; investigation, S.A.A., L.M., M.P., I.G. and E.H.; resources, S.A.A., L.M., M.P., I.G. and E.H.; data curation, S.A.A., L.M., M.P., I.G. and E.H.; writing—original draft preparation, S.A.A., L.M., M.P., I.G. and E.H.; writing—review and editing, S.A.A., L.M., M.P., I.G. and E.H.; visualization, S.A.A., L.M., M.P., I.G. and E.H.; supervision, S.A.A., L.M., M.P., I.G. and E.H.; project administration, S.A.A., L.M., M.P., I.G. and E.H.; funding acquisition, S.A.A., L.M., M.P., I.G. and E.H. All authors have read and agreed to the published version of the manuscript.

Funding: This research received no external funding.

Institutional Review Board Statement: Not applicable.

Informed Consent Statement: Not applicable.

Data Availability Statement: Not applicable.

Conflicts of Interest: The authors declare no conflict of interest.

References

Alfaro, Emigdio, Fei Yu, Naqeeb Ur Rehman, Eglantina Hysa, and Patrice Kandolo Kabeya. 2019. Strategic management of innovation. In *The Routledge Companion to Innovation Management*. London: Routledge, pp. 107–68.

Anghelache, Constantin, Mădălina-Gabriela Anghel, and Ştefan Virgil Iacob. 2021. Economic and social development based on education, research and industry. In *MATEC Web of Conferences*. Petrosani: EDP Sciences, vol. 342.

Arellano, Manuel, and Stephen Bond. 1991. Some tests of specification for panel data: Monte Carlo evidence and an application to employment equations. *The Review of Economic Studies* 58: 277–97. [CrossRef]

Audretsch, David B. 2014. From the entrepreneurial university to the university for the entrepreneurial society. *The Journal of Technology Transfer* 39: 313–21. [CrossRef]

Audretsch, David B., and Michael Fritsch. 1996. Creative destruction: Turbulence and economic growth in Germany. In *Behavioral Norms, Technological Progress, and Economic Dynamics*. Ann Arbor: University of Michigan Press, pp. 137–50.

Audretsch, David B., Max C. Keilbach, and Erik E. Lehmann. 2006. *Entrepreneurship and Economic Growth*. Oxford: Oxford University Press.

Avram, Gabriela, and Eglantina Hysa. 2022. Education, Knowledge and Data in the Context of the Sharing Economy. In *The Sharing Economy in Europe*. Cham: Palgrave Macmillan, pp. 181–206.

Babatunde, Musibau Adetunji, and Rasak Adetunji Adefabi. 2005. Long run relationship between education and economic growth in Nigeria: Evidence from the Johansen's cointegration approach. Paper presented at Regional Conference on Education in West Africa, Dakar, Senegal, November 1–2.

Baumol, William J., and Robert J. Strom. 2007. Entrepreneurship and economic growth. *Strategic Entrepreneurship Journal* 1: 233–37. [CrossRef]

Benos, Nikos, and Stefania Zotou. 2014. Education and economic growth: A meta-regression analysis. *World Development* 64: 669–89. [CrossRef]

Berhani, Riada, and Eglantina Hysa. 2014. The Economy of Albania Today and then: The Drivers to Growth. Paper presented at 4th International Conference on European Studies, Tirana, Albania, November 10; p. 598.

Blessinger, Patrick, and Barbara Cozza. 2016. *University Partnerships for Academic Programs and Professional Development*. Bingley: Emerald Group Publishing.

Blessinger, Patrick, Enakshi Sengupta, and Mandla Makhanya. 2018. Higher education's key role in sustainable development. *University World News*, September 7, vol. 519.

Blessinger, Patrik, Sengupta Enakshi, and Mandla Makhanya. 2019. New higher education literacies for a sustainable future. *University World News*, October 19, vol. 571.

Bosma, Niels, Mark Sanders, and Erik Stam. 2018. Institutions, entrepreneurship, and economic growth in Europe. *Small Business Economics* 51: 483–99. [CrossRef]

Chaganti, Radba, and Patricia G. Greene. 2002. Who are ethnic entrepreneurs? A study of entrepreneursapos; ethnic involvement and business characteristics. *Journal of Small Business Management* 40: 126–43. [CrossRef]

Chang, Ha-Joon. 2010. Poverty, Entrepreneurship, and Development. United Nations University. Available online: https://www.wider.unu.edu/publication/poverty-entrepreneurship-and-development (accessed on 17 June 2022).

Chatterji, Monojit. 1998. Tertiary education and economic growth. *Regional Studies* 32: 349–54. [CrossRef]

Chavis, Larry W., Leora F. Klapper, and Inessa Love. 2011. The Impact of The Business Environment on Young Firm Financing. *The World Bank Economic Review* 25: 486–507. [CrossRef]

Cooper, Sarah Y., and William A. Lucas. 2007. Enhancing self-efficacy for entrepreneurship and innovation: An educational approach. In *University Collaboration for Innovation*. Leiden: Brill Sense, pp. 79–98. Available online: https://researchbank.swinburne.edu.au/file/372a46d8-e7c1-4804-a9b1-72c7d81764e3/1/PDF%20%28Published%20version%29.pdf (accessed on 21 May 2022).

Cozza, Barbara, and Patrick Blessinger. 2017. Novel approaches in university partnerships: An introduction to university partnerships for pre-service and teacher development. In *University Partnerships for Pre-Service and Teacher Development*. Bingley: Emerald Publishing Limited, vol. 10, pp. 3–16.

Cumming, Douglas, Sofia Johan, and Minjie Zhang. 2014. The Economic Impact of Entrepreneurship: Comparing International Datasets. *Corporate Governance: An International Review* 22: 162–78. [CrossRef]

Enache, Roxana, Alina Brezoi, Alina Crişan, and Emil Stan. 2013. *Marketing Educaţional*. Iaşi: Institutul European.

Fairlee, Robert, and Aaron Chatterji. 2013. High-tech entrepreneurship in Silicon Valley: Opportunities and opportunity costs. *Journal of Economics & Management Strategy* 22: 365–89. [CrossRef]

Farrukh, Muhammad, Jason Wai Chow Lee, and Imran Ahmed Shahzad. 2019. Intrapreneurial behavior in higher education institutes of Pakistan: The role of leadership styles and psychological empowerment. *Journal of Applied Research in Higher Education* 11: 273–94. [CrossRef]

Finardi, Ugo. 2013. Clustering research, education, and entrepreneurship: Nanotech innovation at minatec in grenoble. *Research-Technology Management* 56: 16–20. [CrossRef]

Fiore, Eleonora, Giuliano Sansone, and Emilio Paolucci. 2019. Entrepreneurship education in a multidisciplinary environment: Evidence from an entrepreneurship programme held in Turin. *Administrative Sciences* 9: 28. [CrossRef]

Fritsch, Michael. 2013. New Business Formation and Regional Development: A Survey and Assessment of the Evidence. *Foundations and Trends in Entrepreneurship* 9: 249–364. [CrossRef]

Garavan, Thomas N., and O'Cinneide Barra. 1994. Entrepreneurship Education and Training Programmes: A Review and Evaluation–Part 1. *Journal of European Industrial Training* 18: 3–12. [CrossRef]

Gerschenkron, Alexander. 1952. Economic Backwardness in Historical Perspective. In *The Progress of Underdeveloped Areas*. Edited by Berthold Frank Hoselitz. Chicago: University of Chicago Press.

Gigauri, Iza, Mirela Panait, and Maria Palazzo. 2021. Teaching Corporate Social Responsibility and Business Ethics at Economic Programs. *LUMEN Proceedings* 15: 24–37.

Greene, William H. 2003. *Econometric Analysis*, 5th ed. Upper Saddle River: Prentice Hall.

Guerrero, Maribel, and David Urbano. 2012. The development of an entrepreneurial university. *The Journal of Technology Transfer* 37: 43–74. [CrossRef]

Gyimah-Brempong, Kwabena. 2011. Education and economic development in Africa. *African Development Review* 23: 219–36. [CrossRef]

Hanushek, Eric A. 2013. Economic growth in developing countries: The role of human capital. *Economics of Education Review* 37: 204–12. [CrossRef]

Hanushek, Eric A. 2016. Will more higher education improve economic growth? *Oxford Review of Economic Policy* 32: 538–52. [CrossRef]

Hanushek, Eric A., and Ludger Woessmann. 2010. Education and economic growth. *Economics of Education* 60: 67.

Harkema, Saskia J. M., and Henk Schout. 2008. Incorporating student-centred learning in innovation and entrepreneurship education. *European Journal of Education* 43: 513–26. [CrossRef]

Hausman, Jerry A. 1978. Specification tests in econometrics. *Econometrica: Journal of the Econometric Society* 46: 1251–71. [CrossRef]

Hodges, Nancy, Kittichai Watchravesringkan, Jennifer Yurchisin, Jane Hegland, Elena Karpova, Sara Marcketti, and Ruoh-Nan Yan. 2015. Assessing Curriculum Designed to Foster Students' Entrepreneurial Knowledge and Small Business Skills from a Global Perspective. *Family and Consumer Sciences Research Journal* 43: 313–27. [CrossRef]

Holtz-Eakin, Douglas, and Chihwa Kao. 2003. Entrepreneurship and economic growth: The proof is in the productivity. *Syracuse University Center for Policy Research Working Paper N50.* [CrossRef]

Hysa, Eglantina. 2014. Defining a 21st Century Education: Case Study of Development and Growth Course. *Mediterranean Journal of Social Sciences* 5: 41–41. [CrossRef]

Hysa, Eglantina, and Msc Egla Mansi. 2020. The Entrepreneurs' Role in Innovation: Developed versus Developing Countries. Paper presented at XIV IBANESS Congress Series on Economics, Business and Management, Plovdiv, Bulgaria, June 6–7; Available online: http://ibaness.org/conferences/plovdiv_2020/ibaness_plovdiv_proceedings_draft_5.pdf (accessed on 1 May 2022).

Hysa, Eglantina, Alba Kruja, Naqeeb Ur Rehman, and Rafael Laurenti. 2020. Circular economy innovation and environmental sustainability impact on economic growth: An integrated model for sustainable development. *Sustainability* 12: 4831. [CrossRef]

Iamsiraroj, Sasi. 2016. The foreign direct investment–economic growth nexus. *International Review of Economics & Finance* 42: 116–33.

Im, Kyung So, M. Hashem Pesaran, and Yongcheol Shin. 2003. Testing for unit roots in heterogeneous panels. *Journal of Econometrics* 115: 53–74. [CrossRef]

Jiménez, Alfredo, Carmen Palmero-Cámara, María Josefa González-Santos, Jerónimo González-Bernal, and Juan Alfredo Jiménez-Eguizábal. 2015. The impact of educational levels on formal and informal entrepreneurship. *BRQ Business Research Quarterly* 18: 204–12. [CrossRef]

Karimi, Saeid, Harm J. A. Biemans, Karim Naderi Mahdei, Thomas Lans, Mohammad Chizari, and Martin Mulder. 2017. Testing the relationship between personality characteristics, contextual factors and entrepreneurial intentions in a developing country. *International Journal of Psychology* 52: 227–40. [CrossRef] [PubMed]

Klofsten, Magnus, Alain Fayolle, Maribel Guerrero, Sarfraz Mian, David Urbano, and Mike Wright. 2019. The entrepreneurial university as driver for economic growth and social change-Key strategic challenges. *Technological Forecasting and Social Change* 141: 149–58. [CrossRef]

Levin, Andrew, Chien-Fu Lin, and Chia-Shang James Chu. 2002. Unit root tests in panel data: Asymptotic and finite-sample properties. *Journal of Econometrics* 108: 1–24. [CrossRef]

Lewrick, Michael, Maktoba Omar, Robert Raeside, and Klaus Sailer. 2010. Education for entrepreneurship and innovation: "Management capabilities for sustainable growth and success". *World Journal of Entrepreneurship, Management and Sustainable Development* 6: 1–18. [CrossRef]

Li, Jing, Hong Fang, Siran Fang, and Sultana Easmin Siddika. 2018. Investigation of the relationship among university–research institute–industry innovations using a coupling coordination degree model. *Sustainability* 10: 1954. [CrossRef]

Maradana, Rana P., Rudra P. Pradhan, Saurav Dash, Kunal Gaurav, Manju Jayakumar, and Debaleena Chatterjee. 2017. Does innovation promote economic growth? Evidence from European countries. *Journal of Innovation and Entrepreneurship* 6. [CrossRef]

Marcotte, Claude. 2013. Measuring entrepreneurship at the country level: A review and research agenda. *Entrepreneurship & Regional Development* 25: 174–94. [CrossRef]

Matei, Mirela. 2013. *Responsabilitatea Socială a Corporaţiilor şi Instituţiilor şi Dezvoltarea Durabilă a României.* Bucharest: Expert Publishing House.

Morina, Fatbardha, Eglantina Hysa, Uğur Ergün, Mirela Panait, and Marian Catalin Voica. 2020. The effect of exchange rate volatility on economic growth: Case of the CEE countries. *Journal of Risk and Financial Management* 13: 177. [CrossRef]

Naudé, Wim. 2010. Entrepreneurship, developing countries, and development economics: New approaches and insights. *Small Business Economics* 34: 1–12. [CrossRef]

Ndou, Valentina. 2021. Social entrepreneurship education: A combination of knowledge exploitation and exploration processes. *Administrative Sciences* 11: 112. [CrossRef]

Ndou, Valentina, Gioconda Mele, and Pasquale Del Vecchio. 2019. Entrepreneurship education in tourism: An investigation among European Universities. *Journal of Hospitality, Leisure, Sport & Tourism Education* 25: 100175.

Nowak, A. Z., and Gangadhar Dahal. 2016. The contribution of education to economic growth: Evidence from Nepal. *International Journal of Economic Sciences* 5: 22–41. [CrossRef]

Panait, Mirela, Eglantina Hysa, Marius Gabriel Petrescu, and Hailong Fu. 2022. Universities–Players in the Race for Sustainable Development. In *Higher Education for Sustainable Development Goals.* Edited by Carolina Machado and João Paulo Davim. New York: River Publishers, pp. 23–42.

Panait, Mirela, Marius Petrescu, and Raluca Podasca. 2016. Remodeling the role of universities in the context of sustainable development. Paper presented at International Conference "Economic Scientific Research—Theoretical, Empirical and Practical Approaches"—ESPERA, București, România, December 3–4; vol. 2, pp. 375–83.

Pece, Andreea Maria, Olivera Ecaterina Oros Simona, and Florina Salisteanu. 2015. Innovation and Economic Growth: An Empirical Analysis for CEE Countries. *Procedia Economics and Finance* 26: 461–67. [CrossRef]

Permani, Risti. 2009. The role of education in economic growth in East Asia: A survey. *Asian-Pacific Economic Literature* 23: 1–20. [CrossRef]

Pillay, Pundy. 2011. *Higher Education and Economic Development: Literature Review*. Cape Town: Centre for Higher Education Transformation.

Purwatiningsih, Anggraeni, Kamaludin Kamaludin, and Pudjo Sugito. 2018. The Effect of Constructivistic Learning Models on Entrepreneurial Orientation. *International Journal of Advanced Engineering, Management and Science* 4: 264323. [CrossRef]

Qazi, Wasim, Jawaid Ahmed Qureshi, Syed Ali Raza, Komal Akram Khan, and Muhammad Asif Qureshi. 2020. Impact of personality traits and university green entrepreneurial support on students' green entrepreneurial intentions: The moderating role of environmental values. *Journal of Applied Research in Higher Education* 13: 1154–80. [CrossRef]

Räty, Hannu, Katri Komulainen, Ulla Hytti, Kati Kasanen, Päivi Siivonen, and Inna Kozlinska. 2019. University students' perceptions of their abilities relate to their entrepreneurial intent. *Journal of Applied Research in Higher Education* 11: 897–909. [CrossRef]

Reynolds, Paul D., Michael Hay, and S. Michael Camp. 1999. *Global Entrepreneurship Monitor*. Kansas City: Kauffman Center for Entrepreneurial Leadership.

Rusu, Valentina Diana, and Angela Roman. 2017. Entrepreneurial Activity in the EU: An Empirical Evaluation of Its Determinants. *Sustainability* 9: 1679. [CrossRef]

Saini, Neha, and Monica Singhania. 2018. Determinants of FDI in developed and developing countries: A quantitative analysis using GMM. *Journal of Economic Studies* 45: 348–82. [CrossRef]

Secundo, Giustina, Valentina Ndou, Pasquale Del Vecchio, and Gianluigi De Pascale. 2019. Knowledge management in entrepreneurial universities: A structured literature review and avenue for future research agenda. *Management Decision* 57: 3226–57. [CrossRef]

Sengupta, Enakshi, and Patrick Blessinger. 2019. Integrating Sustainable Development into Teaching and Learning Methods. *International Journal for Education Law and Policy* 15: 111.

Sengupta, Enakshi, Patrick Blessinger, and Tasir Subhi Yamin. 2020. Introduction to university partnerships for sustainable development. In *University Partnerships for Sustainable Development*. Bingley: Emerald Publishing Limited.

Stefani, Lorraine, and Patrick Blessinger, eds. 2017. *Inclusive Leadership in Higher Education: International Perspectives and Approaches*. New York: Routledge.

Stoica, Ovidiu, Angela Roman, and Valentina Diana Rusu. 2020. The Nexus between Entrepreneurship and Economic Growth: A Comparative Analysis on Groups of Countries. *Sustainability* 12: 1186. [CrossRef]

Szabo, Zsuzsanna K., and Emilia Herman. 2012. Innovative Entrepreneurship for Economic Development in EU. *Procedia Economics and Finance* 3: 268–75. [CrossRef]

Thurik, R. 2009. Entreprenomics: Entrepreneurship, Economic Growth, and Policy. In *Entrepreneurship, Growth, and Public Policy*. Edited by Zoltán J. Ács, David B. Audretsch, Robert Strom and Robert J. Strom. Cambridge: Cambridge University Press, pp. 219–49. [CrossRef]

Tsai, Chun-Li, Ming-Cheng Hung, and Kevin Harriott. 2010. Human capital composition and economic growth. *Social Indicators Research* 99: 41–59. [CrossRef]

Tvaronavičienė, Manuela, Elena Tarkhanova, and Nino Durglishvili. 2018. Sustainable economic growth and innovative development of educational systems. *Journal of International Studies* 11: 248–56. [CrossRef]

Ur Rehman, Naqeeb, and Eglantina Hysa. 2021. The effect of financial development and remittances on economic growth. *Cogent Economics & Finance* 9: 1932060.

Vasile, Valentina, Gheorghe Zaman, Steliana Pert, and Felicia Zarojanu. 2007. *Restructuring Romania's Education System Considering the Evolutions from the Domestic Market Perspective and Impact on RDI Progress*. No. 2007, 2. Bucharest: Strategy and Policy Studies (SPOS).

Vodă, Ana Iolanda, and Nelu Florea. 2019. Impact of personality traits and entrepreneurship education on entrepreneurial intentions of business and engineering students. *Sustainability* 11: 1192. [CrossRef]

Wolff, Edward N. 2000. Human capital investment and economic growth: Exploring the cross-country evidence. *Structural Change and Economic Dynamics* 11: 433–72. [CrossRef]

Wong, Poh Kam, Yuen Ping Ho, and Erkko Autio. 2005. Entrepreneurship, innovation and economic growth: Evidence from GEM data. *Small Business Economics* 24: 335–50. [CrossRef]

Wooldridge, Jeffrey M. 2002. *Econometric Analysis of Cross Section and Panel Data*. Cambridge, MA: MIT Press.

Xu, Haiying, Wei-Ling Hsu, Teen-Hang Meen, and Ju Hua Zhu. 2020. Can higher education, economic growth and innovation ability improve each other? *Sustainability* 12: 2515. [CrossRef]

Yusuf, Shahid, and Kaoru Nabeshima, eds. 2007. *How Universities Promote Economic Growth. Directions in Development; Human Development*. Washington, DC: World Bank. Available online: https://openknowledge.worldbank.org/handle/10986/6631 (accessed on 21 May 2022).

administrative sciences

MDPI

Review

Entrepreneurial Education at Universities: A Bibliometric Analysis

Hiranya Dissanayake [1], Anuradha Iddagoda [2,*] and Catalin Popescu [3,*]

1 Department of Accountancy, Faculty of Business Studies and Finance, Wayamba University of Sri Lanka, Lional Jayathilake Mawatha, Kanadulla, Kuliyapitiya 60200, Sri Lanka
2 Management Sciences Unit, Department of Mathematics, Faculty of Applied Sciences, University of Sri Jayewardenepura, Nugegoda 10250, Sri Lanka
3 Department of Business Administration, Petroleum-Gas University of Ploiesti, 100680 Ploiesti, Romania
* Correspondence: anuradhaiddagoda@sjp.ac.lk (A.I.); cpopescu@upg-ploiesti.ro (C.P.)

Abstract: This study aims to identify the primary topics and present dynamics in the field of entrepreneurship education at universities and to make recommendations for future research directions. We conduct a bibliometric analysis on a selection of 447 studies from the Web of Science database to determine the extent of research on entrepreneurship education at universities between 2004 and 2022. In this study, researchers identify the most influential articles and writers based on their citations, publications, and geographical location. Additionally, they assess existing themes, identify bottlenecks to growth in the literature, and recommend future study options. While research on entrepreneurship education at universities happens globally, there is a dearth of collaboration across national borders, particularly between writers from developed and developing countries. Most of the research on entrepreneurship education at universities focuses on a quantitative approach in the analysis of entrepreneurship. Lastly, we conclude by proposing possible avenues for future research.

Keywords: entrepreneurship education; bibliometric analysis; higher education; entrepreneurial self-efficacy; entrepreneurial intention

Citation: Dissanayake, Hiranya, Anuradha Iddagoda, and Catalin Popescu. 2022. Entrepreneurial Education at Universities: A Bibliometric Analysis. *Administrative Sciences* 12: 185. https://doi.org/10.3390/admsci12040185

Received: 26 October 2022
Accepted: 1 December 2022
Published: 6 December 2022

Publisher's Note: MDPI stays neutral with regard to jurisdictional claims in published maps and institutional affiliations.

1. Introduction

The current world is facing challenges after the COVID-19 pandemic faced by all around the world; in this situation the survival of economies is mainly based on successful entrepreneurs. Professor Howard Stevenson defined entrepreneurship thusly: "Entrepreneurship is the pursuit of opportunity without regard to resources currently controlled" (Matei and Voica 2013, p. 3). This study further analyzed the behavior of the entrepreneur in two different scenarios, including here the promoter and the trustee. The promoter feels capable of making the most of the opportunities presented, regardless of the means at hand, while the trustee believes in his or her own ability to make the most of opportunities, regardless of the means at hand.

Essential characteristics of entrepreneurs include the ability to recognize opportunity where others see chaos, contradiction, and confusion; the willingness to take calculated risks with one's time, equity, or career; the ability to form an effective venture team; the creative skill to marshal needed resources; the fundamental skill of building a solid business plan; and, finally, the vision to recognize opportunity where others see chaos, contradiction, and confusion (Kuratko 2016).

In order to cultivate entrepreneurs in the world it is important to enhance entrepreneurship education. One of the most influential contemporary management theorists, Peter Drucker, has said "The entrepreneurial mystique? It's not magic, it's not mysterious, and is has nothing to do with the genes. It's a discipline. And, like any discipline, it can be learned" (Drucker 1985). Accordingly, the behaviors needed by entrepreneurs are mainly cultivated through entrepreneurship education. Entrepreneurship education is defined as 'any pedagogical process of education for entrepreneurial attitudes and skills' (Fayolle et al. 2006, p. 702; Ndou et al. 2018). Numerous studies demonstrate the benefits of entrepreneurship education and the ways in which it can strengthen entrepreneurial motivations (Kariv et al. 2018; Breznitz and Zhang 2021; Ndou et al. 2019). Accordingly, entrepreneurship education plays a crucial role in developing successful entrepreneurs (Ndou 2021; Panait et al. 2022).

In this context, universities play a significant role in developing curriculums and curricula related to entrepreneurship education (Coşkun et al. 2022; Apostu et al. 2022; Fayolle et al. 2006; Avram and Hysa 2022). Institutional support for student businesses can take many forms, including business plan contests, accelerator and incubator programs, intellectual property (IP) services, and entrepreneurship education programs (Lüthje and Franke 2003; Foote and Hysa 2022).

In addition, it is important to boost the relevance of entrepreneurship education in higher education, with a focus on the value of hands-on experience, and encourage the growth of programs that provide both theoretical and practical training in the field (Breznitz and Zhang 2021; Hysa 2014). There are many studies that have investigated entrepreneurial education at universities (Lüthje and Franke 2003; Kariv et al. 2018; Popescu 2019; Breznitz and Zhang 2021).

There are a few studies that have conducted literature reviews on entrepreneurial education, but no study has comprehensively analyzed literature related to entrepreneurial education at universities in the current context (Lüthje and Franke 2003; Kuratko 2017) Nevertheless, according to the researchers' knowledge, there is no study that has conducted a bibliometric analysis in entrepreneurial education at the universities. Accordingly, this paper bridges the gap by conducting comprehensive bibliometric analysis on entrepreneurship education at universities in the period of 1994–2022 and explores the research gaps in this research field.

The following research questions were explored in the paper through the use of bibliometric analysis and content analysis techniques.

RQ 1: What is the trend of publications related to entrepreneurship education at universities?
RQ 2: Who are the most cited pioneer authors in the subject of entrepreneurship education at universities?
RQ 3: Which journals dominate entrepreneurship education at universities?
RQ 4: What is the total number of articles based on countries, and international collaboration in the subject of entrepreneurship education at universities?
RQ 5: Which publications and papers on entrepreneurship education at universities have the most significant citation impact?
RQ 6: What are the relevant author keywords related to the entrepreneurship education at universities?
RQ 7: What are the future research recommendations related to entrepreneurship education at universities?

It is critical to delve into these research questions in order to identify the state of knowledge, trends of research and research requirements in the context of entrepreneurial education at universities. Based on the aforementioned, the current research employs a bibliometric-based evaluation methodology to assess the quality of previous works on the topic of entrepreneurial education at universities.

The following contributions will be demonstrated: First, we will provide a comprehensive overview of the research contribution of scientific journals, authors, and countries; secondly, we will take a closer look at the most-cited works and most-productive authors; third, we qualitatively analyze the highly cited articles in the domain of entrepreneurship education at universities and, finally, we will examine the research agenda's top priorities and any potential structural gaps.

This paper is organized as follows: introduction literature review; empirical results and their discussion; methodology; conclusion; and policy implications.

2. Materials and Methods

2.1. Explanations

In this paper, a bibliometric analysis was used to provide an all-encompassing picture of the current state of scientific production and evaluate the quality of previous studies, providing a wealth of information on a specific topic. Bibliometric analysis has grown significantly, starting with 1998 till 2017 and even after, at an even higher rate (see White and McCain 1998; van Eck and Waltman 2017). As described by statisticians and mathematicians (Garfield 1955), this approach utilizes a wide range of mathematical tools and statistical methodologies to examine and survey published works such as articles and books. Statistical methods shed light on scientific research explanations and disciplinary patterns (De Bakker et al. 2005; Bouyssou and Marchant 2011). Bibliometric analyses tell researchers about the history of a field, illuminate its current state, and suggest new research directions (Durieux and Gevenois 2010; Bilal et al. 2022).

This study of bibliometric analysis concentrated on the field of entrepreneurship education in university research. Our review only included empirical and review articles. We also did not include studies that were not written in English in our analysis. The analysis did not include other forms of literature, such as books, book chapters, or conference proceedings. In fact, this study method of data collection and data analysis is depicted in Figure 1. There are five stages in a typical bibliometric analysis: research design, data gathering, analysis, visualization, and interpretation (Zupic and Čater 2014).

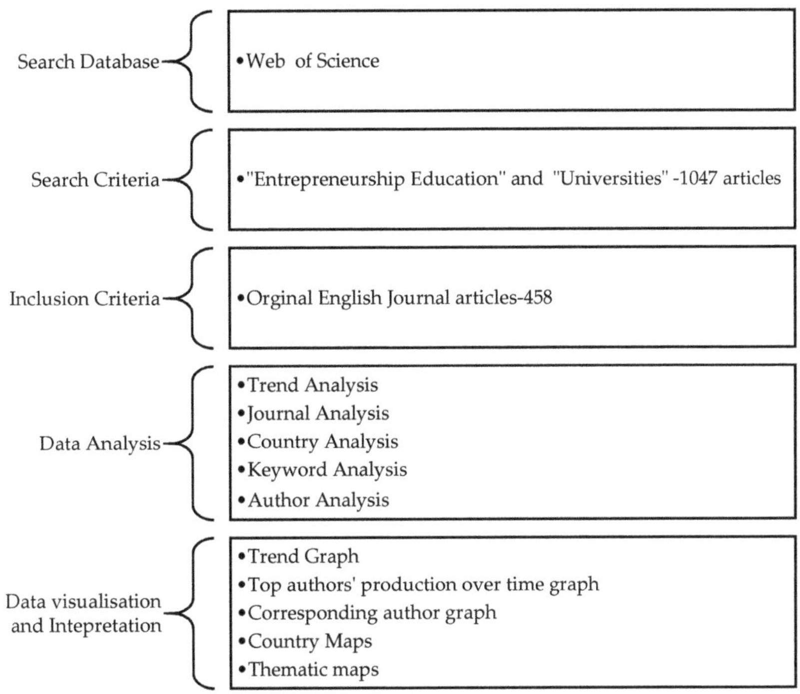

Figure 1. Paper Methodology. Adapted from Zupic and Čater (2014).

2.2. Search Database

When it comes to scientific literature, the Web of Science is unrivalled as the largest and most comprehensive database in existence. More than 11,000 peer-reviewed, high-impact academic journals covering the life and physical sciences, technology, medicine, and other related disciplines are included. The Web of Science database was used for the research in this article.

2.3. Search Criteria

This study used "entrepreneurship education" and "universities" as search terms in the Web of Science database. Initially, 1047 documents were extracted.

2.4. Inclusion Criteria

Next, non-English language articles and non-journal publications (589) were removed in order to obtain a better review. Finally, we analysed 458 English language journal publications.

Web of Science found 1047 publications covering entrepreneurship education at universities; of these, 458 (43.74%) were original research articles, 558 (53.30%) were conference proceedings, 13 (1.24%) were review articles, and 40 (3.82%) were other types of publications like book reviews, meeting abstracts, etc., while only one paper accepted to be published in 2023 was removed. Approximately 1016 papers (97%) were published in English. Finally, this study selected 458 articles for further analysis.

2.5. Data Analysis

In this paper, we used the Biblioshiny program to examine and depict the current state and future directions of entrepreneurship education in university research. Massimo Aria created the Biblioshiny software with the Shiny package written in the R programming language (Aria and Cuccurullo 2017).

Next, this study used bibliometric analysis techniques to explore the trend of publications, source analysis, country analysis, author analysis and keyword analysis. The author' journal productivity is analyzed using h-index, g-index, m-index and total citations. The h-index is a non-dimensional measure of an author's scholarly influence based on the frequency with which their own work has been cited by other scholars in the field. According to the definition of the h-index provided by Bornmann and Daniel (2007) and Choudhri et al. (2015), an h-index author has published at least h articles that have been cited at least h times. Similar to the h-index, the m-quotient (or m-index) is calculated by dividing an author's h-index by the number of years since their first publication. The g-index is the middle value of the number of citations (or the frequency with which an article has been referenced) for the top 'g' articles.

2.6. Data Visualisation and Interpretation

Finally, data is visualized using tables and figures including trend graphs, top authors' production over the graph, corresponding author country figure, country map, and thematic map. Next, these tables and figures are interpreted to derive meaningful conclusions.

3. Results

This section explains trend analysis, author analysis, source analysis, country analysis, and keyword analysis.

3.1. Trend Analysis

This study used time series analysis and stages of development analysis to explain the evolution of trends in entrepreneurship education at universities. A time series analysis allows for a year-by-year look at the evolution of development by the overall situation, and research trends are reflected in the yearly distribution of documents. Next, the articles can be broken down into discrete phases, and the features of the overall trend are displayed through the description of various stages of development. For the analysis of the articles, 10-year periods were used (i.e., well-defined decades). Figure 2 illustrates three time periods, including 1994–2003, 2004–2013 and 2014–2022.

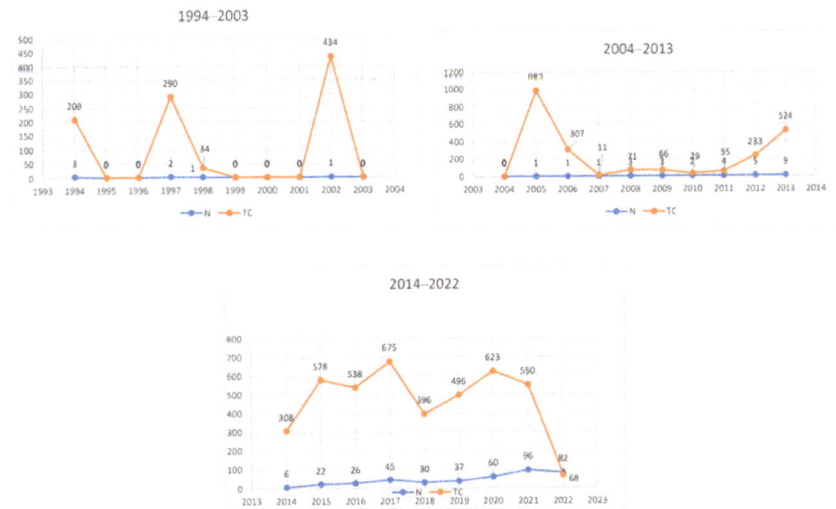

Figure 2. Trend Analysis. Source: Constructed by the authors.

In the period 1994–2013, a maximum of 5 publications per year were recorded in most years, and in some years there was not even one publication (according to the blue line in Figure 2). However, this small number of published articles have each received more than 200 citations, meaning that they can be considered seminal papers. The first paper in the Web of Science database that was published, "Experiments in Entrepreneurship Education—Successes and Failures", by Gartner and Vesper, received 143 citations (Gartner and Vesper 1994). "In pursuit of a new 'enterprise' and 'entrepreneurship' paradigm for learning: creative destruction, new values, new ways of doing things and new combinations of knowledge" by Gibbs (1993) was the highest cited paper in this period.

The next period, 2004 to 2013, illustrates steady growth in the number of publications. It was observed that all the years except 2004 saw the publishing of at least one article, and all articles received at least 11 citations. The highest number of citations was received by an article published in 2005 titled "The emergence of entrepreneurship education: Development, trends, and challenges", which was authored by Kuratko (2017).

The last period, 2014–2022, shown in Figure 2, illustrated an upward growth in the number of publications. In 2021, the highest number of publications was recorded (96), while in all other years, at least 20 papers were published (except for 2014). The highest number of citations received in the year 2017 was 675 citations for 45 papers. In this period the highest number of citations was received for "The Impact of Entrepreneurship Education: A Study of Iranian Students' Entrepreneurial Intentions and Opportunity Identification", which was authored by Karimi, Biemans and Mulder, which received 176 citations (Karimi et al. 2016). The newest paper in the Web of Science database was "Model Construction of College Students' Entrepreneurial Ability Cultivation in Mental Health Education Environment", authored by Huang (2022).

3.2. *Author Analysis*

There were a total of 1096 authors in the study, with 987 contributing one paper, 28 contributing two or more, and 12 contributing four or more. Table 1 shows that Kuratko DF, Gibb A, Gartner WB, Vesper KH, Rasmussen EA and Solheim R are the highest-cited authors who received more than 300 citations. Secundo G, an Italian scholar, has published the highest number of university-education articles on entrepreneurship education. Secundo G has an h-index of 5, a g-index of 7, and a total citation count of 111. Secundo G is well-respected in the study of entrepreneurship education due to the high quality of the many publications she has authored and published on the topic. Secundo G began publishing papers in 2016, as shown in Figure 3 (the size of the circle in the Figure represents the number of documents, and the shade of the colour represents the number of citations), with the most published documents and the highest frequency of average citations per item occurring in 2021.

Table 1. Most relevant authors.

Element	h_index	g_index	m_index	TC	NP	PY_start
Kuratko DF	1	1	0.056	982	1	2005
Gibb A	1	1	0.048	434	1	2002
Gartner WB	3	3	0.103	397	3	1994
Vesper KH	2	2	0.069	369	2	1994
Rasmussen EA	1	1	0.059	307	1	2006
Sorheim R	1	1	0.059	307	1	2006
Sanchez JC	1	1	0.1	241	1	2013
Cloodt M	1	1	0.111	218	1	2014
Duysters G	1	1	0.111	218	1	2014
Zhang Y	1	1	0.111	218	1	2014

Source: Constructed based on Biblioshiny Software.

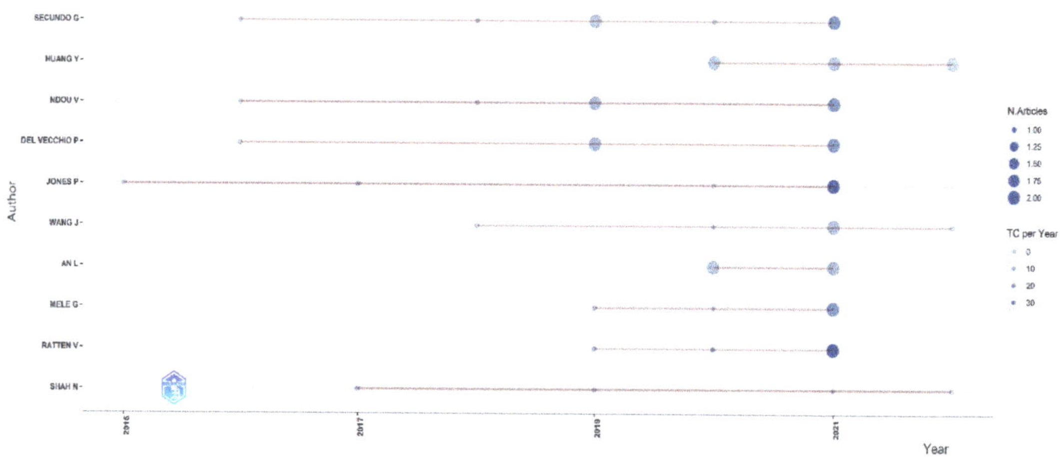

Figure 3. Top authors' production over time. Source: constructed based on Biblioshiny software.

As an illustration, G. Secundo et al. (2021) article in the Technol Forecast Soc Change journal, titled "Threat or opportunity? A case study of the digital-enabled redesign of entrepreneurship education in the COVID-19 emergency", was cited 37 times (Secundo et al. 2021). The research employs a mixed-method approach to enumerate the accomplishments of the University of Salento's Contamination Lab (CLab@Salento), an entrepreneurship education program focusing on innovative and technology-based entrepreneurship. Through digital technology, this study demonstrates a novel method for entrepreneurial education through storytelling, pitching, and business planning and development.

Kuratko was the highest cited author, and received 982 citations for one paper (see more in Table 1). This article discusses contemporary issues and developments in the field of entrepreneurship education. Accordingly, Ndou had the second highest number of publications and the h index and g index 5 and 6, respectively. She had her first publication in the year 2017 and the most recent publication in 2021. The highest cited article was co-authored with Secundo G. in 2021, which is discussed above. Jones P had the third highest number of publications, and the h index and g index was 5, respectively. His first publication was in 2016 and his most recent was in 2021, and he published articles related to entrepreneurship education. His highest cited article was titled "COVID-19 and entrepreneurship education: Implications for advancing research and practice", published and co-authored by Ratten and Jones in 2021.

3.3. Source Analysis

There was a total of 173 sources as part of the study, with eight journals considered core journals producing 148 papers, 38 journals (the middle zone) producing 148 papers and zone three, which had 128 journals (a selection of most relevant journals is presented in Table 2). Table 3 shows that *Entrepreneurship Theory and Practice, The Journal of Small Business Management, Education and Training, The International Journal of Management Reviews* and *The Journal of Business Venturing* are the highest cited journals which received more than 4000 citations. *Entrepreneurship Theory and Practice,* a Q1 journal, has published the highest cited articles on entrepreneurship education at universities. This journal has an h-index of 1, a g-index of 1, and a total citation count of 982.

Table 2. Most relevant journals.

Element	h_index	g_index	m_index	TC	NP
Education and Training	14	21	1.75	478	28
Frontiers in Psychology	6	9	2	103	19
International Journal of Management Education	11	15	1.375	326	15
Sustainability	6	12	1.2	156	13
Industry and Higher Education	4	6	0.5	50	11
International Journal of Entrepreneurial Behavior & Research	7	10	0.538	127	10
Studies in Higher Education	7	9	0.875	99	10
Journal of Small Business Management	7	8	0.7	844	8
Journal of Technology Transfer	7	8	0.778	263	8
Entrepreneurship and Regional Development	6	6	0.545	252	6

Table 3. Most Cited Sources.

Journal	h_index	g_index	m_index	TC	NP
Entrepreneurship Theory and Practice	1	1	0.056	982	1
Journal of Small Business Management	7	8	0.7	844	8
Education and Training	14	21	1.75	478	28
International Journal of Management Reviews	1	1	0.048	434	1
Journal of Business Venturing	3	3	0.103	433	3
Technovation	3	3	0.12	387	3
International Journal of Management Education	11	15	1.375	326	15
International Entrepreneurship and Management Journal	4	4	0.444	287	4
Journal of Technology Transfer	7	8	0.778	263	8
Entrepreneurship and Regional Development	6	6	0.545	252	6

Source: Constructed based on Biblioshiny Software.

The *Education and Training* journal published the highest number of publications with an h-index and g-index of 14 and 21, respectively, and is well respected in the study of entrepreneurship education due to the high quality of the many articles published on the topic. This journal began publishing papers in 2015, with the highest frequency of average citations per item occurring in the same year. The highest cited article in this journal is titled "Beyond intentions—what makes a student start a firm?" co-authored by Joensuu-Salo et al. (2015).

3.4. Country Analysis

It is possible that a country's prominence and sway in the study of entrepreneurship education in universities can be gauged by the number of papers published there on a particular topic. Between 1994 and 2022, authors from 78 different nations and regions published their research. Table 4 shows the top ten cited countries. Only China is in Asia; seven countries are in Europe (the United Kingdom, the Netherlands, Italy, Spain, Norway, Germany and Portugal). Two are in the Americas (the United States, Brazil), while China, the USA, the UK, Italy, and Spain are the top five countries in total documents, with the order reflecting decreasing importance.

Table 4. Most Cited Countries.

Country	Total Citations	Total Publications
USA	1990	121
United Kingdom	796	119
Netherlands	604	30
Italy	478	83
Spain	465	76
Norway	389	15
China	315	321
Germany	233	33
Portugal	171	38
Brazil	116	35

Source: Constructed based on Biblioshiny Software.

Table 4 shows that developed regions, like Europe and North America, are where most research papers on entrepreneurship education at universities are published. These findings suggest that these regions are driving the field. A more significant theoretical impact on developing countries could result from studying entrepreneurial education, but academic research focuses primarily on developed countries. There are several factors at play here. Most developing regions receive inadequate investment in entrepreneurial education, making it difficult to support more academic research.

China performs exceptionally well when working with other countries, as shown in Figure 4. At least 14 studies have involved collaboration from many countries. The United States, United Kingdom, Spain and Italy all frequently collaborate, with rates of 7, 10, 7, and 7 times each year, respectively. While China has published on entrepreneurship education at universities more than any other country, most of these studies have been conducted independently. The country has only worked with Malaysia, United Arab Emirates, Nigeria, Pakistan, Saudi Arabia, Kazakhstan, and Bahrain. Figure 5 displays the collaboration statistics of sample countries on land degradation, showing that 85.8% of China's papers are written independently. While countries like the Netherlands, Australia, Pakistan, Croatia, and Uganda are engaged in international collaboration, which is a greater than 70% multiple-country collaboration, the vast majority of nations research on their own. There are more publications involving only domestic authors than those from other countries.

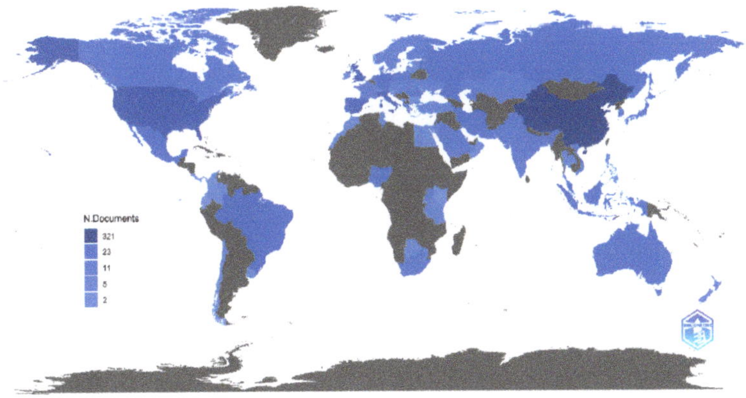

Figure 4. Country Scientific Production. Source: Constructed based on Biblioshiny Software.

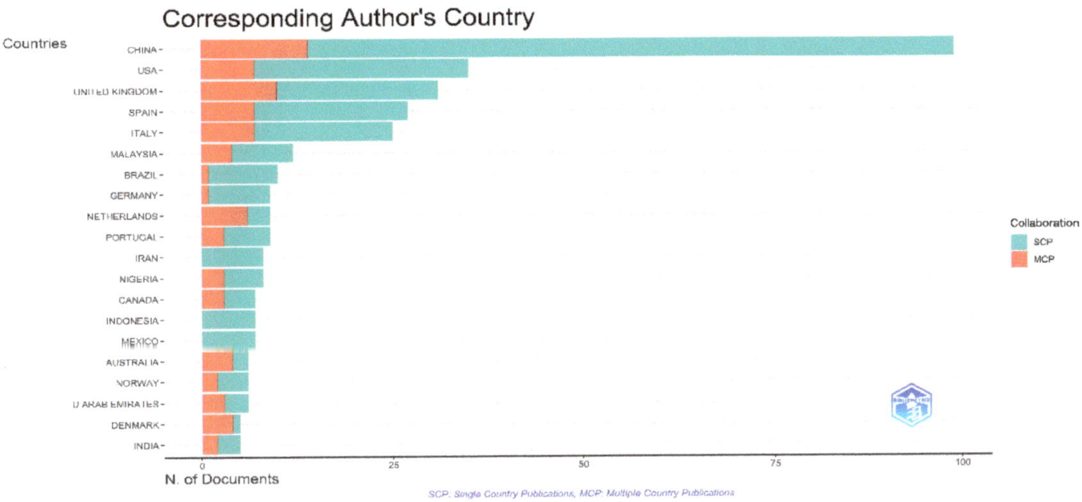

Figure 5. Corresponding author's Country. Source: Constructed based on Biblioshiny Software.

3.5. Keyword Analysis

The article's core is summarized and refined at a high level in the keywords (Xie et al. 2020). The highly frequent keywords used in this study, including cluster and multiple correspondence analysis, clearly and intuitively convey the article's concept and writing style in university entrepreneurial education. The software program Biblioshiny does data mining and statistical analysis of the high-frequency keywords of the research publications. Keywords with a word frequency of more than or equal to 10 are chosen and displayed as a word cloud (see Figure 6) using Biblioshiny to do data mining and statistical analysis on the high-frequency keywords of the research papers. Entrepreneurship education, entrepreneurship, entrepreneurial intention and education are the most commonly used keywords in entrepreneurship education at universities, appearing in 26.466%, 15.021%, 7.725% and 7.582%, from the total number of the keywords analysed from the literature review, respectively, accordingly to Figure 6 (See also Table 5). Social entrepreneurship and student entrepreneurship have also been discussed by a few authors (Apostu et al. 2022; Matei and Voica 2013; Secundo et al. 2021).

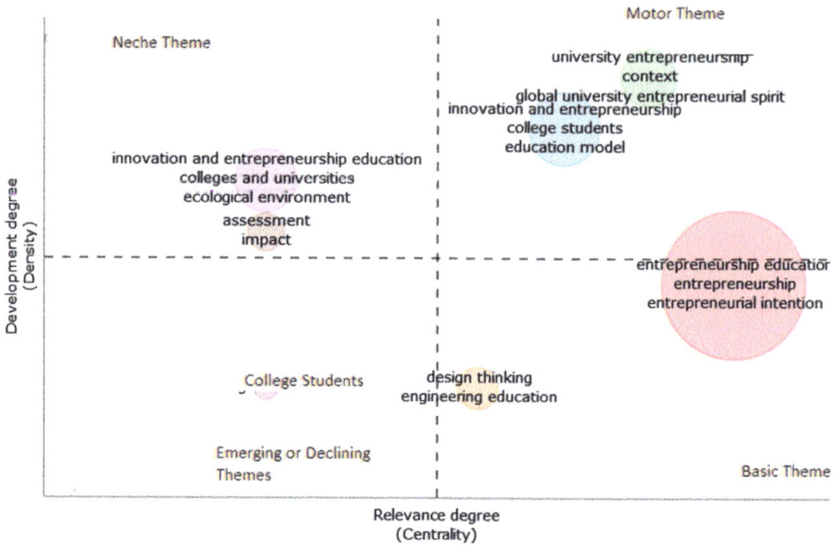

Figure 6. Thematic Map. Source: Constructed based on Biblioshiny Software.

Table 5. Frequency Analysis of Keywords.

Words	Occurrences	Percentage of Occurrence
entrepreneurship education	185	26.466
entrepreneurship	105	15.021
entrepreneurial intention	54	7.725
education	53	7.582
entrepreneurial	40	5.722
higher education	29	4.149
university	29	4.149
entrepreneurial university	24	3.433
innovation	24	3.433
entrepreneurial education	22	3.147
students	20	2.861
universities	19	2.718
entrepreneurial intentions	15	2.146
entrepreneurial self-efficacy	13	1.860
intention	13	1.860
self-efficacy	13	1.860
gender	12	1.717
social entrepreneurship	10	1.431
student entrepreneurship	10	1.431

Figure 6 exhibits the thematic map related to entrepreneurship education at universities. It provides a pictorial presentation of the trending themes in this area of research.

bottom-right part of the map shows the basic themes representing the well-established research issues in this area. The main themes are entrepreneurship education, entrepreneurship, and entrepreneurial intention, which will be discussed jointly. In addition, engineering education considers design thinking, which affects entrepreneurial education. The themes gaining importance in the recent past are presented in the top-right part of the map. They mainly include two research issues: university entrepreneurship and innovation and entrepreneurship. Keyword analysis also shows that significantly less research has been done in these areas. It is important to discuss about global universities, which have to include models of education, innovation and entrepreneurship in research. College students and motivation can be considered as declining themes. The niche themes include innovation and entrepreneurship, ecological environment, non-linear models, and quality evaluation (Hoxhaj and Hysa 2015; Hysa and Foote 2022). Additionally, assessment and impact analysis also considered niche themes (Hysa and Rehman 2019). Entrepreneurship education at universities is analyzed throughout the research process using a thematic evolution map, and the theme's course through evolution is determined by looking at the evolution trend (Figure 7). Comparing the evolutionary path map to the evolutionary state of each era reveals that land degradation research is still in its formative stages; it has not yet reached its full potential. There is clear evidence of differentiation, integration, transfer, and regeneration of themes, as demonstrated by the wide range of study themes across periods and the complexity of thematic evolution interactions. Evolutionary change is a highly unpredictable process. Since this study's inception, sixteen different evolutionary lines have emerged from two distinct origins.

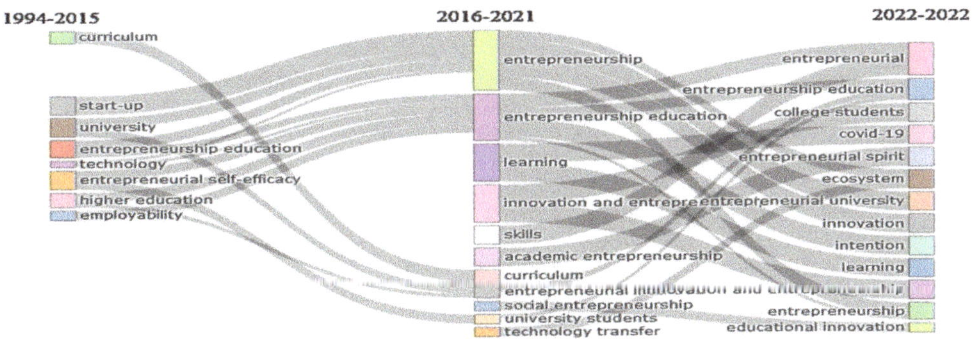

Figure 7. Thematic Evolution. Source: Constructed based on Biblioshiny Software.

The education research curriculum aspect in entrepreneurship started to be discussed with 1994. In fact, from 1994 to 2016, business startups were discussed, analyzed and researched, while from 2016 to 2021 the concept of entrepreneurship was elaborated and developed, under multiple aspects (including education). Starting with 2021, topics related to entrepreneurial universities were mentioned more and more often. The subjects or themes related to entrepreneurship proposed and discussed in the period 1994–2016 were modified by approaches in 2021 when they were massively integrated in the education sector, especially in universities (under the concept of entrepreneurial universities). If starting with 1994 we were discussing entrepreneurial self-efficacy, in 2016 the major theme related to entrepreneurship started to be entrepreneurial education. In the last years the discussions related to graduate people moved from employability to generate skills for the entrepreneurship (Hysa and Mansi 2020). On the other hand, the approach initiated in 2016 under the name of social entrepreneurship has in mind especially educational innovation and the new ecosystem structure, in the light of the new global challenges.

3.6. Qualitative Analysis of Highly Cited Articles

This section reviews the 10 highly cited articles related to entrepreneurship education at universities explained in chronological order. Table A1 in the Appendix A included the information related to the highly cited papers. Gartner and Vesper (1994) conducted a longitudinal survey conducted over 20 years and found a plethora of interactions between all of the different aspects of an entrepreneurship course when conducting a pedagogical experiment, such that modifications to one aspect of a course influence and are influenced by modifications to other aspects. In addition, this study assists readers who are attempting new things in their entrepreneurship classes to make an effort to understand the contextual factors that may determine the ultimate success or failure of their ventures.

In addition, In the latter part of 1994, a mail survey was conducted by Vesper and Gartner (1997). Over 940 business school presidents in the United States, 42 in Canada, and 270 elsewhere in the world were asked to fill out this survey. Course availability, faculty publications, community impact, alumni accomplishments, innovations, alumni start-ups, and scholarly outreach were cited as the top seven criteria for ranking entrepreneurship programs. In the programs surveyed, students could take classes on topics including "entrepreneurship or starting new firms", "small business management", "field projects/venture consulting", "starting and running a firm", "venture plan writing", and "venture finance", among others. This study suggests that the Malcolm Baldrige National Quality Award (MBNQA) evaluation is a comprehensive and robust method. In addition, this study suggests that more debate and dialogue among academics, administrators, students, and other stakeholders must be encouraged to prepare criteria for evaluating entrepreneurship education at universities.

Specifically, Gibb (2002) examines the political necessity of fostering an "enterprise culture" in Europe, which is largely attributable to the need to boost international competitiveness. Following this analysis of the educational response, several recent surveys are used to review some of the most pressing concerns surrounding the growth of entrepreneurship education at universities across the United Kingdom and Europe. The second section makes some attempts to address the imperative conceptually. The degree of uncertainty and complexity in the task and broader environment, as well as the desire of an individual in search of an opportunity or problem solution, are thought to play a role in inspiring entrepreneurial behaviour.

Amidst this massive growth, the obstacle of full academic legitimacy for entrepreneurship persists (Kuratko 2005). There is a case to be made that entrepreneurship education has finally arrived at a level of legitimacy, but significant obstacles remain. Entrepreneurship is cutting-edge; it requires constant originality. It's the way of the future for MBA programs, so it needs to start taking the reins. Words like "dream", "create", "explore", "invent", "pioneer", and "imagine" are now commonly used to describe the new innovation regime of the 21st century. Teachers of entrepreneurship should demonstrate the same creative zeal as their students.

Moreover, Rasmussen and Sørheim (2006) detailed several action-based entrepreneurship education initiatives currently underway at five different Swedish universities. These examples demonstrate that entrepreneurship education places less emphasis on lecturing to isolated students and more on participatory learning in teams and online communities. Several programs aim to do more than one thing at once, such as teach aspiring businesspeople or launch innovative companies or make university research available to the public. Constructing an action-oriented entrepreneurship education program has implications for the future of the field.

To confirm (or disprove) the common belief that entrepreneurship education increases the intention to start a business, Sánchez's (2013) study aims to highlight the crucial role played by an EE program on the entrepreneurial competencies and intentions of secondary school students. We adopted a quasi-experimental design consisting of a series of tests taken before and after the intervention. The findings corroborate our hypotheses, showing that students in the "experimental" group improved their skills and motivation for self-employment, while students in the "control" group did not. The results add to the literature on EE and the theory of planned behaviour by illuminating the impact of the program's individual benefits on the students.

Afterwards, Zhang et al. (2014) use the Entrepreneurial Cognition Theory, Ajzen's Theory of Planned Behavior, the Shapero Entrepreneurial Event Model, and previous research on entrepreneurship education, exposure, perceived desirability, and feasibility to examine the relationship between these variables and university students' entrepreneurial intentions (EI). Our sample size was 10 universities, and we were able to collect 494 valid responses. Using probit estimation, we found that people's opinions of a thing's desirability have a substantial effect on EI, while people's opinions of a thing's feasibility have none. While it may come as a surprise, exposure has a major negative effect, while entrepreneurship education has a major positive one. Higher levels of EI can be found among males and those educated at technologically focused institutions or who come from technologically focused backgrounds. In addition, the correlation between entrepreneurship education and EI is significantly strengthened by the positive interactive effects of gender, institution type, and field of study.

Moreover, Saeed et al. (2015) proposed and tested an integrative, multi-perspective framework. We have hypothesized that the three dimensions of university support, that is, perceived educational support, concept development support, and business development support, together with institutional support, shape students' entrepreneurial self-efficacy. In turn, entrepreneurial self-efficacy and individual motivations constitute the fundamental elements of the intention to start a business. A sample of 805 university students took part in the study and data were analyzed using structural equation modelling. Our findings showed that perceived educational support exerted the highest influence on entrepreneurial self-efficacy, followed by concept development support, business development support, and institutional support. Self-efficacy in turn had a significant effect on entrepreneurial intention. Individual motivations such as self-realization, recognition, and role had an additional impact on intention. However, the intention was not related to financial success, innovation, and independence. The findings suggest that a holistic perspective provides a more meaningful understanding of the role of perceived university support in the formation of students' entrepreneurial intention.

Karimi et al. (2016) used a pre-and post-survey to compare the effects of required and elective entrepreneurship education programs (EEPs) on students' entrepreneurial motivation and ability to spot new business prospects, drawing on insights from the theory of planned behaviour. In total, 205 students from six different Iranian universities filled out the questionnaires used to collect the data. In both types of EEPs, students reported significant improvements in their subjective norms and perceived behavioural control. The results also showed that the entrepreneurial aspirations of students were significantly raised by the elective EEPs but not by the required EEPs. This research adds to our understanding of planned behaviour and may influence how EEPs are developed and delivered.

Importantly, Wright et al. (2017) constructed an eco-system framework to start-ups by university students. This framework takes into account the following factors: the nature of the university environment and the external context; the involvement of different types of entrepreneurs, support actors, and investors; the evolution of these factors over time; and university mechanisms to facilitate student entrepreneurship, including a continuum of involvement from pre-accelerators through accelerators. Methods of financial support are also discussed.

4. Discussion

According to the bibliometric analysis using quantitative and qualitative analysis of highly cited documents, the following research gaps and future recommendations can be identified. Overall, researches on entrepreneurship education at universities is still in its early phases, as seen by the extant pieces of literature, and future studies should focus on expanding on the following fronts:

1. It is better to research the relationship between innovation and entrepreneurship education at universities and entrepreneurship performance in detail for future researchers (Hernández-Sánchez et al. 2019). Accordingly, questionnaires and in-depth interviews can explore using quantitative and qualitative research. Previous scholars consider intellectual property and partnership, but no model has been constructed to examine this relationship (Schmitz et al. 2017).
2. In addition, entrepreneurship education and intention should be tested longitudinally by studying university students whose knowledge must be verified at the level of each program and study cycle (Sherkat and Chenari 2020; Pascucci et al. 2022). In particular, it is better to conduct studies in different academic programs at universities (Management, Engineering, Arts, etc.) and to different levels of students. Moreover, conducting cross-country studies with questionnaire surveys on these themes is better integrated with in-depth interviews.
3. Moreover, spirituality and entrepreneurship education have not been examined in detail in the existing literature. Combining the spirit with the entrepreneurial skills aims at a behavioral and attitudinal transformation in order to generate sustainable businesses. It is essential to consider the human values in entrepreneurship, including mindfulness, compassion, a meaningful life, and a sense of community. Universities should consider the spiritual values incorporated in their curriculum. Spirituality can be considered a moderator in the relationship between entrepreneurship education and entrepreneurship or mediation between entrepreneurship education and entrepreneurship. Nevertheless, no study has been conducted with comprehensive surveys or mixed approaches. It is better to conduct a sequential exploratory study to examine this relationship.
4. Entrepreneurship education and the sustainability of entrepreneurs is also a topic that needs more investigation in the future (Pascucci et al. 2021). There is a need to consider how the required sustainability practices introduced by the universities in their curriculum can improve the sustainability of entrepreneurs. It is better to conduct case studies, in-depth interviews, and mixed-method research in different countries and contexts.
5. A university-based entrepreneurial education ecosystem (Liu et al. 2021) has been elaborated upon, but only considered the views of university executives to build the model. It is better to consider future researchers to obtain stakeholder opinions, including students, government, industry, and communities, to expand the model to evaluation. This model can be tested using case studies, in-depth interviews, and mixed-method research in different countries and contexts.

5. Conclusions

Pieces of literature in the field of entrepreneurial education were retrieved from the Web of Science database for 1994–2022 and then analyzed with the help of the Biblioshiny software package. The study of entrepreneurial education at universities displays the following traits and methodological rigor (Rejeb et al. 2022):

(1) The first research question is analyzed using trend analysis by observing the changes of publications from the period of 1994 to 2022. According to an analysis of publication patterns, the number of published works addressing entrepreneurial education at universities has been steadily increasing since 2004. The first part of the analysis related to the first research question had three distinct phases: the initial, low publication stage; the intermediate, sprouting stage and the expansive, higher publication stage. According to citation counts, research into entrepreneurship education saw the most growth between 2014 to 2022. The number of people concerned about entrepreneurship education at universities and the number of academics working on this issue has grown over time.

(2) The second research question analyzed the use of the number of publications and citations per author. Accordingly, the highest cited author is Kuratko DF, who is The

Jack M. Gill Distinguished Chair of Entrepreneurship attached to Indiana University, USA. The second most highly cited author was Professor Allan Gibb who is attached to Durham University in the UK.

(3) The third research question was analysed using the number of publications and citations per journal. Accordingly, highly cited journals including *Entrepreneurship Theory and Practice*, and *The Journal of Small Business Management* are Q1 journals. Therefore, future scholars should direct their publications to these journals to receive a greater number of citations.

(4) The fourth research question was analyzed using the number of publications but also the citations of the authors assigned to their countries. In this regard, China has more clout in the field of research in recent years than most countries in the world, with the possible exception of the USA and Great Britain. As a major developed nation, the United Kingdom also has significant research conducted in this area. An analysis of published works reveals infrequent international collaboration and a preponderance of solo research efforts. While scientific research is becoming increasingly globalized, this trend is counterproductive.

(5) The fifth research question was analysed using the critical review of highly cited papers related to entrepreneurship education at universities. These studies considered the evaluation of entrepreneurship education, entrepreneurship culture, action-based entrepreneurship education, entrepreneurial intention and university support, the effectiveness of entrepreneurship education and eco-system framework to start-ups by university students.

(6) The sixth research question was analysed using word clouds and thematic maps. The most frequently used keywords are entrepreneurial education, entrepreneurship, and entrepreneurship intention. A thematic analysis was conducted and identified future research implications.

(7) The final question suggests future research areas to be considered by researchers including innovation and entrepreneurship, entrepreneurial education and intention, spirituality and entrepreneurship education, entrepreneurial education and sustainability, and entrepreneurial eco-systems using comprehensive (mixed, longitudinal) studies.

These findings from the synthesis improve our familiarity with entrepreneurship education with regard to universities' research and trends, but there are still some gaps in our knowledge that need to be filled by additional research. Our analysis does not specifically investigate the factors related to the longitudinal shifts in the choice of topics, co-authorships, and journal citations that are indicated by our study, nor do we delve into the causes that accelerated the rise in entrepreneurship education-related publications. Other researchers can investigate the causes behind shifts and conduct analyses on co-citations and bibliographic coupling.

Only articles published in journals with strict peer review were included in this study. Therefore, other sources of information, such as entrepreneurship education-related conference proceedings, books, and chapters were not.

Our reliance on the Web of Science database alone was also a weakness. While the Web of Science is the best place to find bibliometric-related articles, we may have missed some important ones by focusing on this database exclusively (Rejeb et al. 2022). Future researchers can consider SCOPUS, a Google Scholar database, for data collection. On the other hand, focusing on only English-language articles may have overlooked significant contributions from publications and networks that employ other languages.

In addition, we avoided studying qualitative indicators in favour of focusing exclusively on quantitative ones in this study. Including qualitative indicators in the future can open up new avenues of inquiry and shed light on previously unknown phenomena. Journal performance is measured by looking at how many times each article has been cited by other works (Rejeb et al. 2022). Therefore, future research can rely on article-level metrics (Altmetrics) (according to Luc et al. 2021) and other journal performance indicators that take entrepreneurship education mentions into account. More information about the strengths and weaknesses of a journal, as well as its relative reach, can be gleaned by utilizing alternative indicators (Bang et al. 2019).

Although we were able to successfully investigate and map entrepreneurship education-based global scholarly studies, our results did not identify the main drivers behind the explosive growth of this literature over time. As a result, we can move forward with studies that shed light on the driving forces behind entrepreneurship education research's rapid development.

Even with these limitations, this study contributes to the synthesis of the literature on entrepreneurship education at universities, which will be important for the research initiators and research scholars with regard to identifying trends and future research recommendations.

Author Contributions: Conceptualization, H.D., A.I. and C.P.; Data curation, H.D., A.I.; Formal analysis, H.D., A.I. and C.P.; Investigation, H.D., A.I. and C.P.; Methodology, H.D., A.I.; Supervi-sion, H.D., A.I. and C.P.; Validation, H.D., A.I. and C.P; Visualization, H.D., A.I. and C.P.; Writing—original draft, H H.D., A.I. and C.P.; Writing—review & editing, H.D., A.I. and C.P. All authors have read and agreed to the published version of the manuscript.

Funding: This research received no external funding.

Institutional Review Board Statement: Not applicable.

Informed Consent Statement: Not applicable.

Conflicts of Interest: The authors declare that they have no conflict of interest.

Appendix A

Table A1. Highly Cited Documents.

Authors	Title of the Paper	Findings	Total Citations
Kuratko (2005)	The emergence of entrepreneurship education: development, trends, and challenges	Identifies issues and developments in 21st-century entrepreneurial education Avoid paradigm paralysis Entrepreneurship educators must have innovative drive Consider about spirituality	982
Gibb (2002)	In pursuit of a new 'enterprise' and 'entrepreneurship' paradigm for learning: creative destruction, new values, new ways of doing things and new combinations of knowledge	If there is to be a sufficient response, it is necessary to apply the Schumpeterian idea of creative destruction to the higher education sector in order to find innovation (new ways of doing things) and new combinations of knowledge.	434
Rasmussen and Sørheim (2006)	Action-based entrepreneurship education	Entrepreneurship education places more of an emphasis on learning-by-doing activities in a network context than it does on teaching individuals in a traditional classroom setting. Several programs aim to educate entrepreneurs, launch new businesses, and commercialize academic research, among other things.	307
Sánchez (2013)	The impact of an entrepreneurship education program on entrepreneurial competencies and intention	With a number of entrepreneurially related competencies and intentions, entrepreneurial education has positive and significant relationships.	241
Vesper and Gartner (1997)	Measuring progress in entrepreneurship education	The top seven criteria suggested for ranking entrepreneurship programs are: - courses offered - faculty publications - impact on community - alumni exploit - innovations - alumni start-ups, and - outreach to scholars	226

Table A1. *Cont.*

Authors	Title of the Paper	Findings	Total Citations
Zhang et al. (2014)	The role of entrepreneurship education as a predictor of university students' entrepreneurial intention	Entrepreneurial Intention (EI) is higher in men and people with technological backgrounds and/or universities than in women and people with other backgrounds and universities. The relationship between entrepreneurship education and EI is also significantly influenced by - the gender - type of university, and - study major.	218
Karimi et al. (2016)	The impact of entrepreneurship education: A study of iranian students' entrepreneurial intentions and opportunity identification	Entrepreneurial education programs had minimal effects on students' attitudes toward entrepreneurship and their perceptions of opportunity identification. Entrepreneurship education programs significantly influenced subjective norms and perceived behavioral control.	176
Gartner and Vesper (1994)	Experiments in entrepreneurship education: successes and failures.	When conducting a pedagogical experiment, it appears that there are numerous interactions between every aspect of an entrepreneurship course, such that changes made to one aspect have an impact on, and are in turn influenced by, other aspects.	143
Saeed et al. (2015)	The role of perceived university support in the formation of students' entrepreneurial intention	A holistic viewpoint offers a more insightful understanding of the part that students' perceptions of university support play in the development of their entrepreneurial intentions.	
Wright et al. (2017)	An emerging ecosystem for student start-ups	Eco-system framework should include - continuum of involvement from pre-accelerators to accelerators - university mechanisms to support student entrepreneurship - the participation of a range of entrepreneurs - support actors, and investors, - the unique characteristics of the university environment and - the external context, and their evolution over time	

References

Apostu, Simona Andreea, Lindita Mukli, Mirela Panait, Iza Gigauri, and Eglantina Hysa. 2022. Economic Growth through the Lenses of Education, Entrepreneurship, and Innovation. *Administrative Sciences* 12: 74. Available online: https://www.mdpi.com/2076-3387/12/3/74/htm (accessed on 14 October 2022). [CrossRef]

Aria, Massimo, and Corrado Cuccurullo. 2017. bibliometrix: An R-tool for comprehensive science mapping analysis. *Journal of Informetrics* 11: 959–75. [CrossRef]

Avram, Gabriela, and Eglantina Hysa. 2022. Education, Knowledge and Data in the Context of the Sharing Economy. In *The Sharing Economy in Europe*. Edited by Vida Česnuitytė, Andrzej Klimczuk, Cristina Miguel and Gabriela Avram. Cham: Palgrave Macmillan. [CrossRef]

Bang, Chang Seok, Jae Jun Lee, and Gwang Ho Baik. 2019. The Most Influential Articles in Helicobacter PYLORI RESearch: A Bibliometric Analysis. *Helicobacter* 24: e12589. Available online: https://onlinelibrary.wiley.com/doi/full/10.1111/hel.12589 (accessed on 16 October 2022). [CrossRef] [PubMed]

Bilal, Bilal, Eglantina Hysa, Ahsan Akbar, Fakhra Yasmin, Atteeq ur Rahman, and Shengbing Li. 2022. Virtual Learning During the COVID-19 Pandemic: A Bibliometric Review and Future Research Agenda. *Risk Management and Healthcare Policy* 15: 1353. [CrossRef]

Bornmann, Lutz, and Hans-Dieter Daniel. 2007. What Do We Know About the h Index? *Journal of the American Society for Information Science and Technology* 58: 1381–85. Available online: https://onlinelibrary.wiley.com/doi/full/10.1002/asi.20609 (accessed on 18 September 2022). [CrossRef]

Bouyssou, Denis, and Thierry Marchant. 2011. Ranking Scientists and Departments in a Consistent Manner. *Journal of the American Society for Information Science and Technology* 62: 1761–69. Available online: http://onlinelibrary.wiley.com/doi/10.1002/asi.22883/abstract (accessed on 11 September 2020).

Breznitz, Shiri M., and Qiantao Zhang. 2021. Entrepreneurship Education and Firm Creation. *Regional Studies* 56: 940–55. Available online: https://www.tandfonline.com/doi/abs/10.1080/00343404.2021.1878127 (accessed on 15 October 2022). [CrossRef]

Choudhri, Asim F., Adeel Siddiqui, Nickalus R. Khan, and Haris L. Cohen. 2015. Understanding Bibliometric Parameters and Analysis. *Radiographics* 35: 736–46. Available online: www.rsna.org/rsnarights (accessed on 18 September 2022). [CrossRef]

Coşkun, Hüseyin Emre, Catalin Popescu, Deniz Şahin Samaraz, Akif Tabak, and Bulent Akkaya. 2022. Entrepreneurial University Concept Review from the Perspective of Academicians: A Mixed Method Research Analysis. *Sustainability* 14. 10110. Available online: https://www.mdpi.com/2071-1050/14/16/10110/htm (accessed on 14 October 2022).

De Bakker, Frank G. A., Peter Groenewegen, and Frank Den Hond. 2005. A bibliometric analysis of 30 years of research and theory on corporate social responsibility and corporate social performance. *Business & Society* 44: 283–317.

Drucker, Peter F. 1985. *Innovation and Entrepreneurship: Practice and Principles*. Available online: https://book4you.org/book/1177680/78db76 (accessed on 15 October 2022).

Durieux, Valerie, and Pierre Alain Gevenois. 2010. Bibliometric idicators: Quality masurements of sientific publication. *Radiology* 255: 342–51. [CrossRef]

Fayolle, Alain, Benoît Gailly, and Narjisse Lassas-Clerc. 2006. Assessing the impact of entrepreneurship education programmes: A new methodology. *Journal of European Industrial Training* 30: 701–20. [CrossRef]

Foote, Russell, and Eglantina Hysa. 2022. Improving Operational–Developmental Connections: Creating Positive "Butterfly Effects" in Universities. In *Governance and Management in Higher Education (Innovations in Higher Education Teaching and Learning)*. Edited by Sengupta Enakshi, Patrick Blessinger and Nasiruddin Nezaami. Bingley: Emerald Publishing Limited, vol. 43, pp. 57–70. [CrossRef]

Garfield, Eugene. 1955. Citation Indexes for Science. *Sciences* 122: 108–11. [CrossRef] [PubMed]

Gartner, William B., and Karl H. Vesper. 1994. Experiments in entrepreneurship education: Successes and failures. *Journal of Business Venturing* 9: 179–87. [CrossRef]

Gibb, Allan. 2002. In Pursuit of a New 'Enterprise' and 'Entrepreneurship' Paradigm for Learning: Creative Destruction, New Values, New Ways of Doing Things and New Combinations of Knowledge. *International Journal of Management Reviews* 4: 233–69. Available online: https://onlinelibrary.wiley.com/doi/full/10.1111/1468-2370.00086 (accessed on 10 October 2022). [CrossRef]

Gibbs, Philip A. 1993. Determinants of Corporate Restructuring: The Relative Importance of Corporate Governance, Takeover Threat, and Free Cash Flow. *Strategic Management Journal* 14: 51–68. Available online: https://www.scopus.com/inward/record.uri?eid=2-s2.0-84989040574&doi=10.1002%2Fsmj.4250140906&partnerID=40&md5=f78b658c138cb631a4c5b4b5067fe7d2 (accessed on 8 March 2017). [CrossRef]

Hernández-Sánchez, Brizeida Raquel, José Carlos Sánchez-García, and Alexander Ward Mayens. 2019. Impact of Entrepreneurial Education Programs on Total Entrepreneurial Activity: The Case of Spain. *Administrative Sciences* 9: 25. [CrossRef]

Hoxhaj, Jona, and Eglantina Hysa. 2015. Comparing ENQA, British, German & Albanian Standards of Quality in Higher Education. *European Journal of Sustainable Development* 4: 243. [CrossRef]

Huang, Mugian. 2022. Model Construction of College Students' Entrepreneurial Ability Cultivation in Mental Health Education Environment. *International Journal of Environmental Research and Public Health* 2022: 1–9. [CrossRef]

Hysa, Eglantina. 2014. Defining a 21st Century Education: Case Study of Development and Growth Course. *Mediterranean Journal of Social Sciences* 5: 41. Available online: https://www.richtmann.org/journal/index.php/mjss/article/view/1957 (accessed on 10 October 2022). [CrossRef]

Hysa, Eglantina, and Egla Mansi. 2020. Integrating teaching and learning in graduate studies: Economic development course. *Technology Transfer: Innovative Solutions in Social Sciences and Humanities*, 61–64. [CrossRef]

Hysa, Eglantina, and Naqeeb Ur Rehman. 2019. Assessing the Teaching Quality of Economics Programme: Instructor Course Evaluations. *Integration of Education* 23: 556–67. [CrossRef]

Hysa, Eglantina, and Russell Foote, eds. 2022. *New Perspectives on Using Accreditation to Improve Higher Education*. Hershey: IGI Global.

Joensuu-Salo, Sanna, Elina Varamäki, and Anmari Viljamaa. 2015. Beyond intentions—What makes a student start a firm? *Education + Training* 57: 853–73. [CrossRef]

Karimi, Saeid, Harm J. A. Biemans, Thomas Lans, Mohammad Chizari, and Martin Mulder. 2016. The Impact of Entrepreneurship Education: A Study of Iranian Students' Entrepreneurial Intentions and Opportunity Identification. *Journal of Small Business Management* 54: 187–209. [CrossRef]

Kariv, Dafna, Luis Cisneros, and Mihai Ibanescu. 2018. The Role of Entrepreneurial Education and Support in Business Growth Intentions: The Case of Canadian Entrepreneurs. *Journal of Small Business & Entrepreneurship* 31: 433–60. Available online: https://www.tandfonline.com/doi/abs/10.1080/08276331.2018.1468974 (accessed on 15 October 2022).

Kuratko, Donald F. 2005. The Emergence of Entreprenerusship Education—Luratko, 2005. *Entrepreneurship Theory and Practice* 29: 577–97. Available online: https://journals.sagepub.com/doi/abs/10.1111/j.1540-6520.2005.00099.x?journalCode=etpb (accessed on 5 June 2019). [CrossRef]

Kuratko, Donald F. 2016. *Entrepreneurship: Theory, Process, Practice*, 10th ed. Boston: Cengage. 465p.

Kuratko, Donald F. 2017. The Emergence of Entrepreneurship Education: Development, Trends, and Challenges. *Entrepreneurship Theory and Practice* 29: 577–98. Available online: https://journals.sagepub.com/doi/10.1111/j.1540-6520.2005.00099.x (accessed on 15 October 2022). [CrossRef]

Liu, Haibin, Sadan Kulturel-Konak, and Abdullah Konak. 2021. Key Elements and Their Roles in Entrepreneurship Education Ecosystem: Comparative Review and Suggestions for Sustainability. *Sustainability* 13: 10648. Available online: https://www.mdpi.com/2071-1050/13/19/10648/htm (accessed on 25 September 2022). [CrossRef]

Luc, Jessica G. Y., Michael A. Archer, Rakesh C. Arora, Edward M. Bender, Arie Blitz, David T. Cooke, Tamara Ni Hlci, Biniam Kidane, Maral Ouzounian, Thomas K. Varghese, and et al. 2021. Does Tweeting Improve Citations? One Year Results from the TSSMN Prospective Randomized Trial. *The Annals of Thoracic Surgery* 111: 296–300. Available online: https://pubmed.ncbi.nlm.nih.gov/32504611 (accessed on 16 October 2022). [CrossRef]

Luthje, Christian, and Nikolaus Franke. 2003. The 'Making' of an Entrepreneur: Testing a Model of Entrepreneurial Intent among Engineering Students at MIT. *R&D Management* 33: 135–47. Available online: https://onlinelibrary.wiley.com/doi/full/10.1111/1467-9310.00288 (accessed on 15 October 2022).

Matei, Mirela, and Marian Catalin Voica. 2013. Some Considerations Regarding the Social Responsibility of Entrepreneurs in the European Union. *International Journal of Sustainable Economies Management* 2: 26–35. Available online: https://www.igi-global.com/article/some-considerations-regarding-social-responsibility/77340 (accessed on 14 October 2022). [CrossRef]

Ndou, Valentina. 2021. Social Entrepreneurship Education: A Combination of Knowledge Exploitation and Exploration Processes. *Administrative Sciences* 11: 112. [CrossRef]

Ndou, Valentina, Gioconda Mele, and Paquale Del Vecchio. 2019. Entrepreneurship education in tourism: An investigation among European Universities. *Journal of Hospitality, Leisure, Sport & Tourism Education* 25: 100175.

Ndou, Valentina, Giustina Secundo, Giovanni Schiuma, and Giuseppina Passiante. 2018. Insights for shaping entrepreneurship education: Evidence from the European entrepreneurship centers. *Sustainability* 10: 4323. [CrossRef]

Panait, Mirela, Eglantina Hysa, Marius Gabriel Petrescu, and Hailong Fu. 2022. Universities–Players in the Race for Sustainable Development. In *Higher Education for Sustainable Development Goals*. Edited by Carolina Machado and João Paulo Davim. New York: River Publishers, pp. 23–42.

Pascucci, Tancredi, Brizeida Raquel Hernández-Sánchez, and José Carlos Sánchez-García. 2021. Cooperation and Environmental Responsibility as Positive Factors for Entrepreneurial Resilience. *Sustainability* 14: 424. [CrossRef]

Pascucci, Tancredi, Giuseppina Maria Cardella, Brizeida Hernàndez-Sànchez, and Jose Carlos Sànchez-Garcìa. 2022. Environmental Sensitivity to Form a Sustainable Entrepreneurial Intention. *Sustainability* 14: 10398. [CrossRef]

Popescu, Cătălin. 2019. Adaptive Sustainable Academic Management Practices. In *Sustainable Management Practices*. Edited by Muddassar Sarfraz, Muhammad Ibrahim Adbullah, Abdul Rauf and Syed Ghulam Meram Shah. London: IntechOpen. [CrossRef]

Rasmussen, Einar A., and Roger Sørheim. 2006. Action-based entrepreneurship education. *Technovation* 26: 185–94. [CrossRef]

Rejeb, Abderahman, Karim Rejeb, Alireza Abdollahi, and Horst Treiblmaier. 2022. The big picture on Instagram research: Insights from a bibliometric analysis. *Telemat Informatics* 73: 101876. [CrossRef]

Saeed, Saadat, Shumaila Y. Yousafzai, Mirella Yani-De-Soriano, and Moreno Muffatto. 2015. The Role of Perceived University Support in the Formation of Students' Entrepreneurial Intention. *Journal of Small Business Management* 53: 1127–45. [CrossRef]

Sánchez, José C. 2013. The impact of an entrepreneurship education program on entrepreneurial competencies and intention. *Journal of Small Business Management* 51: 447–65. [CrossRef]

Schmitz, Ademar, David Urbano, Gertrudes Aparecida Dandolini, João Artur de Souza, and Maribel Guerrero. 2017. Innovation and Entrepreneurship in the Academic Setting: A Systematic Literature Review. *International Entrepreneurship and Management Journal* 13: 369–95. Available online: https://link.springer.com/article/10.1007/s11365-016-0401-z (accessed on 25 September 2022). [CrossRef]

Secundo, Giustina, Gioconda Mele, Pasquale Del Vecchio, Gianluca Elia, Alessandro Margherita, and Valentina Ndou. 2021. Threat or opportunity? A case study of digital-enabled redesign of entrepreneurship education in the COVID-19 emergency. *Technological Forecasting and Social Change* 166: 120565. [CrossRef]

Sherkat, Anoosheh, and Alireza Chenari. 2020. Assessing the Effectiveness of Entrepreneurship Education in the Universities of Tehran Province Based on an Entrepreneurial Intention Model. *Studies in Higher Education* 47: 97–115. Available online: https://www.tandfonline.com/doi/abs/10.1080/03075079.2020.1732906 (accessed on 25 September 2022). [CrossRef]

van Eck, Nees Jan, and Lutman Waltman. 2017. Citation-based clustering of publications using CitNetExplorer and VOSviewer. *Scientometrics* 111: 1053–70. [CrossRef]

Vesper, Karl H., and William B. Gartner. 1997. Measuring progress in entrepreneurship education. *Journal of Business Venturing* 12: 403–21. [CrossRef]

White, Howard D., and Katherine W. McCain. 1998. Visualizing a discipline: An author co-citation analysis of information science, 1972–1995. *Journal of the Association for Information Science and Technology* 49: 327–55.

Wright, Mike, Donald S. Siegel, and Philippe Mustar. 2017. An emerging ecosystem for student start-ups. *The Journal of Technology Transfer* 42: 909–22. [CrossRef]

Xie, Hualin, Yanwei Zhang, Zhilong Wu, and Tiangui Lv. 2020. A bibliometric analysis on land degradation: Current status, development, and future directions. *Land* 9: 28. [CrossRef]

Zhang, Ying, Geert Duysters, and Myriam Cloodt. 2014. The role of entrepreneurship education as a predictor of university students' entrepreneurial intention. *International Entrepreneurship and Management Journal* 10: 623–41. [CrossRef]

Zupic, Ivan, and Tomaž Čater. 2014. Bibliometric Methods in Management and Organization. *Organizational Research Methods* 18: 429–72. Available online: https://journals.sagepub.com/doi/10.1177/1094428114562629 (accessed on 26 September 2022). [CrossRef]

administrative sciences

MDPI

Concept Paper

Emergent Strategy in Higher Education: Postmodern Digital and the Future?

Mohamed Ashmel Mohamed Hashim [1], Issam Tlemsani [2], Robin Matthews [3], Rachel Mason-Jones [1] and Vera Ndrecaj [1,*]

1 Cardiff School of Management, Cardiff Metropolitan University, Western Avenue, Cardiff CF5 2YB, UK
2 The Centre for International Business, TCIB, Omega House, New Malden, London KT3 6DR, UK
3 London School of Commerce, Chaucer House, White Hart Yard, London SE1 1NX, UK
* Correspondence: vndrecaj@cardiffmet.ac.uk

Abstract: Mintzberg's version of emergent strategy is based on the idea that strategies are contingent on circumstances that change from time to time often very rapidly and therefore papers focused on strategy and detailed planning are limited in their practical application. The word strategy as far as Mintzberg is concerned is anathema, therefore, introducing a concept that has a misconception embedded in it. This paper claims that education for sustainable development and higher education institutions' survival depends on adopting postmodern thinking, in other words, digital transformation. This conceptual paper proposes a blueprint of a process for developing a series of agile potentially short-term conceptual solutions thereby embracing the expectation that the rate of change in societies is accelerating. This paper scrutinizes (a) the applicability of emergent strategy/strategic approach to higher education institutions, (b) how postmodernism influences higher education institutions to become digital hubs of commoditization of knowledge and (c) how the integrated capabilities of digital transformation build sustainability in education delivery. Structural Equation Methodology is proposed to examine the impact of postmodernism on the sustainable delivery of education in higher education institutions, and the need to foster relevant emergent strategies is also justified. The paper also develops new research propositions and managerial implications for driving optimistic digital education. Ultimately, it offers a framework for spear-leading effective and leading post-modernistic digital transformation. Emerging education technology, sustainable digital transformation and advanced use of robotic-human cognitive collaboration are experiencing a significant transformation. Universities play a vital role in enhancing engagement within higher education. One of the managerial implications of the results and discussion is the need for higher education institutions to provide taught leadership and planning in emergent strategy formulation and implementation. The findings confirm the significant importance of linking the Structural Equation Method and the postmodern strategic context in which we argue that higher education institutions require emerging rethinking.

Keywords: postmodernism; digital transformation; DAO and emerging university strategy; sustainability

Citation: Hashim, Mohamed Ashmel Mohamed, Issam Tlemsani, Robin Matthews, Rachel Mason-Jones, and Vera Ndrecaj. 2022. Emergent Strategy in Higher Education: Postmodern Digital and the Future? *Administrative Sciences* 12: 196. https://doi.org/10.3390/admsci12040196

Received: 8 November 2022
Accepted: 7 December 2022
Published: 15 December 2022

1. Introduction

Higher education institutions in the 21st century face a competitive landscape that has changed entirely in the last 40 years or so. Indeed, the adjective 'competitive' has shifted from the ancient rivalry based on academic reputation to peaks to a correlated landscape in multiple versions.

Higher education institutions have become distributed autonomous organizations (DAOs), partly by design but mainly because of the fragmentation of the competitive landscape and their own almost involuntary internal adaptive digital processes. The existing narrative no longer captures their fragmented competitive landscape (Cabrera et al.

2022; Garrod 2016; Wang et al. 2019; Bellavitis et al. 2022). Fragmentation has occurred because of a variety of contributing factors such as the emergence of Massive Open Online Courses (MOOCs), the trend towards blended learning, reinforced by COVID-19, information exchange, commoditization of knowledge, and strategic decision making, which is distributed among internal and external influences: the ingredients of a postmodern situation (Player et al. 2020; Raviolo 2013; Manning 2012). Higher education institutions tend to invest in digital transformation strategies in order to be prepared for the pressing challenges of globalized education (Mohamed Hashim et al. 2021). Radical postmodern changes in global education have enabled higher education institutions to develop sustainable digital transformation strategies to stay competitive. What does it mean to stay competitive in global education? Staying competitive in the global education landscape demands long-term strategies to coup-up with the postmodern challenges. How can higher education institutions use sustainable digital transformation strategies to attain sustainability in education delivery? There is a trade-off between cost implications and various facets of achieving sustainability (Grenčíková et al. 2021). We set ourselves principal tasks in the paper. First, to elucidate the current situation higher education institutions are in and second, to explore and identify a scheme or framework for adaptation processes.

Postmodernism is a phenomenon that has developed a stronghold in higher education. The depth of critical exploration is found in the writing of established scholars (Lyotard 1984; Clark 2006; Richardson and Jencks 1989; Kahraman 2015; Lyotard 1984). This phenomenon has considerably influenced the sustainable delivery of global education and its landscape and is closely associated with the unique phenomenon of sustainable digital transformation. Digital transformation in the global higher education industry determines the future roadmap to a sustainable education management strategy. Thus, there is a need for higher education to develop emergent education strategies integrated with the forces of postmodernism (Vica Olariu et al. 2020; El Kamel and Rigaux-Bricmont 2011; Kahraman 2015). Table 1 presents a detailed examination of literature and influence voices exploring postmodern forces of education.

Table 1. Key features of postmodern society and digital transformation. Source: Based on dwellcc.org (2020), Richardson and Jencks (1989), Kahraman (2015), Lyotard (1984).

Key Features	Features Education of Modern Society	Changes in Features of Postmodern Society	Role of Digital Transformation	The Impact on Education Sustainable Delivery
The form of Knowledge	Essentially is controlled by an authoritative mechanism and unbiased knowledge.	Biased knowledge and the higher education institutions' educators are the architects of biased/ new knowledge.	Academic Program Manage-ment/Review/Monitor and control.	Regulate the academic delivery.
Spending pattern.	Spending is controlled and approved by the state.	Independent of spending priorities, but justification is required.	Virtual planning, communication, and coordination of the academic programmes.	Track the delivery progress/gain visibility of the key global changes in education.
Commoditization of education.	Education is fixed—time, place, and cost.	Types of choices for selecting higher education institutions—virtual, online, distributed and distance learning.	Offers students various options to follow and complete the course through information technology education tools.	Virtual, online, distributed and distance learning.

Table 1. *Cont.*

Key Features	Features Education of Modern Society	Changes in Features of Postmodern Society	Role of Digital Transformation	The Impact on Education Sustainable Delivery
Change.	Lecturer/teacher lead.	Independent learning, teachers are there to guide and set up the challenges.	Cases, sums, and challenges are posted to the students online	Offer the opportunity for blended learning/ new pedagogy
Culture and Values.	Unique, students are expected to learn the culture; it also can be viewed as a barrier. Attempt to be value neutral.	Unified learning society and education build diverse personal values.	Digitalization promotes unity among students. Thus, unity is based on the dominant digital culture in education.	Digital delivery creates/develops an equal culture.
Student nature and the curriculum.	The objective is to meet the national curriculum.	Complex, it needs to meet the needs of globalization of education. Responding to meet global, social, economic, and political pressure. Move from the national to the global context.	It has become the common platform for delivering globalized/commoditization of education.	Enable higher education institutions to deliver the courses according to global delivery standards such as AQA and AACSB.

This conceptual paper aims to develop a conceptual model for implementing postmodernistic digital transformation in higher education. The model advocates how digital transformation can act as an enabling force to develop competitive advantages for higher education institutions in the context of postmodern education (Morze and Strutynska 2021). Building competitive advantage is a relative, evolving, and important concept in strategy formulation. In recent years, specifically in the education industry, the notion of building competitive advantage has been challenged by global phenomena such as digital transformation, globalization, information exchange, digitization, and social media in most global industries. These phenomena have collectively made the process of building a competitive advantage in a rapidly changing, short-term landscape (Abad-Segura et al. 2020; Akhmetshin et al. 2020).

The emergent strategy/approach has become increasingly important in global education because of its ability to deal with inevitable changes such as the impact of postmodernism, digital transformation, and sustainability of delivery (Foss et al. 2021; Mirabeau and Maguire 2013; Davies and Walters 2004; Kahraman 2015). The successful implementation of an emergent strategy is broadly recognized in global industries. Seasoned education strategists and entrepreneurs can teach us a lot about the need for emergent strategy and how to best approach it. The global education industry is evolving, it is typified by key features such as innovation, transformation, and agility. Thus, higher education institutions face various challenges in establishing a model to build an emergent strategy while systematically integrating the influence of postmodernism and digital transformation. On this notion, we propose a unique emergent approach for education strategy- an emergent strategy for education using design thinking. Thus, this paper critically reviews (a) the need for emergent strategy, (b) the integration of postmodernism-digital transformation and (c) the sustainability of the digital delivery of education. Higher education strategy endures a prime responsibility for establishing competitiveness, economic performance, and shaping graduates' futures. Education strategists embark on formulating emergent strategies for higher education institutions to cope with the changing global education landscape/market conditions (Mintzberg and Waters 1985). Thus, there is a significant need to establish a practical approach to emergent strategy (Fixson and Rao 2014).

Mintzberg (1978), the chief architect of the emergent strategy, argues that the intended strategy does not necessarily come into a realisation. Thus, it becomes an unrealised strategy. Hence, there is a need to understand the realised strategy using an empirical approach while highlighting outside the planning activities/process. In this context, the realised strategy results in patterns of activities the management does not anticipate (Vica Olariu et al. 2020; Hernández-Betancur et al. 2017), this resulting response approach is titled an emergent strategy. Figure 1 illustrates the emergent strategy process. There are four factors that define the current situation that higher education institutions find themselves having to respond to, these being (a) the eruption of accelerating technological change, (b) expanding range of product attributes, (c) internationalisation and conflicting government policies and (d) that define the current state of higher education institutions. This paper explores is the role emergence of digital technologies and the information revolution on the resulting fragmented postmodern landscape.

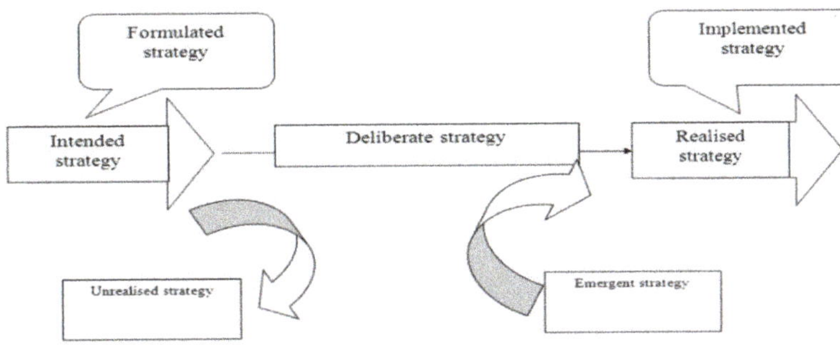

Figure 1. Pattern in strategy formation. Source: Based on H. Mintzberg: "Management Science May 198; 24, P.945".

Higher education institutions can explore new business opportunities by going beyond the traditional approach to strategies, tools, and changing market conditions. Specifically, the influence of postmodernism, globalization, digital transformation, and information exchange have rapidly changed global education. Thus, the significant need for an emergent strategy/strategic approach is realised. Using design thinking, the simplistic approach to emergent strategy adopts an incremental mechanism, which is demonstrated below in Figure 2 the incremental-act model.

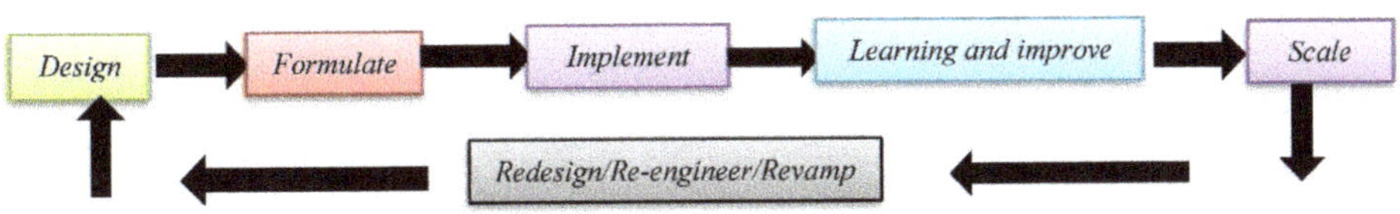

Figure 2. Fundamental model of emergent approach to education strategy. Source: Based on Fixson and Rao (2014).

The higher education institutions that adopt an emergent strategy approach do not only/constrain themselves to critically analysing historical data to predict the future/via conventional forecasting. Instead, they capitalize on market opportunities based on predicated changes by taking calculated risks using reliable and scalable steps using an experimental approach, examining and evaluating the outcome of each step.

Adopting a robust lesson learnt approach allows a structured approach to formulating measured actions. Meaning each step forward reveals the previously covered challenges and baseline for the next step- thereby ensuring the notion underlying the emergent strategy. We argue that one of the key problems of the prescriptive/analytical approach to business strategy is not the mechanism but rather the unreliability of tools predicting uncertainties specifically in terms of application and understanding potentially good and bad scenarios. For example, statistical analysis of return on investment, discounted cash flow provides stakeholders with false interpretation certainty about the naturally uncertain condition. On this notion, we claim the emergent approach becomes a necessity for education strategy amid the agile changes of postmodernism and digital transformation.

Postmodernists argue that increasingly societies are characterised by consumerism and choices. The influence of postmodernism challenges global education to explore beyond the conventional operations of higher education institutions and conventional education delivery, which favours liberal education. We claim that we still live in a postmodern society/postmodern age which typified five major characteristics but not limited to (a) diversity of individuals, (b) better fluidity in identity and appearance, (c) emergence of cross-culture, (d) globalization of education, (f) commoditization of knowledge and (f) media-saturated life (El Kamel and Rigaux-Bricmont 2011; Emerick 2007; Nielsen 2006; Hassard 2003).

What do these changes mean to global education? How does it impact the delivery of university education? Why does digital transformation become inevitable in the delivery of education? What impact would it have on the sustainability of education? These are critical questions in the age of information exchange.

The key characteristics of postmodern education society are shown in Figure 3, although its significance and impact are unclear, specifically in global education. Relatively it is under investigation, and a paucity of knowledge is evident in the literature. We favour the conventional of wisdom postmodernism to enrich contemporary education society, verify the compatibility of its key features and validate the need to establish rigorous organizational research. Global education is closely associated with the information revolution powered by digitalization and technologies. Almost by definition, digitalization leads to parallel computing, which leads to the emergence of DAO's fragmentation and hence postmodernism.

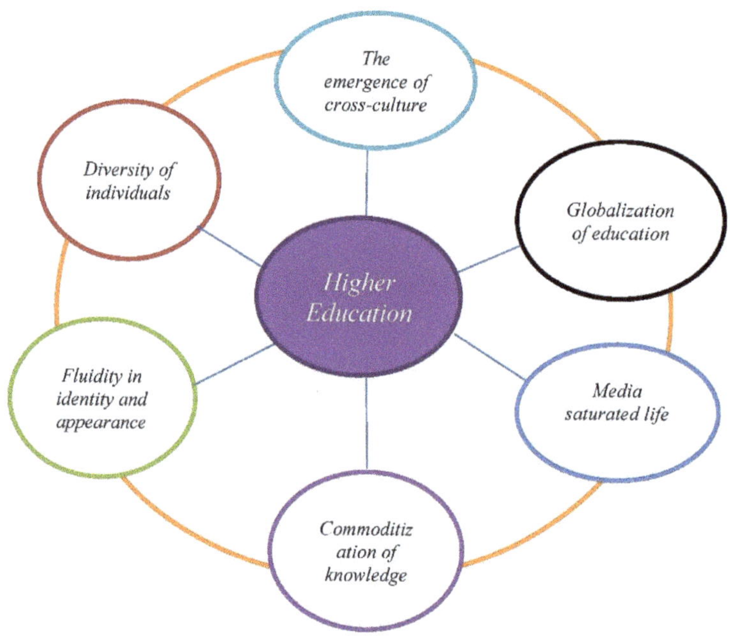

Figure 3. The key characteristic of postmodern education society. Source: Based on Lyotard (1984), Richardson and Jencks (1989), Kahraman (2015), and Authors' proposal (2022).

Figure 4 indicates that the emergent strategic approach to higher education becomes a necessity because of the significant influence of real-world phenomena, namely (a) the influence of postmodernism and (b) digital transformation shaping the sustainable delivery of education. Despite the collective impact of postmodernism and digital transformation on education and its wide acknowledgement, there is limited of knowledge regarding how it puts pressure on higher education institutions to deliver education sustainably by amalgamating delivery models. This body of literature states that digital transformation act as a proxy between the way university integrates the impact of postmodernism into their digital transformation strategy, in turn, its impact on the delivery of sustainable education/its delivery (Usher and Edwards 1994; Klimski 2018; Vica Olariu et al. 2020; El Kamel and Rigaux-Bricmont 2011; Kahraman 2015). To fill this gap in this literature, this research explores (a) how postmodernism impact sustainable digital transformation and (b) how digital transformation influences sustainable education of higher education institutions.

Figure 4. The fundamental model of postmodernism on sustainable digital transformation. Source: Based on the author's proposal (2022).

How should higher education institutions shape and reshape their education delivery amid the pressing changes of postmodernism and digital transformation to foster sustainable education delivery while coping with the growing demand of the age of globalization? Digital transformation, agility in education delivery, the resilience of higher education institutions, blended education, and relevant affiliation are continued to be the vital elements of the postmodernism of higher education and the higher education digital future.

At this juncture, it is critical to examine the key changes in education from a modern society to a postmodern society, particularly how it evolves. What does it mean to the future of education, and how higher education institutions can make sense of it? What is the role of digital association? How is it intrinsically interrelated with postmodern education society? In order to answer these questions, first, it is essential to identify and determine the key changes caused by postmodernism and how it is closely associated with the process of digital transformation as far as university education is concerned (Usher and Edwards 1994; Klimski 2018). Table 1 presents an overview of the key features identified within the literature.

We argue, comparatively, that there is a paucity of knowledge in terms of how postmodernism changes to sustainable digital transformation-based education globally. Further, how this change leads to sustainable education delivery at the university level has had limited exploration in the literature at the university level. It is therefore, posited that the relevance of Postmodernism on sustainable digital transformation and how it impacts the need for sustainable education requires significant investigation both globally and locally (Kia 1988).

The higher education institutions need to examine how the impact of most modernism influenced emerging changes in the delivery of sustainable education. The demand for the globalized sustainable delivery of education has enabled higher education institutions worldwide to demonstrate critical features such as (a) flexibility, (b) differentiation, (c) agile, (d) mobility and (e) decentralization while design-developing and delivering portfolios of educational courses (Mohamed Hashim et al. 2021, 2022; Lozano et al. 2015) Meaning, as a community, we are transitioning into a new educational era, and inevitably, the phenomena-impact of postmodernism and sustainable digital transformation- are collectively changing our educational balance. Within this examination we need to explore a couple of key questions, such as: Are we living in a new type of educational landscape? If so, what kind of landscape is it? In order to explore these questions, we require constructive dialogue to enable effective critical exploration.

This paper attempts to bridge an existing research gap by (a) developing the existing body of knowledge about postmodern education to the next level, (b) critically examining the close associations between postmodern education, digital transformation, and the need for emergent strategy and (c) developed practical conceptual models for assessing the realistic impact of postmodernism on sustainable education delivery in higher education institutions. As stated, this paper aims to offer a relatively implementable model for post-modernistic digital transformation in higher education.

The authors have developed the following key research questions to achieve the aim.

(a) How postmodernism of education impacts the sustainable delivery transformation of higher education institutions.
(b) What forces influence the postmodernism of education and sustainable delivery of education?
(c) How international collaboration integrates with the impact of postmodernism on sustainable education delivery.

2. Literature

The literature review drew upon the existing literature exploring (the emergent strategy process based on Quinn (1980), Senge (1990), Argyris (2014), Mintzberg (1987) and Lynch (2018). Whittington et al. (2020) empirically demonstrate how organizations seize incremental, adaptive, flexible, experimental, and learning and development while design-developing, implementing, and re-engineering business strategies.

This conceptual paper claims that developing a strategic emergent approach that interacts with postmodernism and digital transformation capabilities (Figure 5), in turn, enables higher education institutions to gain sustainability in the delivery of education. It needs to be highlighted that in the age of globalization, the critical success factors of university education are increasingly standardised. Thus, the notion of building competitive advantage requires thinking beyond conventional structures and resourcing approaches within the education industry, it requires unique models, processes, design thinking and selective integration capabilities. The subsequent sections critically examine the formation/approach to emergent strategy for education, the importance of integrating the global influence of postmodernism and the utility of digital transformation on university education and how it enables building sustainable delivery models for education. Specifically, the phenomenon- Emergent Strategy in Higher Education: Postmodern Digital and the Future? Require thorough examination/investigation to use knowledge in the educational society and strategic management of education as to how to build, retain and protect the process of building competitive advantages (de S. Oliveira and de Souza 2021; Teece 2020; Halliday 2020; Lamichhane and Wagley 2013).

2.1. *The Emergent Approach of Education Strategy*

The robustness of strategy tends to be thought of as forceful, flexible, interactive and based on learning and development. Therefore, the concept of enabling the emergent strategy is increasingly becoming important in the rapidly changing landscape of global education. However, the application and integration of the emergent strategy approach lack significant theoretical underpinning, specifically in terms of how to interact in the dynamic education environment and leverage it to create superior value. The emergent strategy approach offers education a way to achieve sustainable education delivery. We provide a theoretical foundation and a unique approach by amalgamating postmodernism and digital transformation educational changes. An emergent strategic approach is wide spreading, specifically in the global education industry. Specifically, education stakeholders are comfortable with the notion/logic the emergent strategy underpins. However, it can be shown there is limited understanding of the approach and especially its practical application within managerial levels. One could argue that the formulation of an emergent strategy requires the identification and determination of uncertainty within the macro and micro environment as the first step in the current education industry (globalised and post-modernised industry). Thus, the stakeholders who closely interact with strategy formulation and implementation require sensing skills, corporate games knowledge, and simulation practices.

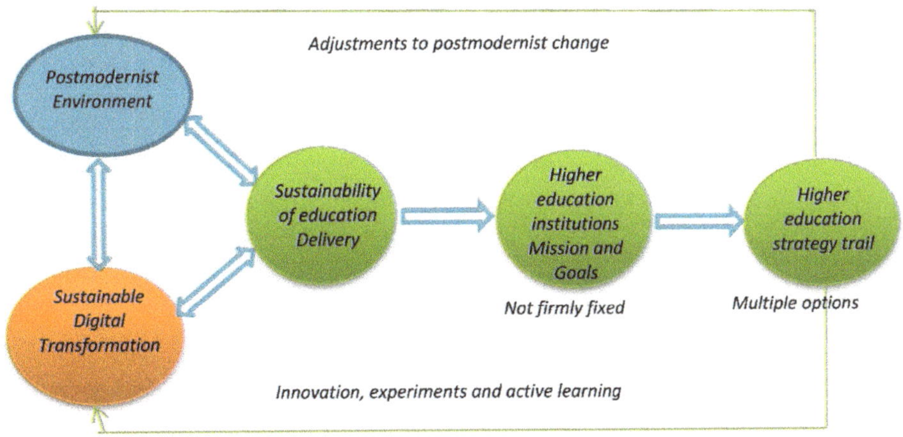

Figure 5. The postmodern-digital future. Source: Based on Argyris (2014), Mintzberg (1987), Quinn (1980), Senge (1990), Lynch (2018) and Whittington et al. (2020).

The application of an emergent approach to education requires (a) continuous knowledge acquisition, (b) the use of a reliable method, and (c) an organizational mindset to strive in global education. The emergent strategy implementation empowers higher education institutions to develop a portfolio of educational offerings at various levels through experiential learning. Further, the implementation of an emergent strategy also enables the lecturers, education administrators and students to learn from the unsuccessful experiments (Morze and Strutynska 2021).

2.2. Postmodernism

Lyotard (1984) on postmodernism claimed that postmodernism is due to people's wariness and disbelief regarding the metanarratives- therefore no longer believing there is only one truth or mechanistic solution. Postmodern society stops blind belief in the big stories and is more open to an exploration of contextual understanding and adaption. Instead, individuals are more open to develop their unique perspectives on events and indeed understanding the challenges of variation of interpretation to move towards a collective understanding of the wider narrative. This feature of postmodernism has impacted the design and delivery of education.

Today higher education institutions are under increasing pressure to impart one-to-one engagement mechanisms and indeed experience while delivering courses. Irrespective of the generic course delivery, stimulating one-to-one engagement experience is becoming a fundamental need for higher education institutions, digital transformation platforms this experience primarily because of its customisable ability to generate various solutions (Leaning 2014; Pivovarova et al. 2020; Lamichhane and Wagley 2013; Zhu 2009; Done and Knowler 2013).

So, what do postmodernists believe in? We assess that postmodernism is typified by the distrust of educational experts, meaning that students in the modern era believe in more than one truth in selecting and pursuing a course while still understanding some core truths that exist. They validate the authenticity through various measures. In other words, students are potentially more engaged with the concepts of critical discussion. For example, understanding phenomena such as COVID-19 and climate change crises cannot be distilled down to one universal solution but is recognised to have a variety of complex contributors and requirements in terms of effective response. It is also a noticeable fact that while pursuing education, students are willing to remain as global citizens, and their structural identities, such as nationalities, tribes, classes, and ethnicities, are becoming less important.

Usher et al. (1997) state that education in the postmodern society is explained and shaped by diversity. Meaning education delivery must provide the learners with lots of choices. Robin Usher and Richard Edwards were the pioneers who studied the empirical relationship between postmodernism and education in a global context. Specifically, they stated that postmodernism and its influence are significantly important for university-level education.

Education institutions should offer courses that suit the learners changing needs, which may influence by the changing economic, social and political situations. It should mean that global education reflects the postmodern economy or service economy. This phenomenon has created desirable opportunities for higher education institutions to offer education in the forms of distance learning, virtual learning and blended learning. Thus, the digital transformation of education becomes inevitable. Particularly, it has become a centric feature of postmodernist education life. It is important to highlight the key features of postmodernist education at this juncture.

Genosko (2001) quoted Baudrillard's view on postmodernism in the year 2021 stating that postmodernism prioritises individualism (choices and opinions) relative to socialism. Further, some scholars argue the birth of postmodernism leads to the death of socialism. What does this mean to the higher education institutions' education system? Increasingly, higher education institutions' signs and symbols have increasing importance. It is argued there are no standard key performance indicators to validate if the required values (or indeed achieved) are reflected in the delivery of education. Value t is quite hard to differentiate between reality and hyper-reality.

We argue that higher education institutions are increasingly building international collaboration and gaining accreditation to maximise the educational values communicated by the sign, symbols, and badges. Often collaboration is used to power the accreditation of degrees, and higher education institutions use it as a shield to attract revenue and fill any potential revenue gap. In the global education landscape, the standard of education accreditation impacts the delivery models and hence potential profitability. When higher education institutions gain multiple accreditations, it enables them to tailor the education delivery while building the flexibility to actively engage students in pursuing their expected learning outcomes and career aspirations (Mohamed Hashim et al. 2021).

The other key distinguishing feature of postmodernist education is hyper-social media saturation, which builds to hyper-reality. The images and logos of educational institutions often describe the values/experience delivered to the students. We view that as it is an illusion of reality-simulacrum. Today, what is unreal is perceived as real because of social media reality. Education institutions excessively use simulacra to which students believe with the stimuli until they no longer believe in the reality of education. At this juncture, a key critical question is what the role of social media in global education continues to be unclear but complex.

This is most evident across Facebook, YouTube, Instagram and Twitter. Social media personalities/superstars are increasingly used for branding/attracting students for various education portfolios. Specifically, this is excessively evident in fashion design courses. At a high-level, images of celebrities are presented selectively in social media to paint the illusion of reality. As part of their strategy, most of the university's practices distinguish the reality of students and what they perceive from the social media paints about the delivery of education. The illusion built via social media fails to materialise real meaning to students' education life and experience.

Students choose courses based on industrial demand. Thus, there is a danger that the philosophy of wider education choice may become fragmented/unstable if universities focus too much attention on current trends. This phenomenon has imposed noticeable volatility in identifying and determining the portfolio of courses to be offered. Thus, higher education institutions are under considerable pressure to understand the key changing education trends. To be very specific, there is no fixed formula to determine the portfolio of courses. As factually justified, we live in a society where the influence of postmodernism is immense. It has broken down the education society into individual narratives recognising multiple identities, volatility, complexities, and potential confusion (Lyotard 1984).

The rationality of postmodernism is important to transformative education. In the age of globalization, it is achieved via the digital transformation of education in higher education institutions, as it establishes the notion of theoretical basis, the foundation of digitalization of university education/learning experience, importantly; this is relevant to the digitalization of sustainable education/education delivery-an increasingly popular subject in university education. This paper engages with the future of university education as an interdisciplinary with postmodernism and digital transformation. We argue, what could be viewed as important of postmodernism in gaining academic respectability? (Holsberry 1981; Lyotard 1984).

In the age of postmodernism, the delivery of global education is characterised by but not limited to (a) digitalization of education, (b) education as a service, (c) commoditisation of education, (d) privatisation of education (e) marketization of education (f) virtual learning, (g) independent learning, (h) decentralised learning (i) social media-saturated education, (j) scientific thinking, (k) accreditation is influenced by distributed education, (l) standard of education and (m) quality of education. The collective and serious changes of postmodernism problematize and disrupt the deep-rooted assumptions of university education, specifically teaching, learning, and delivery. This conceptual paper concludes what university education might become due to post-modernistic disruption and turbulence.

2.2.1. Digitalization of Education

Higher education is globally undergoing significant changes, which are primarily influenced by the impact of postmodernism and technological advancements. Particularly, the influence of postmodernism has enabled higher education institutions to adjust their educational deliverables using innovative delivery models. The deliverables are tailored according to the knowledge of economic changes, the digital transformation is used as a tool/platform to create value the educational delivery (Bican and Brem 2020; Benavides et al. 2020; Abad-Segura et al. 2020; Bogdandy et al. 2020; Akhmetshin et al. 2020; Iivari et al. 2020).

This approach enables higher education institutions to take a sustainable approach to global education delivery. We argue that the influence of the postmodern digital transformation and how it impacts education sustainability should be viewed as a key change in the socio-economic education system. Further, other factors such as globalization and information exchange are fuelling the key characteristics of global education ((a) digitalization of education, (b) education as service, (c) commoditisation of education, (d) privatisation of education (e) marketization of education (f) virtual learning, (g) independent learning, (h) decentralised learning (i) social media-saturated education, (j) scientific thinking, (k) accreditation is influenced by distributed education, (l) standard of education and (m) quality of education).

Higher education institutions worldwide use digital transformation as an alternative to fill the student enrolment gap. This is a common feature of the digital transformation strategy, relatively influenced by the shock of postmodernism. This unique education phenomenon has enabled higher education institutions to examine the sustainability of their education delivery. However, this realm is relatively still in the embryonic stage, drastically different in scope, and requires rigorous investigation. We view that the empirical insight of this research leads higher education institutions to build entrepreneurial capabilities, which can be translated into developing competitive advantage in the long run.

2.2.2. Commoditisation of Education and Virtual Learning

Commoditisation of education has provided a strategic option for higher education institutions to equalize work experiences to degree-level standards. As stated, the commoditisation of education, which is one of the key features of postmodernism, has aided higher education institutions in attracting qualified faculty members in relevant and demanded fields of education.

The pressure of commoditisation of education in higher education institutions' planning also potentially fueled resources to be viewed as a commodity to fulfil the delivery commitment of education. This situation diminishes the strategic importance of faculties in the role of delivery. Thus, higher education institutions explore alternative delivery methods and unique delivery models to meet the demands of the commoditisation of education.

In the postmodern environment, we argue that the average life span of education delivery (actual delivery) is shrinking. The power of digitalization also reduces and optimizes the lifecycle times or patterns. Thus, faculties go through specialised training to cope with the need for the digital transformation of education. We argue that education's virtual commoditization has already set a new standard for faculties and students. Utilizing digital transformation capabilities to meet delivery needs goes beyond technical understanding and requires a deeper understanding of pedagogical preparation in the digital environment (Carter et al. 2020; Chambers 2016). The benefits of Virtual learning are widely acknowledged. As indicated by Figure 6 it attempts to enhance collaborative and engaging learning. A standardised virtual environment is typified by three distinct features, namely (a) physicality, (b) interactivity and (c) persistence. The participation and engagement of the students are represented by digital/graphical representation (Bican and Brem 2020).

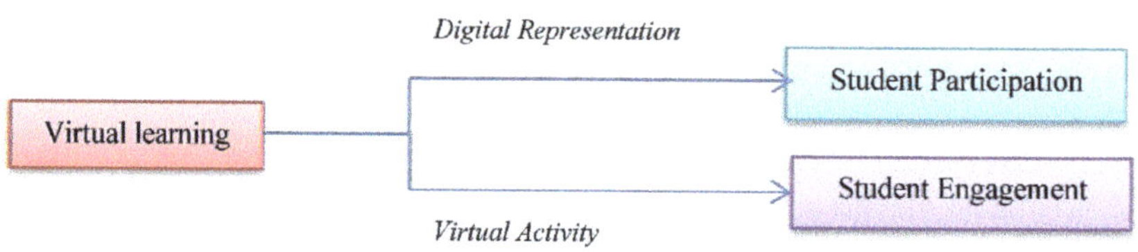

Figure 6. Key components of virtual learning. Source: Based on Al-Azzam et al. (2020), Johnson and Blitzer (2020), and Authors proposal (2022).

Virtual learning is a significant component of student learning approach pedagogies and has been fueled by the response to COVID-19. Virtual Learning (VL) enables students to access content/videos/presentations anytime via multiple communication channels. Modern-day virtual learning tools' innovative features and capabilities assist learners in engaging in profound interactions and have close engagement experiences; otherwise, they would have in classical face-to-face learning. Arguably, virtual learning is taking the centric orientation of blended learning. The use of digital space is a desirable feature (Henseruk and Martyniuk 2020).

Inevitably, virtual learning has become a propelling force for higher education institutions, particularly because of COVID-19 (Al-Azzam et al. 2020; Johnson and Blitzer 2020). Thus, the authors argue that higher education institutions must extend the utility of virtual learning to instrumentalize the sustainable delivery of education. Therefore, higher education institutions must reconnoiter how to combine humanistic qualities with virtual learning to guarantee student collaboration and engagement as well as face-to-face learning (Powell and McGuigan 2021).

Virtual learning also provides a competitive opportunity for higher education institutions to gain inclusiveness by catering their courses to a wide range of international students.

This also enables higher education institutions to pursue competitive brand positioning (Chatzoglou and Chatzoudes 2018). Virtual learning is utilized to create opportunities for transnational students. However, increasingly higher education institutions are using VL as a tool to engage disabled students who prefer limited movements (Gerrard 2007). There is a growing emphasis on developing a process model for building the space architecture, aiming at adequate student engagement.

2.2.3. Independent Learning

The modern education system recognises the importance of independent learning as part of its pedagogy. The COVID-19 pandemic further pushed higher education institutions to use independent learning as one of their main tools to overcome the challenges associated with not being able to utilise conventional learning environments. Utilising and encouraging independent learning practices has become necessary to cope with the rapidly changing current state of both scientific and information technology bodies of knowledge. Additionally, to stay abreast with the research and development needs.

The purpose of independent learning in a university environment is to enable students to (a) develop content-based knowledge, (b) gain technical know-how and (c) other capabilities (Sudirtha et al. 2021). In a postmodern society, one could argue that higher education institutions should design and develop content for independent learning and tailor the learning process according to the rapid changes both in the scientific and information technology disciplines- the industrialized world. Thus, we argue that developing an empirical model to establish a uniquely changing pattern becomes a fundamental necessity.

Independent learning among university students promotes entrepreneurial spirit (Tan 2013; Beeson 2016). Thus, there is a greater emphasis on this phenomenon (however, to establish independent learning among students effectively, higher education institutions require the right combination of tools and techniques, most importantly, its integration with the digital transformation strategy/blueprint (Carter et al. 2020). Notably, independent learning primarily happens through online channels in the current era; thus, it offers the luxury for the students to selectively utilize the tools and the sources (Lemmetty and Collin 2019). However, disseminating information and information exchange is the key to independent learning (Sudirtha et al. 2021).

2.2.4. Social Media-Saturated Education

The increasing use of social media-based education creates a new gap: how knowledge collaboration occurs in traditional society versus knowledge collaboration in social media saturated postmodern society (Abney et al. 2018). The integration of social media has become one of the key elements of the digital transformation of higher education institutions. It is viewed as a platform to engage students to develop a positive attitude about the globalised world (Carrigan and Jordan 2021). Faculties in higher education institutions also utilize social media for instruction and lecturing purposes using various technologies closely integrated with social media (García-Peñalvo 2021).

Specifically, there is a vibrant movement towards Facebook and LinkedIn (West et al. 2015). The gravitational movement is logical, given the incredible numbers of subscribers/users on Facebook; however, different results have been found for the usefulness, learning and development and engagement (Heiberger and Harper 2008; Kirschner and Karpinski 2010; Kolek and Saunders 2008). Higher education institutions have started to explore how Twitter might be used to develop engagement opportunities for students, faculty, and education communities (Kassens-Noor 2012; Rinaldo et al. 2011). For example, research indicates that Twitter is assessing a microblogging feature that facilitates educational dialogues delivered in just time/real-time (Junco et al. 2011).

The competitive advantages of social media for higher education institutions are:

a. The competitiveness of the university is dependably exhibited by social media and there is a rising hype around it.
b. Social media networks have produced unique values and benefits

c. The content shared relatively affects the productivity of student groups
d. Influence to build/enhance industrial knowledge/intelligence
e. Collaborative learning is possible
f. Consistent, people-to-people interaction led to convergence and divergence

Postmodernist scholars argue that the saturated use of saturated social media is a vibrant feature of postmodernism. The use of online connectivity and the effective use of social media have collectively enabled the modern generation to pursue knowledge uniquely. This phenomenon has allowed higher education institutions to explore new patterns via social media interactions to stimulate the learning process. Researchers have found that effective use of social media leads to critical thinking and student engagement (García-Peñalvo 2021).

2.3. *Distributed Autonomous Organizations (DAO)*

Higher education institutions have focused on digital transformation strategies to ensure futureproofing and maintenance of competition in global education. This highlights the importance of examining what is required to stay competitive and how does the competitive landscape change? We claim that DAO implementation is critical, and it must be held accountable for developing a digital transformation blueprint/regulating sustainable digital delivery using digital business models. The emergence of DAO leads to new educational opportunities, develop the digital economy, forms digital cooperation among higher education institutions and their distributed operations (Burkov 2020).

The rise of DAO also led to the rapid development of innovative technologies, digitalization of education, enhanced digitalization of society and increased the number of hardware devices connected through the Internet of Things. There is a potential that the connected hardware devices will lead to borderless higher education institutions. It will introduce a diverse range of needed digital education and learning skills and technologies. Forming such DAO is an educational business challenge. Additionally, DAO could be viewed as a direct substitute for the conventional educational delivery of traditional higher education institutions. Despite the criticality of DAO, there is also a need to regulate and optimize several processors due to increasing digital freedom and scaling security-related issues to deal with b both the short-term and long-term postmodern-educational challenges.

The introduction of Blockchain Technology has increased the scope, scale and practicality of DAO. The unique technology that underpins blockchain has broadened the efficient functioning of DAO, specifically in the education sector, where content (digital assets), selection of courses (individualism/consumerism) and the recognition (offering certificate of recognition) are performed over the distributed autonomous network. Thus, it becomes strategically important for higher education institutions to focus on DAO, which holds the accountability to build digital delivery advantage. DAO development in the education industry is achieved by monitoring and controlling using formalized rules. These rules are designed based on real-time performance indicators. Increasingly the process of managing and controlling the DAO is automated using digital transformation capabilities, software, digital technologies and rule engines are used (Kaal 2021; Virovets and Obushnyi 2020).

Scholars argue that DAOs will be the future of many global industries, thus, the education industry cannot be an exception. However, there are many arguments and discussions about the DAOs structure, delivery model and capital growth. It is believed that future educational opportunities will be created via the formation of DAOs, based on their digital transformational capabilities-digital interaction between higher education institutions/students. The efficiency of the functioning of the DAOs is believed to be based on their ability and flexibility to decentralise their governance. However, digitally distributed autonomous organisations' security, governance, regulation, and legislation are still underdeveloped areas (Yan et al. 2013).

2.4. The Fundamental Model of Postmodern Education Delivery

The derived fundamental model shown in Figure 7, highlights several propositions. The model also simplifies the approach and underlying notion of the research. It draws on the current literature and attempts to illustrate the methodology, analysis and conclusions through (a) forces of postmodern education parameters estimated (b) quantifying its impact on education and (c) the close association sustainable digital transformation enabling the delivery of education. Thus, it offers a unique perspective on the postmodern education phenomenon and sustainable digital transformation, which is strategically important for university education worldwide. Global education has proved to be noteworthy in the evolution of education and it is highly likely to be even more significant in the future amid globalization (Dlačić et al. 2013; Bagci and Celik 2018; Alalwan et al. 2021).

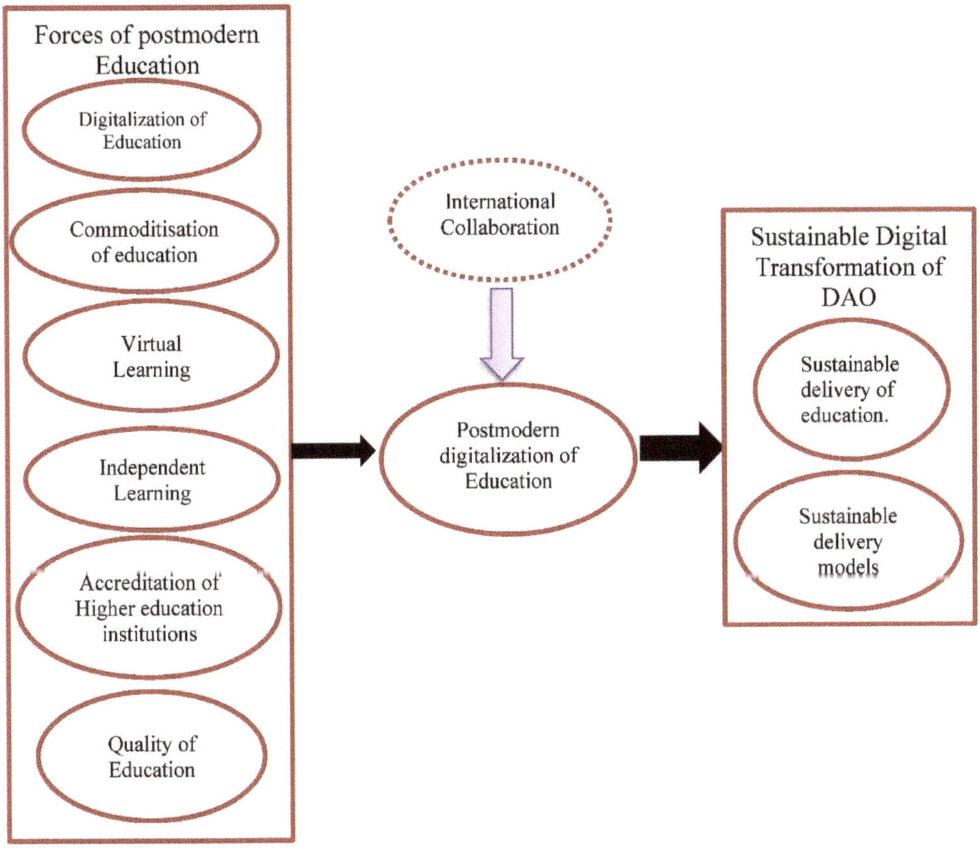

Figure 7. The nested fundamental model derived from the literature review. Source: Based on Dlačić et al. (2013); Bagci and Celik (2018) and Alalwan et al. (2021).

3. Methodology

Structural Equation Methodology (SEM) has been proposed to examine the impact of postmodernism and the university's ability to transform sustainable delivery of education using digital transformation capabilities. SEM offers a robust approach to examine the accumulated influence of postmodern educational forces (latent exogenous variables), the impact of postmodernism on education (proxy) and its impact on sustainable delivery capabilities (latent endogenous). Thus, the emphasis on understanding the accumulated impact using a quantitative approach, SEM using Confirmatory Factor Analysis, is recommended.

The SEM approach enables higher education institutions to examine the accumulated impact of postmodernism on the sustainable delivery of education via digital transformation. It is a robust statistical framework that is increasingly used across organizational research. It specifically enabled the researchers to test (a) latent variables, (b) measured variables and (c) the direct and indirect relationship in a structural model. The model identified (postmodern-digital) encapsulates both measurement and structural models. Thus, the need for the SEM method becomes inevitable. Figure 8 shows the authors' perspective regarding the examination of the impacts of postmodern digitalization. It is possible for future researchers to adopt either Confirmatory Factor analysis (CFA) or Exploratory Factor Analysis as the data analysis techniques on SEM. Thus, the interpretation of the data analysis is limited by the standard fit indices (model fit the data/data fit the model).

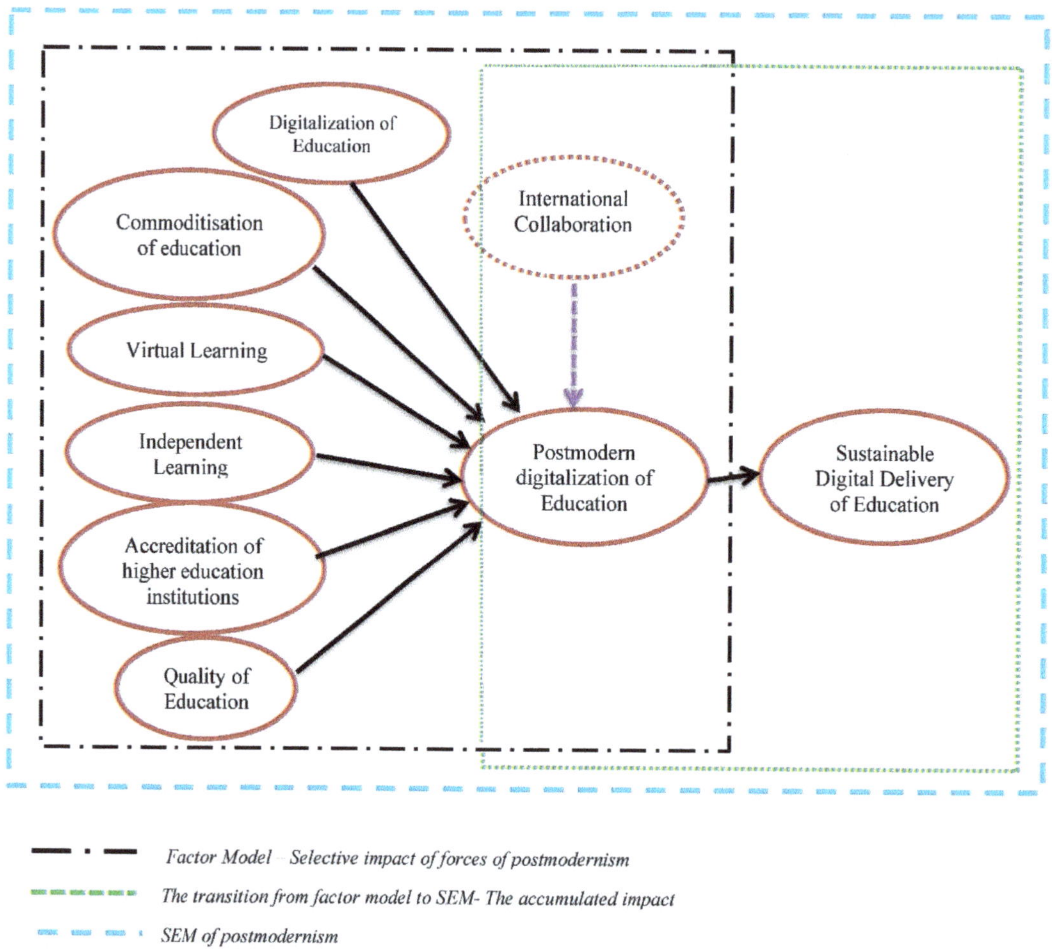

Figure 8. The measurements models and structural model of postmodern digital. Source: Based on Authors variable view.

This approach enables an organization to quantify (a) the individual impact of postmodern forces and (b) its collective impact on sustainable delivery; hence, transforming the same approach as a potential management information system becomes easy. Further, SEM also enables capturing the indirect effects among the variables (path models, mediation, and moderation). This insight is critical to developing the narrative form of the knowledge of postmodernism.

The conceptual framework of the research (Figure 7) shows the projected association and relationship of postmodern forces of education. We view those as the antecedents of postmodernism of education. Then, it is linked to the influence of postmodernism on education, and as an outcome, how the sustainable delivery of education is impacted is examined by the conceptual model. Further, how the international collaboration of higher education institutions further amplifies the sustainable delivery of education is an important feature of the digital delivery of education. In the postmodern view, this could be viewed as a contextual parameter. Thus, the mediating mechanism is incorporated as one of the key elements of the conceptual framework.

4. Analysis

One of the primary tasks of higher education institutions is to provide informed knowledge to meet the demand in the wider community, engage in research that contributes to the body knowledge and economic factors such as making a profit. The process is significant, complex and evolving. Higher education institutions worldwide generally establish knowledge production and delivery by retaining recognized intellectuals and a dependable process. Digitalization, demand for industrialised knowledge, the dominance of information technology, commoditization, and the influence of postmodernism have enabled higher education institutions to re-engineer their education strategy to adapt to the rapidly changing environment by adopting innovative capabilities. Specifically, in the current era for a university to be competitive, we constructively argue that they require the integration of (a) an emergent strategy, (b) the integration of postmodern-digitalization and (c) developing a mechanism for sustainable delivery of education. We examine how integrating an emergent strategic approach with postmodern digitalization can improve the sustainable delivery of education. Scientific approaches can be formulated to test and validate this proposition. This is the centric argument/value addition of this paper. We believe that this approach would enable us to discover the systematic changes/differences triggered by postmodern digitalization in higher education institutions in a wider context (Holsberry 1981; Lyotard 1984; Arends 2014; Czainska 2009; Audebrand 2010).

Higher education institutions have adopted the emergent strategy approach to enable agile responsiveness and establish competitive advantages. It is creative, requires design thinking, dynamic and analytical in approach. The use of the emergent strategy is widely acknowledged. The characteristics of the emergent strategy enabled higher education institutions to gain distinct benefits, primarily; it enables higher education institutions to be flexible and agile in their approach to the delivery of education while fostering continuous learning and improvement. We argue that demonstrating agility for higher education institutions has become necessary to meet the need for the postmodern digitalization of education. Meaning the integration of postmodern digitalization needs an emergent approach to the portfolio of courses, particularly its design and delivery. As far as this research is concerned, it offers the luxury to the higher education institutions to deal with the business environment with agility (absorbing the influence of postmodernism) based on using competitive resources (digital transformation capabilities) to meet the purpose (sustainable delivery of education) (Fixson and Rao 2014; Du Toit and Verhoef 2018).

We propose using an emergent approach (Figure 9) for education strategy is innovative, incremental and complex to describe but a necessity. The intended strategy predicts the future based on the patterns of business activities from the past (Mintzberg 1987). Industrial knowledge is increasingly becoming a commodity. Particularly, information technology has gained accelerated growth as a service. The commoditized knowledge is forming new industries; this process has become a new trend among higher education institutions. By developing new knowledge, higher education institutions gain power and build competitive advantages. The mechanisation of knowledge is bound to affect higher education institutions when it becomes irrelevant or of no use (Lyotard 1984; Duvnjak et al. 2020).

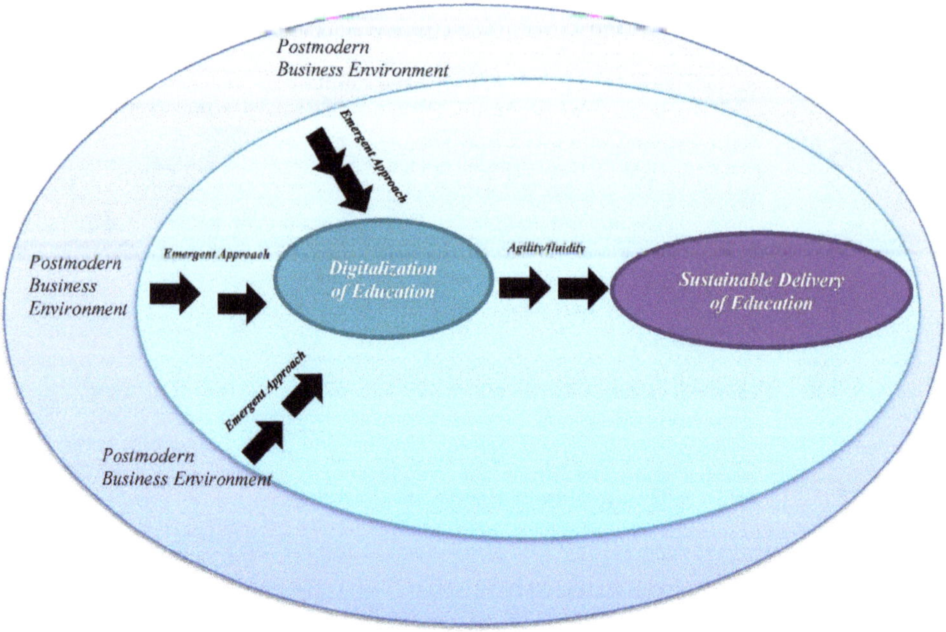

Figure 9. An emergent education strategy approach amid postmodernisms. Source: Based on Fixson and Rao (2014), Du Toit and Verhoef (2018) and Authors proposal (2022).

Higher education institutions' perspectives on the learning patterns and delivery of courses fall within the boundaries of the national education agenda, the national educational framework, and society's brain. Educational communication and transparency are directly related to the potential commercialisation of education. The higher education institutions' economic performance becomes the point of imperilling stability or standard. The gained capital and the monetary advantages are invested or pumped into new channels of multinationals. It could be observed that learning is going through, designed and retained in the same channels as money (Kane 2017).

Modern technology permits education to scale via a more agile approach, constantly producing scientific, technological and research driven knowledge, a typified feature of a postmodern society. It is noticeable that although higher education institutions show dissonance that scientific knowledge does not characterize the totality of knowledge production, it is also important to ensure a high emphasis on industrial research knowledge because of the demand conditions of global education. What will happen to the narrative form of knowledge production in higher education institutions continues to be a grey area. Despite our dissonance that the narrative form of knowledge cannot supersede the research-science-technology knowledge, the narrative form of knowledge is necessary to develop an implementable system body of knowledge (Lyotard 1984; Lacan 2019).

To cope effectively with the postmodern influence, higher education institutions have explored the case of education delivery. The power of computing to magnify the knowledge base of late digital transformation is used to amplify the sustainable delivery of education worldwide. The use of pedagogy-blended learning, independent learning, and virtual learning enhances the students' high use of digital education platforms. They have simply oversaturated education life. In a postmodern society, the bargaining power of the students relevant to selecting a course and pursuing it is high because of the interpretation and perception power of the customer. Additionally, many higher education institutions are more open about their commercial performance; the sustainability of the delivery is determined by how many unities of a commodity (portfolio of courses) are sold to the students through digital channels.

What is a core (set of) proposition (s)? Sudden (late 20th or early 21st century) eruptions have fragmented the competitive landscape of higher education institutions. From a gradual, relatively smooth evolution landscape of them over the centuries, which Lyotard, for example, describes as modernistic, an era, described as postmodernity, arrived as a shock to businesses and governments. As it is often the case, the early heralds appeared in the arts: literature, criticism the graphic and performance arts (Lyotard 1984).

Interestingly, postmodernity was fostered, particularly within the humanities departments (i.e., working with sophisticated digital transformation technologies has become the new normal, business as usual). Academics fostered postmodern themes in the humanities both in action and reaction: that is, embracing postmodernism as an outlook and resisting it, and so creating a dialogue, on the other hand, promoters of postmodernism and the other opposes. Action and reaction create a dialogue that raises postmodernism to eminence with supporters and refusers. Early refusers were in the arts, and postmodernist trends became visible in the nano and digital technologies. The new technology erupted through the economics of scope and scale and network effects, which diffused the new technologies exponentially, simultaneously the variety of new products increased, their cost fell, and productivity in high-tech industries accelerated.

Again, ironically, higher education institutions were central to all these trends: central to its development and adoption. By adopting digital technologies operationally, they became networks. Strategically, though they remained hierarchies, governed from within by managerial hierarchies and governed from the outside by direct central governments and central government quangos and other regulatory bodies decided upon by higher education institutions themselves. Thus, we describe aspects of postmodernity: fragmentation and distribution system, operationally, flattened/consoles by the internal and external hierarchies, hierarchies that inhabited adaptive strategies.

Postmodernism relatively has provoked architectural layers and features in education. The influence of the postmodern is a phenomenological experience, representing the core thinking on central issues of postmodernity (Foss et al. 2021). Postmodernism rejects universality/single universal science base. It brings the advantage of producing multiple methodological approaches to the research process. In this background, if we want to discuss/approach the production/representation of knowledge, what methodologies will we select and apply? We believe SEM as a methodology offers a broad option to test the integrated influence of postmodernism in digital delivery. SEM integrates both exogenous and endogenous latent variables (forces of postmodern education and its outcome), universities' existence and how they can transform these rapid changes to their advantage and its observables using a statistical framework that is needed to test the accumulated impact of a postmodern variable on digital delivery. CFA on SEM enables the researchers to quantify with the high-loading observables, thus determining the factors that empirically influence postmodern-digital reality formation. Further, the use of CFA on SEM enables higher education institutions to smoothly transform the postmodern-digital reality into a management information system, using its core features (Bagci and Celik 2018; Alalwan et al. 2021). A theoretic approach postmodern-digital approach can improve the strategy formulation and implementation in education (Figure 10).

Pursuing sustainability in digital delivery amid modern, postmodern challenges is complex; it requires an incremental approach to delivery but identifying and determining the key elements of the delivery becomes critical to impart sustainable digital systems in higher education institutions. Thus, we suggest using the model of the Deming Cycle (Plan-Do-Check-Act) incrementally. Further, Glavič and Lukman (2007) developed an incremental model capturing the key elements aligned in Figure 11.

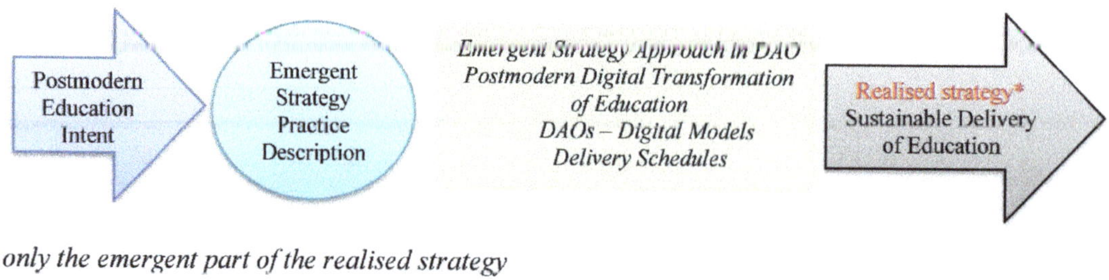

Figure 10. Emergent Strategy is Displayed as the Realised Strategy. Source: Based on (Foss et al. 2021).

Figure 11. Process and Elements of a Sustainable University. Source: Based on Glavič and Lukman (2007). Source: www.google.com (accessed on 1 December 2022) based on Glavič and Lukman (2007).

We also view the incremental approach to gaining sustainable delivery as requiring time, tools and technique. However, defining an approach that discusses the variables and performance indicators is key to success. Based on the facts derived from the literature, we would like to suggest the approach shown in Figure 12 for sustainable postmodern-digital delivery.

The research findings aid the evolution of sustainable digital transformation practices in higher education by producing empirical insights into determining the impact of postmodernism and its association with sustainable education. It also highlights the strategic importance of using a sustainable digital transformation to generate and regulate sustainable education programmes (Demenko and Savina 2019). The paper also delivers fresh insight into the impactful postmodernism changes affecting higher education institutions' existences and how they can transform these rapid changes to their advantage. A different sense of education life creates a unique thinking style and attitude. Postmodernism and its influence have shaped education through the power of digital delivery. In turn, it has impacted sizeably the higher education life of global students. The main elements of postmodern culture and its influence, the importance of viewing through the lens of digital transformation, are integrated with the digital delivery of the educational programme in higher education institutions.

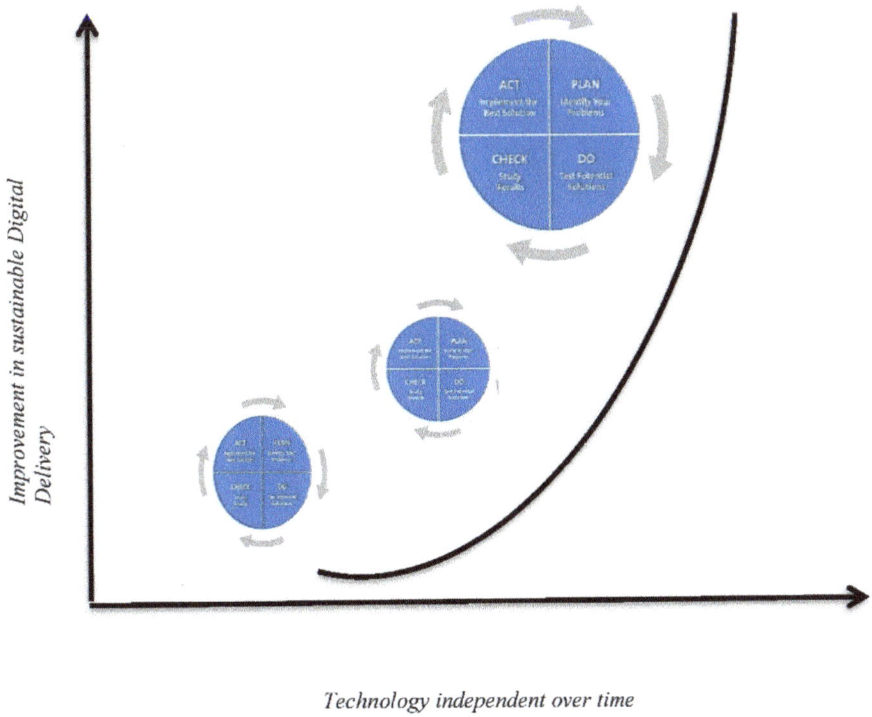

Figure 12. Incremental Improvement of Sustainable Digital Delivery. Source: Based on Demenko and Savina (2019).

5. Conclusions

This concept paper attempts to develop a theoretical model that embarks on how sustainable digital transformation as an educational force could be better utilized to achieve sustainability in higher education amid post-modernism. We propose a deductive research approach to examine this niche phenomenon using the SEM on CFA/EFA as the research strategy/data analysis technique. We claim that the traditional/analytical approach can produce a significant debacle because the approach is conventional, leaner, and premeditated. It also does not include the vibrant feature readily concerning the changing market condition. Emergent strategy produces a practical sequence of logic. We argue emergent approach to education strategy is essential to stay competitive in the rapidly changing globalized education sector.

The implementations of post-modernistic-emergent strategy models lead to new sustainable digital transformation capabilities in higher education and new education technologies, portfolios of courses, and policies. Universities have a role to play in the wider community and ultimately, they have a responsibility to lead the formation of a digital-oriented society. It implies that sustainable digital transformation in post-modernistic education is a necessity/unavoidable but not a luxury. This unique phenomenon put pressure on the applicability of the emergent strategy. It also means utilizing the sustainable digital transformation as a global platform in education to provide equal access to deliver quality education service is a significant challenge primarily because of the digital divide. Although this paper has discussed the significance of those phenomena to a greater extent, it is restricted by several limitations such as (a) it limits the applications of complex theoretical ideas (postmodernism/emergent strategy/digital transformation) to the education industry and (b) it focusses primarily on the technological, human and sustainable drivers in the higher education institutions.

The findings of this research answer our research questions on how postmodern digitalization impacts the sustainable delivery of education in emergent strategy. Our findings (literature/analysis) show that the intent for postmodern education is critical for formulating an emergent approach, explicating the integration of key activities of digital transformation is inevitable to enable the sustainable delivery of education. By cautiously integrating the connection between the influence of postmodernism on education, its digital enablement and sustainable delivery of education, our research makes a notable contribution by developing (a) a measurements model (b) a structural model and (c) offering a data analysis technique confirmatory factor analysis on SEM.

The implementation of emergent strategy requires postmodern education intent and autonomous strategic behaviour. Its integration with digital transformation to resolve the sustainable delivery of education is an approach to fill the global revenue gap as far as higher education institutions are concerned. Higher education institutions are under increasing pressure to regulate their research activities and scholarship to react to these changes.

Future researchers may focus on applying the suggested conceptual models in this paper to validate the practicality of generating beneficial outcomes for higher education institutions. Further, they also may examine how the implementation differs in two distinguished contexts to alter the impact of postmodernism on education delivery, specifically the moderating/mediating effect of contextual parameters. Demonstrating a positive attitude in adopting postmodernist changes in a sustainable digital transformation journey for higher education institutions is an absolute necessity to be future-equipped and stay competitive in education delivery. Such an integrated view of postmodernism on sustainable digital transformation offers a wide range of education programmes worldwide. However, there is a paucity and lack of clarity among educational institutions on how (a) postmodernism and (b) sustainable digital transformation impact, shape, and continuously improve educational delivery.

Author Contributions: All authors have contributed equally and substantially to the manuscript's concept and design. The team conceived the presented idea, developed the theory, verified the analytical methods, supervised the findings, discussed the results, and contributed to the final manuscript. All authors have read and agreed to the published version of the manuscript.

Funding: This research has no external funding.

Institutional Review Board Statement: Not applicable.

Informed Consent Statement: Not applicable.

Data Availability Statement: Not applicable.

Conflicts of Interest: The authors declare no conflict of interest.

References

Abad-Segura, Emilio, Mariana-Daniela González-Zamar, Juan C. Infante-Moro, and Germán Ruipérez García. 2020. Sustainable Management of Digital Transformation in Higher Education: Global Research Trends. *Sustainability* 12: 2107. [CrossRef]

Abney, Alexandra K., Laurel A. Cook, Alexa K. Fox, and Jenifer Stevens. 2018. Intercollegiate Social Media Education Ecosystem. *Journal of Marketing Education* 41: 254–69. [CrossRef]

Akhmetshin, Elvir, Karine Barmuta, Vladimir Vasilev, Hilary Okagbue, and Ogochukwu Ijezie. 2020. Principal Directions of Digital Transformation of Higher Education System in Sustainable Education. *E3S Web of Conferences* 208. [CrossRef]

Alalwan, Ali Abdallah, Abdullah M. Baabdullah, Yogesh K. Dwivedi, Nripendra P. Rana, Banita Lal, and Ramakrishnan Raman. 2021. Et-moone and marketing relationship governance: The effect of digital transformation and ICT during the COVID-19 pandemic. *Industrial Marketing Management* 98: 241–54. [CrossRef]

Al-Azzam, Nosayba, Lina Elsalem, and Farai Gombedza. 2020. A cross-sectional study to determine factors affecting dental and medical students' preference for virtual learning during the COVID-19 outbreak. *Heliyon* 6: e05704. [CrossRef]

Arends, Jessica. 2014. The Role of Rationality in Transformative Education. *Journal of Transformative Education* 12: 356–67. [CrossRef]

Argyris, Chris. 2014. Double Loop Learning in Organizations. Harvard Business Review. Available online: https://hbr.org/1977/09/double-loop-learning-in-organizations (accessed on 1 December 2022).

Audebrand, Luck. 2010. Sustainability in strategic management education: The quest for new root metaphors. *Academy of Management Learning & Education* 9: 413–20.

Bagci, Kübra, and Eray H. Celik. 2018. Examination of Factors Affecting Continuance Intention to use Web-Based Distance Learning System via Structural Equation Modelling. *Eurasian Journal of Educational Research* 18: 1–21. [CrossRef]

Beeson, Helena. 2016. Book review: Emerging strategies for supporting student learning by Barbara Allan. *Journal of Learning Development in Higher Education 11*. [CrossRef]

Bellavitis, Cristiano, Christian Fisch, and Paul P. Momtaz. 2022. The Rise of Decentralized Autonomous Organizations (DAOs): A First Empirical Glimpse. *SSRN Electronic Journal*. Available online: https://www.researchgate.net/publication/359712137 (accessed on 1 December 2022).

Benavides, Lina María Castro, Johnny Alexander Tamayo Arias, Martín Darío Arango Serna, John William Branch Bedoya, and Daniel Burgos. 2020. Digital Transformation in Higher Education Institutions: A Systematic Literature Review. *Sensors* 20: 3291. [CrossRef]

Bican, Peter M., and Alexander Brem. 2020. Digital Business Model, Digital Transformation, Digital Entrepreneurship: Is There a Sustainable "Digital". *Sustainability* 12: 5239. [CrossRef]

Bogdandy, Bence, Judit Tamas, and Zsolt Toth. 2020. Digital Transformation in Education during COVID-19: A Case Study. Paper presented at the 11th IEEE International Conference on Cognitive Infocommunications (CogInfoCom), Mariehamn, Finland, September 23–25.

Burkov, Aleksey V. 2020. Application of Technologies of Distributed Ledgers (Blockchain) in the Management of A Decentralized Autonomous Organization (DAO). *Journal of Advanced Research in Dynamical and Control Systems* 12: 1337–45. [CrossRef]

Cabrera, Daniel, Christopher P. Nickson, Damian Roland, Elissa Hall, and Felix Ankel. 2022. Distributed Autonomous Organization of Learning: Future Structure for Health Professions Education Institutions. *JMIR Medical Education* 8: e28770. [CrossRef] [PubMed]

Carrigan, Mark, and Katy Jordan. 2021. Platforms and Institutions in the Post-Pandemic University: A Case Study of Social Media and the Impact Agenda. *Postdigital Science and Education* 4: 354–72. [CrossRef]

Carter, Jr, Richard Allen, Mary Rice, Sohyun Yang, and Haidee A. Jackson. 2020. Self-regulated learning in online learning environments: Strategies for remote learning. *Information and Learning Science* 121: 311–19. [CrossRef]

Chambers, Matthew. 2016. Fundamental Education: UNESCO and American Post-War Modernism.

Chatzoglou, Prodromos, and Dimitrios Chatzoudes. 2018. The role of innovation in building competitive advantages: An empirical investigation. *European Journal of Innovation Management* 21: 44–69. [CrossRef]

Clark, John A. 2006. Michael Peters' Lyotardian Account of Postmodernism and Education: Some epistemic problems and naturalistic solutions. *Educational Philosophy and Theory* 38: 391–405. [CrossRef]

Czainska, Katarzyna. 2009. Internationalisation as an Alternative Strategy for University Development Case Study of Poznan University College of Business and Foreign Language. Available online: https://cor.sgh.waw.pl/handle/20.500.12182/340?show=full (accessed on 1 December 2022).

Davies, Howard, and Peter Walters. 2004. Emergent patterns of strategy, environment and performance in a transition economy. *Strategic Management Journal* 25: 347–64.

de S. Oliveira, Katyeudo K., and Ricardo A. C. de Souza. 2021. Digital Transformation towards Education 4.0. *Informatics in Education* 21: 283–309. [CrossRef]

Demenko, Olga G., and Natalia P. Savina. 2019. Program for the Development of the Digital Economy: Challenges and Prospects. *Programa Para el Desarrollo de la Economía Digital: Retos y Perspectivas* 40: 25. Available online: https://www.revistaespacios.com/a19v40n25/a19v40n25p25.pdf (accessed on 1 December 2021).

Dlačić, Jasmina, Maja Arslanagić, Selma Kadić-Maglajlić, Suzana Marković, and Sanja Raspor. 2013. Exploring perceived service quality, perceived value, and repurchase intention in higher education using structural equation modelling. *Total Quality Management & Business Excellence* 25: 141–57.

Done, Elizabeth, and Helen Knowler. 2013. Features of a post-identitarian pedagogy (with reference to postgraduate student writing and the continuing professional development of teachers). *Studies in Higher Education* 38: 1319–33. [CrossRef]

Du Toit, Jean, and Annè Hendrik Verhoef. 2018. Embodied digital technology and transformation in higher education. *Transformation in Higher Education* 3: a52. [CrossRef]

Duvnjak, Kristina, Maja Gregorić, and Mateja Gorše. 2020. Sustainable development—An artificial intelligence approach. *Management Research and Practice* 12: 18–28.

El Kamel, Leila, and Benny Rigaux-Bricmont. 2011. The Contributions of Postmodernism to the Analysis of Virtual Worlds as a Consumption Experience. *The Case of Second Life* 26: 71–92. [CrossRef]

Emerick, Chris. 2007. Who's Afraid of Postmodernism? Taking Derrida, Lyotard, and Foucault to Church? James K. A. Smith. *Religious Studies Review* 33: 51. [CrossRef]

Fixson, Sebastian K., and Jay Rao. 2014. Learning Emergent Strategies Through Design Thinking. *Design Management Review* 25: 46–53.

Foss, Nicolai J., Matthew C. McCaffrey, and Carmen-Elena Dorobat. 2021. "When Henry Met Fritz": Rules as Organizational Frameworks for Emergent Strategy Process. *Journal of Management Inquiry 31*. [CrossRef]

García-Peñalvo, Francisco Jose. 2021. Avoiding the Dark Side of Digital Transformation in Teaching. An Institutional Reference Framework for eLearning in Higher Education. *Sustainability* 13: 2023. [CrossRef]

Garrod, Joel Z. 2016. The Real World of the Decentralized Autonomous Society. Triple C: Communication, Capitalism & Critique. *Open Access Journal for a Global Sustainable Information Society* 14: 62–77. [CrossRef]

Genosko, Gary. 2001. The Uncollected Baudrillard, SAGE Publications, Limited, London. Available online: https://www.worldcat.org/title/uncollected-baudrillard/oclc/70773980 (accessed on 1 December 2022).

Gerrard, Catherine. 2007. Virtual learning environments: Enhancing the learning experience for students with disabilities. *Campus-Wide Information Systems* 24: 199–206. [CrossRef]

Glavič, Peter, and Rebeka Lukman. 2007. Review of sustainability terms and their definitions. *Journal of Cleaner Production* 15: 1875–85. [CrossRef]

Grenčíková, Adriana, Marcel Kordoš, and Valentinas Navickas. 2021. The Impact of Industry 4.0 on Education Contents. *Business: Theory and Practice* 22: 29–38. [CrossRef]

Halliday, John. 2020. Postmodernism and Post-compulsory Education. *Paideusis* 14: 31–47. [CrossRef]

Hassard, John. 2003. Postmodernism, philosophy and management: Concepts and controversies. *International Journal of Management Reviews* 1: 171–95. [CrossRef]

Heiberger, Greg, and Ruth Harper. 2008. Have you Facebooked Astin lately? Using technology to increase student involvement. *New Directions for Student Services* 2008: 19–35. [CrossRef]

Henseruk, Halyna, and Sergii Martyniuk. 2020. Digital Technologies of Formative Assessment. *Innovate Pedagogy* 2. [CrossRef]

Hernández-Betancur, Juan Esteban, Iván Montoya-Restrepo, and Luz Alexandra Montoya-Restrepo. 2017. *The Moment of Decision from the Perspective of Deliberate and Emergent Strategies*. AD-Minister, No. 31. Medellín: Escuela de Administración de la Universidad EAFIT. [CrossRef]

Holsberry, Carmen W. 1981. Faulkner, Fitzgerald, and Pynchon: Modernism and Postmodernism in Secondary School. *The English Journal* 70: 25–30. [CrossRef]

Iivari, Netta, Sumita Sharma, and Leena Ventä-Olkkonen. 2020. Digital transformation of everyday life–How COVID-19 pandemic transformed the basic education of the young generation and why information management research should care? *International Journal of Information Management* 55: 102183. [CrossRef]

Johnson, William Rainey, and David Blitzer. 2020. Residents' Perspectives on Graduate Medical Education during the COVID-19 Pandemic and Beyond. *MedEdPublish* 9: 77. [CrossRef]

Junco, Reynol, Greg Heiberger, and Eric Loken. 2011. The effect of Twitter on college student engagement and grades. *Journal of Computer Assisted Learning* 27: 119–32. [CrossRef]

Kaal, Wulf. 2021. A Decentralized Autonomous Organization (DAO) of DAOs. *SSRN Electronic Journal*. Available online: https://papers.ssrn.com/sol3/cf_dev/AbsByAuth.cfm?per_id=460345 (accessed on 1 December 2022). [CrossRef]

Kahraman, Ayse Derya. 2015. Relationship of Modernism, Postmodernism and Reflections of it on education. *Procedia—Social and Behavioral Sciences* 174: 3991–96. [CrossRef]

Kane, Gerald C. 2017. MetLife Centers Its Strategy on Digital Transformation. *MIT Sloan Management Review* 59: 88.

Kassens-Noor, Eva. 2012. Twitter as a teaching practice to enhance active and informal learning in higher education: The case of sustainable tweets. *Active Learning in Higher Education 13*. [CrossRef]

Kia, Ardeshir. 1988. The Transition from Modernism to Post-Modernism and Its Problematic Impact on Art Education Curriculum. *Marilyn Zurmuehlen Working Papers in Art Education* 7: 113–19. [CrossRef]

Kirschner, Paul A., and Aryn C. Karpinski. 2010. Facebook and academic performance. *Computers in Human Behavior 26*. [CrossRef]

Klimski, Marcin. 2018. Call For High Quality Education In The Context of the Postmodern Society Conception. Scientific Papers of Silesian University of Technology. *Organization and Management Series 2018*. [CrossRef]

Kolek, Ethan, and Daniel Saunders. 2008. Online disclosure: An empirical examination of undergraduate Facebook profiles. *Journal of Student Affairs Research and Practice* 45: 1–25. [CrossRef]

Lacan, Arnaud. 2019. Think tank: Meaningful management in a postmodern society. *Global Business and Organizational Excellence 38*. [CrossRef]

Lamichhane, Shreeram P., and Mana P. Wagley. 2013. Postmodernism and Nepal's Education. *Journal of Education and Research* 1: 9–12. [CrossRef]

Leaning, Marcus. 2014. Educating the postmodern child: The struggle for learning in a world of virtual realities. By Fiachra Long. *British Journal of Educational Studies* 62: 77–78. [CrossRef]

Lemmetty, Soila, and Kaija Collin. 2019. Self-Directed Learning as a Practice of Workplace Learning: Interpretative Repertoires of Self-Directed Learning in ICT Work. *Vocations and Learning* 13: 47–70. [CrossRef]

Lozano, Rodrigo, Kim Ceulemans, and Carol Scarff Seatter. 2015. Teaching organisational change management for sustainability: Designing and delivering a course at the University of Leeds to better prepare future sustainability change agents. *Journal of Cleaner Production* 106: 205–15. [CrossRef]

Lynch, Richard. 2018. *Strategic Management*, 8th ed. Pearson: London.

Lyotard, Jean-François. 1984. *The Postmodern Condition: A Report on Knowledge*. Minneapolis: University of Minnesota Press.

Manning, Stephan. 2012. New Silicon Valleys or a New Species? *Commoditization of Knowledge Work and the Rise of Knowledge Services Clusters. Research Policy* 42: 379–90. [CrossRef]

Mintzberg, Henry. 1978. 1978 Patterns in Strategy Formation. *Management Science* 24. [CrossRef]

Mintzberg, Henry. 1987. The Strategy Concept I: Five Ps for Strategy. *California Management Review* 30: 11–24. [CrossRef]

Mintzberg, Henry, and James A. Waters. 1985. Of strategies, deliberate and emergent. *Strategic Management Journal* 6: 257–72. [CrossRef]
Mirabeau, Laurent, and Steve Maguire. 2013. From autonomous strategic behavior to emergent strategy. *Strategic Management Journal* 35: 1202–29. [CrossRef]
Mohamed Hashim, Mohamed Ashmel, Isaam Tlemsani, and Robin Matthews. 2021. Higher education strategy in digital transformation. *Education and Information Technologies* 27: 3171–95. [CrossRef]
Mohamed Hashim, Mohamed Ashmel, Isaam Tlemsani, and Robin Duncan Matthews. 2022. A sustainable University: Digital Transformation and Beyond. *Education and Information Technologies* 27: 8961–96. [CrossRef] [PubMed]
Morze, Nataliia, and Oksana Strutynska. 2021. Digital transformation in society: Key aspects for model development. *Journal of Physics: Conference Series* 1946: 012021. [CrossRef]
Nielsen, Cynthia. 2006. Who's Afraid of Postmodernism? *American Catholic Philosophical Quarterly* 80: 642–46. [CrossRef]
Pivovarova, Margarita, Jeanne M. Powers, and Gustavo E. Fischman. 2020. Moving Beyond the Paradigm Wars: Emergent Approaches for Education Research. *Review of Research in Education* 44: vii–xvi. [CrossRef]
Player, Emily, Alice Shiner, Nicholas Steel, and Veena Rodrigues. 2020. Massive open online courses for continuing professional development of GPs. *InnovAiT: Education and Inspiration for General Practice 13*. [CrossRef]
Powell, Lisa, and Nicholas McGuigan. 2021. Teaching, virtually: A critical reflection. *Accounting Research Journal* 34: 335–44. [CrossRef]
Quinn, James Brian. 1980. Strategies for Change: Logical incrementalism Homewood. Illinois: Irwin. *Academy of Management Review* 7: 324–25. [CrossRef]
Raviolo, Paolo. 2013. Continuing education e Massive Online Open Courses. *Educational Reflective Practices* I: 113–17. [CrossRef]
Richardson, John Adkins, and Charles Jencks. 1989. "Postmodernism". *Journal of Aesthetic Education* 23: 114. [CrossRef]
Rinaldo, Shannon B., Suzanne Tapp, and Debra A. Laverie. 2011. Learning by tweeting: Using Twitter as a pedagogical tool. *Journal of Marketing Education* 33: 193–203. [CrossRef]
Senge, Peter M. 1990. *The Fifth Discipline: The Art and Practice of the Learning Organization.* New York: Doubleday/Currency.
Sudirtha, Gede, Ni Ketut Widiartini, and Made Diah Angendari. 2021. Development of 21st century skill learning designs through the application of the concept of independent learning in the vocational field. *Journal of Physics: Conference Series* 1810: 1–11. [CrossRef]
Tan, Elaine. 2013. Informal learning on YouTube: Exploring digital literacy in independent online learning. *Learning, Media and Technology* 38: 463–77. [CrossRef]
Teece, David. 2020. Fundamental Issues in Strategy: Time to Reassess? *Strategic Management Review* 1: 103–44. [CrossRef]
Usher, Robin, and Richard Edwards. 1994. *Postmodernism and Education.* London: Routledge.
Usher, Robin, Ian Bryant, and Rennie Johnstone. 1997. *Adult Education and the Postmodern Challenge: Learning Beyond the Limits.* London: Routledge.
Vica Olariu, Gabriela, Stelian Brad, and Mircea Fulea. 2020. The sustainable university in the new economic context. *FAIMA Business & Management Journal* 8: 5–18.
Virovets, Denys, and Sergiy Obushnyi. 2020. Decentralized Autonomous Organizations As the New Form of Economic Cooperation In Digital World. *The USV Annals of Economics and Public Administration, Stefan cel Mare University of Suceava, Romania, Faculty of Economics and Public Administration* 20: 41–52. Available online: https://www.econbiz.de/Record/the-usv-annals-economics-public-administration-scientific-journal-faculty-economics-public-administration-%C5%9Ftefan-cel-mare-university-suceava-romania/10009725935 (accessed on 1 December 2022).
Wang, Shuai, Wenwen Ding, Yong Yuan, and Juanjuan Li. 2019. Decentralized Autonomous Organizations: Concept, Model, and Applications. *IEEE Transactions on Computational Social Systems* 6: 870–78. [CrossRef]
West, Bettina, Hélén Moore, and Ben Barry. 2015. Beyond the tweet: Using Twitter to enhance engagement, learning, and success among first-year students. *Journal of Marketing Education 37*. [CrossRef]
Whittington, Richard, Patrick Regner, and Gerry Johnson. 2020. *Exploring Strategy*, 12th ed. London: Pearson.
Yan, Feng, Shreyas Sundaram, S. V. N. Vishwanathan, and Yuan Qi. 2013. Distributed Autonomous Online Learning: Regrets and Intrinsic Privacy-Preserving Properties. *IEEE Transactions on Knowledge and Data Engineering* 25: 2483–93. [CrossRef]
Zhu, Qin. 2009. Pragmatism as post-postmodernism. *Techné: Research in Philosophy and Technology* 13: 172–74. [CrossRef]

MDPI AG
Grosspeteranlage 5
4052 Basel
Switzerland
Tel.: +41 61 683 77 34

Administrative Sciences Editorial Office
E-mail: admsci@mdpi.com
www.mdpi.com/journal/admsci

www.ingramcontent.com/pod-product-compliance
Lightning Source LLC
Chambersburg PA
CBHW061511210926
51850CB00008B/1883
9783725814114